Professor Dr. JOSEPH ZUND

New Mexico State University
Department of Mathematical Sciences
Las Cruces, NM 88003-0003
USA

ISBN-13:978-3-642-79189-5 e-ISBN-13:978-3-642-79187-1
DOI: 10.1007/978-3-642-79187-1

Library of Congress Cataloging-in-Puplication Data. Zund, Joseph, 1939- Foundations of the differential geodesy / Joseph Zund. p. cm. ISBN-13:978-3-642-79189-5
1. Geodesy. 2. Generalized spaces. 3. Coordinates. 4. Geodesy - Mathematics. I. Title. QB283.Z85 1994 526'. 1--dc20 94-23055

Softcover reprint of the hardcover 1st edition 1994

Cover design: E. Kirchner, Heidelberg
Typesetting: Camera ready by author

32/3130-5 4 3 2 1 0 - Printed on acid-free paper

To my mentors
Bernard Chovitz and
Charles Whitten †

Preface

Introduction

This monograph is intended to be both a graduate level textbook and a research book. Although primarily addressed to theoretical geodesists, it may be of interest to those working in any field of geophysics in which one is concerned with the geometric properties of a family of equipotential surfaces, e.g. dynamical meteorology/oceanography. Moreover, the mathematical formalism which is developed — the so called leg calculus — also provides an efficient methodology for investigation of Gaussian differential geometry.

Differential geodesy may be formally regarded as the application of differential-geometric techniques to the geometric properties of the geopotential field. The term 'differential geodesy' is an obvious one, and although of somewhat recent vintage, it was for a time seriously considered by Martin Hotine (1898-1968) as a tentative title for his celebrated treatise. The original theory was created in 1947 by Antonio Marussi (1908-1984); however, the *Marussi-Hotine theory*, as subsequently developed by them, is but *one possible formulation* of differential geodesy. The aim of the present exposition is to reveal the *fundamental structural aspects* of their theory without imposing the various coordinate conditions which essentially particularize their approach. In our view, this more abstract structural approach has several advantages: it reveals not only the attributes, but the limitations of the Marussi-Hotine theory; and it paves the way for a new theory, which we call the *generalized Marussi-Hotine theory.* The latter is in its infancy, and it is our hope that the present study will serve as a stimulus for its development.

Note on Conventions

In this monograph we employ the following conventions and notation.

References are printed in all capital letters, e.g. HOTINE (1969), and to retain their chronological integrity, normally we will quote the original date of an item. The citation in the Bibliography will indicate various translations and more recent editions. Note that the monographs MARUSSI (1985) and HOTINE (1991) contain collections of many of their original papers.

HOTINE (1969) is his *Mathematical Geodesy*, and it will be frequently referred to as 'Hotine's treatise' or simply the 'treatise'. Equation, page, and chapter numbers from the treatise will be enclosed in square brackets, e.g. [1.22], [page 7] and [Chapter 1].

Our chapters will be denoted by Latin numerals, and

SECTION I-2 means SECTION 2 of CHAPTER I;

PROBLEM I.3 denotes PROBLEM 3 in CHAPTER I (collected at the end of the chapter); and

I-(2.6) is equation 6 in SECTION 2 of CHAPTER I.

Within a given chapter, the Latin numeral is omitted, however, it is included for references to items appearing in other chapters.

Some equations, or systems of equations, are of such importance that it is convenient to denote them by a special notation which is easier to remember than a particular equation number. These are

($\mathcal{A}$): the *Marussi Ansatz*, VI-(3.1);

($\mathcal{B}$): the *basic gradient equations*, VI-(3.2);

($\mathcal{L}_{\mathrm{I}}$),...($\mathcal{L}_{\mathrm{X}}$): the *Lamé equations/identities*, IV-(9.10-.18);

[F_{I}], [F_{II}], [$\mathrm{F}_{\mathrm{III}}$]: the *commutators/integrability conditions* of a smooth function F, (these occur in CHAPTER VI for the geopotential N and the local gravity n).

($\mathcal{HM}$): the *Hotine-Marussi equations* in SECTION VI.6.

On occasion, the entire system of equations indicated above are collectively denoted by $\{\mathcal{L}\}$, $\{\mathcal{S}\}$ and $\{F\}$ respectively.

Finally, we denote by

Property (T): the condition IV-(5.3).

Guide for the Reader

Due to the dual character of this monograph it is inevitable that the material is of uneven complexity and difficulty. Although the chapters should be read in successive order, one can make the following distinction between *core material* (required for a general view of differential geodesy) and *optional/supplementary material* (primarily for research purposes). The former consists of

CHAPTERS I and X — a *first* and *last* look at the role of coordinates,

CHAPTERS IV and V — the leg calculus and its application to Gaussian differential geometry,

CHAPTERS VI and VIII — the basic equations and structural features of differential geodesy;

while the latter includes:

CHAPTERS II and III — the geometric and analytic formulations of the leg calculus,

CHAPTERS VII and IX — topics mainly of research interest.

One could go further relative to CHAPTERS II and III, but the present coverage should suffice for our immediate needs. CHAPTER VII is essentially concerned with the global structure of the geopotential field and what can be expected from it. It is only a sketch of the material — much of which can be made more precise — and as such deserves a book length exposition, e.g. see PIZZETTI (1913). CHAPTER IX on conformal geodesy deals with a favorite topic of both Marussi and Hotine. So far their hopes have been only partially fulfilled, but again only future work will reveal its relevance. CHAPTER V is admittedly long, since in effect it is a short course on Gaussian differential geometry from a leg-theoretic viewpoint, and SECTIONS V.1-6 are of basic importance. We suggest a serious study of the core chapters, and upon a first reading, or in connection with a formal course, less emphasis on the optional/supplementary material.

Ideally the reader should have immediate access to HOTINE (1969), MARUSSI (1985) and HOTINE (1991). Much of our presentation is an extensive reworking and extension of topics from Hotine's treatise, and we strongly urge that it be consulted *after* study of our presentation: his eleven chapters of [Part I] can be read as desired/required; but [Part II] should be consulted only *after* reading CHAPTER X. The two reprint collections are probably most useful relative to our problem sets and for historical perspectives.

Each chapter concludes with ten problems. These have several purposes: first to test the reader's comprehension of the material by requesting a routine calculation, while others suggest extensions of the discussion given in the text. The problems are of uneven difficulty, and the purpose of citing a reference in the bibliography is to indicate a source where a complete discussion of the question/topic can be found. The inclusion of a citation does not necessarily indicate that a problem is difficult, or that the solution requires going to the literature.

Acknowledgements

The writing of any book necessarily places the author in debt to both colleagues, friends, and organizations for their encouragement, kindness, and tangible support. This book is no exception, and in the present circumstances I have been singularly fortunate. Bernard Chovitz and Charles Whitten have been especially patient in sharing with me their personal experiences with Hotine and Marussi, as well as clarifying numerous technical issues. For the last seven years my geodesy research has been supported by the United States Air Force (Geophysics Laboratory, now the Phillips Laboratory, at Hanscom AFB) and my contract managers there: Christopher Jekeli and David Gleason have done double duty not only as managers but instructors. Truly without this contract support, it would have been impossible for me to visit various geodetic centers, attend conferences, and become a member of the geodetic community. Numerous geodesists — too numerous to enumerate — have generously encouraged and shared their expertise and insight with me. In particular, Erik Grafarend has always been a source of both boundless enthusiasm and concrete information for me. The manuscript has been critically reviewed by Erik Grafarend, John Nolton, Gerald Rogers, and James Wilkes. Their comments and suggestions have been of immeasurable assistance to me, although the final responsibility for the manuscript is mine. Finally special thanks are due to my editor at Springer-Verlag (Heidelberg), Wolfgang Engel, and Valerie Reed for her expert preparation of the manuscript, and James Strange for his assistance in proofreading. Permission from Princeton University Press to reprint the quote from DUHEM (1914) on page 15 is gratefully acknowledged.

Finally, I must express my continuing obligation to the memory of Marussi and Hotine. Over the last decade the thoughts of these great men have provided me with an intellectual challenge which has been unique in my research career. I will regard my efforts a success, if this monograph conveys to the reader at least a glimpse of the beauty and profundity of their vision of differential geodesy. To borrow a phrase of Hermann Weyl, their ideas, deserve 'to be thought through to the end.'

Joseph Zund

Contents

Chapter I. Role of Coordinates in Geodesy and Geometry 1

1. Coordinates in Geodesy and Geometry 1
2. Marussi-Hotine Approach to Geodesy 5
3. Classical Formulations of the Tensor Calculus 15
4. Abstract Notion of a Tensor 26
5. Preview of the Leg Calculus 39

Problems for Chapter I 41

Chapter II. Ricci Calculus 45

1. Introduction 45
2. 3-Leg Algebra 46
3. Congruences 50
4. Geometry of Congruences 51
5. Ricci Coefficients 53
6. Calculation of Ricci Coefficients 56
7. Commutativity of Leg Derivatives 57
8. Lie Bracket 60
9. Jacobi and Schouten Identities 61
10. Interpretation of Ricci Coefficients 63
11. Miscellaneous Leg Representations 65
12. Lamé Equations 67

Problems for Chapter II 70

Chapter III. Cartan Calculus 73

1. Introduction 73
2. Exterior Product 74
3. Exterior Differentiation 78
4. Basic Properties of the Exterior Product and Differentiation 81
5. Cartan's Viewpoint 83

6. Structural Equations of Cartan 86
7. Commutators and Schouten Identities 92
8. Cartan Calculus and the Classical Tensor Calculus 93
9. Coordinates in the Cartan and Ricci Calculus 97
10. Conclusions and Summary 99
Problems for Chapter III 100

Chapter IV. General Leg Calculus 105

1. Introduction 105
2. Fundamental Idea of the Leg Calculus 106
3. Identification of the Ricci Coefficients 110
4. Permutation Algorithm 111
5. Interpretation of the Leg Coefficients 113
6. Basic Equations in the Leg Calculus 114
7. Commutators and Lie Brackets 116
8. Laplacian of a Scalar Function 121
9. Lamé Equations and Identities 123
10. Coordinates in the Leg Calculus 126
11. Reconstruction Theorem and General Viewpoint 132
Problems for Chapter IV 136

Chapter V. Gaussian Differential Geometry 139

1. Introduction 139
2. The Framework of Gaussian Differential Geometry 141
3. Gauss Operator and First Basic Form 148
4. Second Basic Form 153
5. Third and Fourth Basic Forms 159
6. Congruence and Surface Eigenstructure Theorems 161
7. Two Digressions 169
8. Basic Equations of the Surface Leg Calculus 172
9. Frenet Equations 175
10. Meusnier and Bonnet Formulas 178

11. Geodesic Curvature and Torsion in an Arbitrary Direction, and Euler's Formula 180
12. Euler's Osculating Paraboloid and Dupin's Indicatrix of Curvature 183
13. Triply-Orthogonal Systems and Dupin's Theorem 186
Problems for Chapter V 192

Chapter VI. Basic Equations for Differential Geodesy **197**

1. Introduction 197
2. Newtonian Gravitation and the Geopotential Field 197
3. Marussi Ansatz and the Hotine 3-Leg 200
4. Hotine Leg Calculus and N-Integrability Conditions 203
5. Marussi Tensor 206
6. Bruns Curvature Equation 211
7. Bruns Equation and Hotine's Theorem 213
8. n-Integrability Conditions 216
9. The Hotine-Marussi Equations 218
10. 3-Dimensional Laplacian of Local Gravity 219
11. 2-Dimensional Laplacian of Local Gravity 221
Problems for Chapter VI 223

Chapter VII. The Fundamental Theorem of Differential Geodesy **227**

1. Introduction 227
2. Basic Assumptions 227
3. The Fundamental Theorem of Differential Geodesy 229
4. Discussion 231
5. Extension Problem 232
6. Two Subsidiary Problems 237
7. The Congruence-Forming Property 238
8. The Surface-Forming Property 244
9. Conclusions 250

Problems for Chapter VII 252

Chapter VIII. Algebraic Theory of the Marussi Tensor 255

1. Introduction 255
2. Eigenstructure of Symmetric Tensors 256
3. Eigenstructure of the Marussi Tensor 261
4. The Singularity Condition 264
5. Conclusions 268

Problems for Chapter VIII 269

Chapter IX. Conformal Differential Geodesy 273

1. Introduction 273
2. Fundamental Notions of Conformal Geometry 274
3. Conformal Structural Equations 278
4. Conformal Ricci Coefficients 283
5. Conformal Surface Theory 285
6. Conformal Curve Theory 290
7. Generalized Conformal Transformations 295
8. Conclusions 300

Problems for Chapter IX 302

Chapter X. Coordinates in Differential Geodesy 303

1. Introduction 303
2. Marussi Hypothesis and Hotine Problem 304
3. Hotine's Hierarchy of Local Coordinate Systems 308
4. The Marussi-Hotine Approach: A Critique 319
5. Generalized Marussi-Hotine Approach: A Proposal 322
6. Conclusion 338

Problems for Chapter X 339

Bibliography 345

Index 363

I

The Role of Coordinates in Geodesy and Geometry

I.1 Coordinates in Geodesy and Geometry

In the application of the tensor calculus and differential geometry to physical problems, coordinates play a ubiquitous role. One is accustomed to think almost automatically in terms of them. However, their meaning — either geometrically or physically speaking — is seldom immediately apparent. Indeed, their mathematical existence, and physical measurability often pose questions of equal difficulty. Such questions underlie the fundamental notions concerning differential geodesy, and in this preliminary section we propose to discuss various viewpoints and suggest what we believe is a realistic approach.

The basic ideas concerning coordinates can be traced back to elementary Euclidean geometry. There one traditionally encounters and learns two different approaches: the *synthetic method* and the *coordinate method.* To properly set the stage for our geodetic applications of geometry it is necessary to have a clear understanding of the distinction between these opposing — yet complementary — approaches.

The synthetic method is an axiomatic one in which the notion of coordinates is not required. One introduces a set of axioms and logically deduces the subject from these axioms. This was the method Euclid used in his celebrated *Elements,* although a logically valid formulation was not perfected until over two millennia later by the efforts of G. Peano, D. Hilbert *et al.* In such a formulation, the basic notions and objects are precisely specified and not obscured by any algebraic/analytic machinery. It is often maintained that this approach, is in the modern parlance, 'coordinate-free' or 'pure', whereas coordinate based methods are 'impure' or 'applied'. Certainly the former is logically superior to analytic, or coordinate, geometry where no axioms — only definitions and identifications — appear. However, such a viewpoint, which is essentially an aesthetic judgement implying that one approach is superior to another, ultimately rests on prejudice and tradition. It overlooks the fact that the classical Greeks who devised the synthetic methodology had no choice in the matter. They had to proceed synthetically since they were in possession of neither the real number system, nor the notion of a coordinate, and had only a rudimentary knowledge of algebra.

In contrast, the coordinate method is an algebraic/analytic approach in which full use is made *ab initio* of the real number system. There are no axioms *per se,* only definitions/identications of algebraic notions with geometric objects. Points are taken to be ordered pairs or triples of numbers, and linear equations correspond to lines and planes. This was the great discovery of Descartes in the seventeenth century and the striking power of its methodology is well-known. In effect, at one stroke it reduces geometry to questions of elementary algebra and analysis. In purely mathematical discussions, *all* coordinate systems — subject only to the proviso that they can be unambiguously related to one another — are admissible and one system is superior to another merely because it is more convenient or simpler in a given situation. While such a circumstance is undeniably pleasing, it is also deceptive! Outside the realm of pure mathematics, one wants to choose coordinates which are physically — or at least geometrically — interpretable and susceptible of being measured. In this regard the difficulties are frequently subtle and quite unexpected.

For example, consider the case of spherical polar coordinates $x^m = (r, \vartheta, \varphi)$ in E_3 where r is a radial coordinate and ϑ, φ are respectively the co-latitude and longitude. Then the line element ds^2 has the form

$$ds^2 = dr^2 + r^2 \left(d\vartheta^2 + \sin^2 \vartheta d\varphi^2\right), \tag{1.1}$$

and the radial distance between a pair of values $r_1 < r_2$ with

(ϑ, φ) being held fixed is given by

$$s = \int_{r_1}^{r_2} dr = r_2 - r_1. \tag{1.2}$$

In general relativity, one seeks an analogue of this ds^2 which holds for a centrally-symmetric gravitational field. This is known as the Schwarzschild Problem, and as coordinates one chooses $x^\mu = (ct, r, \vartheta, \varphi)$ where $x^0 := ct$, with c being the speed of light in vacuo and t the time. Then the immediate generalization of (1.1) is

$$ds^2 = c^2 e^\nu dt^2 - e^\lambda dr^2 - r^2 \left(d\vartheta^2 + \sin^2 \vartheta d\varphi^2\right) \tag{1.3}$$

where ν and λ are arbitrary smooth functions of t and r, and the negative signs occur by virtue of the $(+ - --)$ signature of a relativistic space-time. The choice of (1.3) is usually argued on heuristic grounds, but in fact it can be shown that it follows rigorously by group-theoretic considerations, see WILKES and ZUND (1982). The vacuum Einstein equations then show that

$$e^\nu = e^{-\lambda} = 1 + \frac{\text{constant}}{r} \tag{1.4}$$

and physically the constant is chosen to be equal to $-2km/c^2$, where k is the Newtonian gravitational constant and m is the mass of the body generating the gravitational field. Introducing

$$r_g := 2km/c^2, \tag{1.5}$$

the so-called gravitational (or Schwarzschild) radius, (1.3) becomes

$$ds^2 = (1 - r_g/r)\, c^2 dt^2 - (1 - r_g/r)^{-1} dr^2 - r^2 \left(d\vartheta^2 + \sin^2 \vartheta d\varphi^2\right) \tag{1.6}$$

which is the famous (exterior) Schwarzschild solution. A more concise expression of (1.6) is

$$ds^2 = (1 - r_g/r)\, c^2 dt^2 - d\ell^2 \tag{1.7}$$

where

$$d\ell^2 := (1 - r_g/r)^{-1}\, dr^2 + r^2 \left(d\vartheta^2 + \sin^2\vartheta d\varphi^2\right) \tag{1.8}$$

is the spatial part of ds^2. One may then ask for the corresponding radial distance between a pair of values $r_1 < r_2$ when (ϑ, φ) are held fixed. By analogy with (1.2) this is given by

$$\ell = \int_{r_1}^{r_2} (1 - r_g/r)^{-\frac{1}{2}}\, dr, \tag{1.9}$$

and evaluation of (1.9) yields

$$\begin{aligned} \ell &= (r_2 - r_g)^{\frac{1}{2}} (r_2)^{\frac{1}{2}} - (r_1 - r_g)^{\frac{1}{2}} (r_1)^{\frac{1}{2}} \\ &+ r_g \log\left\{\frac{(r_2-r_g)^{\frac{1}{2}}+(r_2)^{\frac{1}{2}}}{(r_1-r_g)^{\frac{1}{2}}+(r_1)^{\frac{1}{2}}}\right\} \end{aligned} \quad . \tag{1.10}$$

Thus, contrary to our naive expectation in regarding r as a radial coordinate, we are not justified in regarding $r_2 - r_1$ as the distance between a pair of values $r_1 < r_2$! Hence, r is not exactly, but only an approximate, radial coordinate in (1.8). It is true, however, that the discrepancy is small. For example, if the body generating the field is the Sun then its radius is $\simeq 700,000$ km and $r_g \simeq 3$ km. Our conclusion is that the physics of the situation — in this case general relativity — has required a modification of our interpretation of the coordinate r. By inspection of (1.8) there is no change in our interpretation of the angular coordinates (ϑ, φ).

The above discussion has revealed an apparent divergence in the methodology applied in elementary Euclidean geometry. Relative to the synthetic versus coordinate dichotomy, our view is that in reality these approaches are simply complementary and merely designed to meet *different* requirements. Relative to each of their purposes, each is admirable and worthy of study. The distinction is real and must be understood, and the wise alternative is to choose a formalism that best suits the descriptive requirements of the situation. Ultimately we must never forget that the final verdict on differential geodesy will be based on its ability to provide a picture of the everyday needs of geodesy, i.e. the measurement of the Earth and its gravity field. In this regard the comment of L. Pasteur remains relevant:

"In the field of observation, chance only favors those who are prepared, viz. 'those open not closed minds'."

The dichotomy in methodology is especially important in differential geodesy where in effect Marussi tended to think synthetically and sought to base his intrinsic geodesy on general ideas and principles. In contrast, Hotine thought analytically and was concerned with expressing his, and of course Marussi's, ideas in terms of the tensor calculus. There is a degree of irony in these viewpoints: Marussi (trained as a pure mathematician) espousing broad physical principles and Hotine (trained as a military engineer) primarily focusing his attention on erecting a precise mathematical formulation of geodesy. Such a comment and characterization is a sweeping oversimplification of the situation, but nevertheless it involves some measure of truth. It is not surprising that in retrospect neither of them completely succeeded in the task they set for themselves. This is by no means intended as a criticism of them or their work, but rather is an indication of the intricacy of the theories they attempted to formulate and the problems which they tried to solve. Indeed, original work is seldom complete or free from ambiguities and misconceptions. In science especially, ideas and theories, as well as fortune and fashion, wax and wane. The victor becomes the vanquished and vice-versa, and the field is littered with the bones of theories which were too narrowly based or conceived.

I.2 The Marussi-Hotine Approach to Geodesy

In the previous section we have seen how a subject as ostensibly simple as elementary Euclidean geometry has been developed by two different approaches. These opposing methodologies are in fact complementary, and in thinking in geometry it is essentially a matter of taste or experience which of the notions, 'axioms' or 'coordinates' comes to mind first. Moreover, one can convincingly argue that *both* are required to have an understanding of the subject.

This dichotomy in methodology also occurs in differential

geodesy. Although Marussi and Hotine shared the basic aim of creating a new three-dimensional formulation of theoretical geodesy which would employ and exploit the geometric foundations of geodesy, their approaches each reflected the intellectual heritage, style, taste, and perhaps even temperament of these men. However, despite these differences, they were united by the common background of not being armchair, or ivory tower, theorists! Both came to theoretical geodesy after a rich and varied life of practical experience in the field: Marussi in Italy, Ethiopia and Greece; and Hotine in England, Africa and Persia. Moreover, both became seriously involved in differential geodesy in midlife, Marussi was almost forty and Hotine was almost fifty when the former first outlined his vision of intrinsic geodesy in Oslo in 1947. The difference in style and viewpoint can be clearly seen by reading their expository papers MARUSSI (1959) and Hotine (1964), which are respectively reprinted as the first papers in the monographs, MARUSSI (1985), and HOTINE (1991). In addition, the Preface of HOTINE (1969) furnishes valuable insight into his way of thinking, and puts on record his intellectual bond with his friend and coworker, i.e.

> "The author's main source of inspiration in the subject of this book has been Professor Antonio Marussi of the University of Trieste, not only for the range and originality of his ideas but also for continual advice and encouragement."

Further information on the historical and mathematical background of Marussi and Hotine can be found in the essay ZUND (1990a) which contains an extensive set of references.

The difference in methodology is also quite explicit in the mathematics they used. Marussi tended to think synthetically and sought to base his intrinsic geodesy on some general guidelines — which will be discussed in a moment — whereas Hotine thought analytically and was concerned with reducing his, and Marussi's, ideas into concrete mathematical expressions. If it can be fairly said that much of Marussi's work had an aesthetic/conceptual tone to it, then Hotine — at least as far as his writings indicate — had little inclination in this direction and simply preferred to let the equations speak for themselves. Hence, as an indication of the broad dissimilarity of their viewpoints, we find: Marussi arguing and espousing general physical principles — see MARUSSI (1947), (1951a), (1957), and (1988) in particular — and Hotine, see HOTINE (1957a), (1957b), (1959) and (1969) primarily devoting his efforts to producing a general mathematical formalism and formulation of geodesy. This divergence is especially evident in their attitudes toward the use of tensor calculus as the prime mathematical apparatus for differential geodesy. Marussi chose to approach tensors from the viewpoint of the homographic calculus. This was an abstract version of vector analysis which, while it is now recognized as being one of the precursors of the modern theory of tensors, never found general acceptance — even in Italy the country of its birth. It is likely that probably only a handful of people ever mastered its intricacies, and none of the great Italian geometers (who largely created the classical theory of tensors) ever made use of it in their work. Marussi apparently thought in homographic terms and his original version of intrinsic geodesy and the tensor calculus made essential use of it. Although he later employed a somewhat more orthodox version of the tensor calculus — see in particular his lectures, MARUSSI (1988) — in his final major paper on intrinsic geodesy, MARUSSI (1979), he returned to the homographic calculus and its related notions. On the other hand, Hotine exclusively employed the classical tensor calculus, e.g., as exemplified by the textbook of McCONNELL (1931), in all of his work. Whereas Marussi's version of the tensor calculus tended to minimize the use of indices, Hotine might almost be said to think in terms of indices! For example, in HOTINE (1959) he remarked that

> "No apology is offered for the inclusion of vector methods in index notation, which are much easier to acquire than to avoid."

While one might lament the lack of a common mathematical language and general approach to the subject, we are actually richer for having *two* versions of essentially the same theory. One of the primary goals of the present book is to unify these two seemingly different formulations into a single version of differential geodesy which in effect displays their work in a more lucid form.

In formulating this version of their work, we have followed neither the synthetic homographic-based approach of Marussi, nor the purely index notation methodology of Hotine. Our approach is a distillation of the best of both methods into the leg calculus, which will be previewed in the final section of this chapter, and then systematically developed in the next four chapters. Both Marussi and Hotine were aware of certain aspects of this calculus in their work, but neither took the crucial step of codifying these features into a well-defined theory. It is not surprising that the resulting theory bears a closer resemblance to the path followed by Hotine, rather than that employed by Marussi, since in the case of the former we have a treatise, i.e., HOTINE (1969), at our disposal. Indeed, much of our work could be characterized as a critical rethinking of Parts I and II of his *Mathematical Geodesy*, however, we have sought to balance this by making explicit use of the conceptual framework of the theory which was laid down by Marussi.

In his early self-study of geodesy, Marussi found many concepts and relations such as those between the geoid and ellipsoid of reference, to be unclear and unsatisfying. He also felt that the then usual two-dimensional treatment of geodesy was both artificial and unnatural. Guided by his practical experience and pure mathematical training, he set out to remedy these deficiencies by seeking a unified point of view which would provide a rational foundation for geodesy. The term 'rational' is not meant to suggest that the existing theories were 'irrational', but that a general theory — such as in the case of rational mechanics, as analytical/classical mechanics is traditionally called — should be developed. No doubt, Hotine shared these convictions and it was this feeling which fired his enthusiasm for Marussi's intrinsic geodesy in Oslo. Later, in HOTINE (1964), he was to give a passionate plea for geodesists to

> "... get back to a stricter mathematical discipline as the rational basis of our work ..."

and in effect to revitalize geodesy as a branch of mathematical physics as it was in the days of Newton, Lagrange, Laplace and Gauss. As the American theoretical physicist J.W. Gibbs wrote in 1881,

> "One of the principal objects of theoretical research in any department of knowledge is to find the point of view from which the subject appears in its greatest simplicity."

From his love of geometry, Marussi naturally came upon this point of view. He discovered that the traditional Bruns-Helmert conception of geodesy as

> "... the science of the measurement and mapping of the Earth's surface"

could be reformulated by regarding

> "Geodesy is the science which is devoted to the study of the Earth's gravity field,"

e.g. MARUSSI (1951a), (1988). In other words, when the gravity field of the Earth is described in potential-theoretic terms, geodesy is merged with the differential geometry of equipotential surfaces of the Earth's gravity field! However, it is more than merely differential geometry since, ultimately, it involves reconciling purely mathematical quantities with the physical processes of performing geodetic measurements on and between these equipotential surfaces. In effect, such a procedure seeks to enhance the physical description of the gravity field by recasting it in geometric form.

Marussi in formulating his approach to theoretical geodesy made a number of basic assumptions which we state as follows:

(i) the geometry of space is Euclidean and 3-dimensional;

(ii) the equipotential surfaces of the Earth's gravity field are locally isometrically and smoothly imbedded in a 3-dimensional Euclidean space $\mathbf{E}_3$;

(iii) the choice of reference systems should be natural and not contrived;

(iv) the reference systems should involve no additional hypotheses or otherwise impose any loss of generality in our description of the gravity field;

(v) the reference systems employed in the analysis must be susceptible of an immediate physical interpretation;

(vi) the components of all vectors/tensors occurring in the theory should be readily measurable;

(vii) the domain of the reference systems must be sufficiently large to be useful, viz. to allow one to make measurements and give a description of the gravity field in a required vicinity;

(viii) the domains of various reference systems should be continuable, in the sense that they are compatible and one can readily pass between neighboring systems of reference.

We have stated these requirements in a form which is slightly more general than that given by MARUSSI (1951a) (1957), (1988) and in a less comprehensive version in MARUSSI (1951b). Originally, he assumed that the reference systems were coordinate systems, i.e., his intrinsic coordinates, but later he extended them to include systems of linearly independent vectors which is what we will later regard as a special type of leg system. The assumption that such coordinates always exist and can be exhibited may seem somewhat naive to us now. However, when he took this for granted, it was a common view in theoretical geodesy, since it was also a belief held by Bruns, Helmert, Jordan et al. No doubt Einstein had similar convictions when he first formulated his theory of general relativity in 1915. To Marussi's credit, when GRAFAREND (1971), (1975) and GROSSMAN

(1974) discovered more general reference systems — the so-called non-holonomic/anholonomic systems — he immediately accepted them and even employed them in MARUSSI (1979). It is less clear that Einstein was so flexible in his thinking!

The requirements (i) and (ii) were implicit in his mathematical formulation of his theory and formally exclude the possibility of relativistic geodesy. This was perfectly natural at the time, and to have included relativistic considerations would have unduly complicated the situation. Even today, when relativistic effects are considered in theoretical geodesy, one might argue that differential geodesy should be completely understood before trying to formulate a relativistic differential geodesy. Neither Marussi nor Hotine ever tried to include relativity — either special or general relativity — into their work. Requirement (iii) is the origin of the term "intrinsic". However, the term intrinsic geometry occurs in Gaussian differential geometry where it refers to properties of a surface that can be described without referring to the surrounding ambient space containing the surface. In this sense, extrinsic geometry, or more explicitly, extrinsic properties of a surface, involve considering the normal vector of the surface which of course lies in the surrounding space. In a somewhat different sense — more in the sense of naturalness of the mode of describing the curves and surfaces — CESÀRO (1896) had used the term "intrinsic geometry", and it is very possible that Marussi knew of this book. Requirements (iv)-(vi) constitute Marussi's basic physical assumptions, while (vii) and (viii) were implicit in the considerations which he hoped to fulfill with his intrinsic geodesy. In practice, not all of these requirements are rigorously satisfiable, but in order of importance — after (i) and (ii) — the requirements (iii) and (iv) were crucial and it was assumed that such systems exist, and that (iii) and (iv) would expedite the fulfillment of (v) and (vi). Requirements (vii) and (viii) are highly desirable, however, they clearly depend on the particular physical situation under consideration. Note that (vii) and (viii) are reminiscent of the conditions which occur for coordinate neighborhoods in the modern theory of differentiable manifolds, except that no differentiability, i.e. smoothness, conditions were imposed. Moreover, when Marussi suggested these conditions, the notion of differentiable manifolds was not widely known except among differential geometers, and it is not likely that he was cognizant of such purely mathematical developments.

Needless to say, Marussi only partially succeeded in demonstrating that, in practice, all these requirements could be rigorously satisfied. Indeed, rather than rigid requirements he probably regarded them as *guidelines* for choosing reference systems. Almost fifty years later no one has done any better in this respect, and few have done as well. More recently, and with full knowledge of the state of existing mathematical theory, GROSSMAN (1979) observed that

> "The predominant view is that space near the Earth is a manifold. The unpleasant fact of life is that no one has ever described a method for coordinatizing that supposed manifold in a way consonant with physical reality."

We shall refer to the requirements (i)-(viii) as the *Marussi Conditions* and regard them as the basic guidelines for formulating differential geodesy. They were designed to yield not only an aesthetically pleasing mathematical theory, but more importantly, a physical theory in which measurements can be made and interpreted. Individually each is obvious, but taken together it is less obvious that apart from Newtonian dynamics, they are ever rigorously satisfiable in practice. In effect they ask for a physical theory which *deals* only with physically measurable quantities! However, if it is granted that Newtonian dynamics comes close to satisfying them, then a merger of it with the Newtonian theory of gravitation (when formulated in potential-theoretic terms) and recast in a geometric setting, viz. by using Gaussian differential geometry, offers us some hope of success. But this was precisely what Marussi offered us in his formulation of intrinsic geodesy, and, in effect, it was a glimpse of what a physical theory could be if we could only achieve it. In this sense, if geometry is regarded as the most perfect of the mathematical sciences, then theoretical geodesy satisfying the Marussi Conditions would play an analogous role in the physical sciences.

Nevertheless, regardless of whether they are weakly or strongly satisfiable, the Marussi Conditions in some form or another are remarkable and a tribute to his genius. Few physical theories are created with a set of guidelines indicating rules for their implementation. Perhaps the only example which readily comes to mind is Newton's famous 'Rules of Reasoning in Philosophy' which occur at the beginning of Book Three of his celebrated *Principia*, NEWTON (1934). However, these are quite general and are to be regarded as rules of thinking and not restricted to a particular physical theory, although at the time there was only *one* such theory and Newton had just created it!

Despite their intuitive and heuristic appeal, the Marussi Conditions are obviously incomplete and imperfect. What is missing, and this is part of what Grossman was alluding to in the previously cited quotation, is a set of instructions for translating geodetic operations into specific geometric terms. The operations of measurability and continuability referred to in (vii) and (viii) are unknown, and it is far from evident that they can be described in geometric terms. Thus, Marussi's theory, like many of the theories of classical physics, requires an explicit and well-defined theory of measurement. This is commonly disregarded in classical physics — and usually regarded as *obvious* by physicists — but it has plagued and spoiled most attempts to organize the theories along axiomatic lines. Curiously, it has been acknowledged and received considerable attention only in quantum theory where it remains an issue of contention and controversy. One naively hopes — as Marussi did — for a theory which deals only with physically measurable quantities. But, is this a realistic possibility, or merely a vainly held expectation? No one knows, but it is possible that it is as much a question of philosophy as it is of physics. Opinions differ and some are genuinely surprising and quite unexpected. In a conversation with the youthful Heisenberg who had just succeeded in setting up his version of quantum mechanics — based exclusively on measurable quantities — Einstein allegedly made the remarkable comment (with italics added for emphasis)

> 'But on principle, it is quite *wrong* to try founding a theory on observable magnitudes *alone*. In reality the very opposite happens. *It is the theory which decides what we can observe.*'

While such a response was obviously intended to apply to the context of quantum theory, which is far from the concerns of practicing geodesists, it nevertheless is food for thought and should make us wonder whether in our quest for a physically realistic differential geodesy we are requiring too much. Hence, keeping Einstein's rejoinder in mind, it is probably sensible merely to try to set out and understand the basic mathematical and physical structure of the theory before subjecting it to demands which it cannot be reasonably expected to meet. In this sense, the Marussi Conditions — regarded purely as guidelines — can still play a meaningful role, even if they set unattainable goals.

In conclusion, the differences between Marussi and Hotine and their thought processes in regard to their formulations of theoretical geodesy can be illuminated and — we believe to a great extent — clarified by perusing the seminal book DUHEM (1914). Duhem (1861-1916) was a bit of a maverick, but also a distinguished nineteenth century physicist and philosopher of science. In Part I, Chapter IV of his book he analyzed in great detail — and perhaps not without a degree of envy — why Maxwell's formulation of electrodynamics was so alien and difficult for European physicists to comprehend and accept. The difference according to Duhem was that European (in particular, French) scientists tended to think in abstract terms while English scientists tended to be more narrowly focused and to leap from particular situations to the general theory *without* a proper concern for general principles. His argument is interesting, and somewhat piquant, but in a sense it is applicable to the approaches taken by Marussi and Hotine (although in the latter's case he preferred to proceed from analytic expressions not mechanical models). Perhaps Duhem's most profound conclusion was his definition of what should be meant and expected from a physical theory, and in effect his entire book is devoted to the explanation and justification of the following assertion:

> "A physical theory is not an explanation. It is a system of mathematical propositions, deduced from a small number of principles which aim to represent as simply, as completely, and as exactly as possible a set of experimental laws."

We do not know whether either Marussi or Hotine ever heard of Duhem or whether they would agree with such a definition. However, it appears to capture — with some measure of truth — the essence of their approach to differential geodesy. As a rule, geodesists are seldom interested in philosophy, however, it is important in theoretical work in that it colors/influences the kind of theory we set up, the questions we ask and the answers which we will accept for these questions. This is frequently encountered in theoretical physics, e.g., in quantum theory, where Weyl and Dirac stressed that the theory be 'beautiful' and 'simple'. In effect, what they meant was that the mathematics being employed was familiar and well-known, viz. it *fit* the physical theory! In this sense, differential geodesy is a beautiful and simple theory since much of it reduces to Gaussian differential geometry.

I.3 The Classical Formulations of the Tensor Calculus

In order to understand what is involved in the leg calculus, which is the principal mathematical formalism employed in our treatment of the Marussi-Hotine approach to differential geodesy, it is necessary to delineate four distinct approaches to the tensor calculus. These are

[i] the classical tensor calculus,

[ii] the Ricci calculus of congruences,

[iii] the Cartan calculus of exterior differential forms, and

[iv] the abstract theory of tensors.

The first three of these will be discussed in this section, while the fourth is given in the next section. We do not include the homographic calculus employed by Marussi in our presentation, since virtually all of Marussi's homographic equations that are of geodetic importance were translated in HOTINE (1969) into classical tensor notation. Finally, in Section 5, we will give a preview of the general leg calculus based on a synthesis of the approaches [i]-[iv]. A detailed development of the Ricci and Cartan theories will then be given in Chapters II and III, and in Chapter IV these are combined into the general leg calculus. Much of our discussion of items [i]-[iii] is historical, but nevertheless it is useful since it gives a clear distinction between the theories and serves to illuminate how and why these approaches were developed.

First, the classical tensor calculus was formalized into a coherent self-contained mathematical discipline by G. Ricci-Curbastro (1853-1925) — more commonly known simply by the surname of Ricci — during the 1880's. Fragments of the theory had previously been given by G.F.B. Riemann, E.B. Christoffel, and R. Lipschitz in Germany some ten to fifteen years earlier, but it fell to Ricci to mold it into a well defined formalism which he called the *Calcolo differenziale assoluto*, i.e., the absolute differential calculus. This terminology is now obsolete and the name tensor calculus was popularized only in the second decade of the twentieth century by Einstein in his work on general relativity. Both the approaches [i] and [ii] are due to Ricci, but it is useful to separate them, since logically they are different theories, although each is subsumed as a special case of the abstract approach [iv]. In Ricci's work, both were developed more or less simultaneously, and this mixture appears in a rudimentary form in Hotine's treatise. Ricci himself was both a geometer and a mathematical physicist, and much of the impetus in creating a new mathematical formalism was derived from his study of electrodynamics and elasticity theory. Much of his original work on tensors, which he called *covariant* and *contravariant systems*, was given in his book RICCI (1898), and also in his undated *Lezioni della teoria matematical della elastistica* RICCI (1956/57). These lectures represent the first systematic application of tensor-theoretic methods to these subjects. The definitive version of his work on tensors was contained in RICCI and LEVI-CIVITÀ (1901) which is regarded as one of the seminal papers on the tensor calculus. This memoir was carefully studied by Einstein as a mathematical preliminary to his formulation of general relativity. With respect to the approach [i] the methods developed by Ricci represents the 'traditional'/orthodox version of tensors which was in vogue for roughly a half-century. The basic idea was to set up a calculus which would operate efficiently in any coordinate system, permit changes of the coordinate system, and be applicable in a space of arbitrary dimension and not necessarily Euclidean. Hence, although there was no physical need for considering spaces of dimensions greater than three at the time, this new formalism was designed *ab initio* to free itself from the limitations inherent in the vector calculus, i.e. three dimensions and rectangular Cartesian coordinate systems. The viewpoint and framework of the new theory was succinctly summarized by LEVI-CIVITÀ and AMALDI (1931) in the following quotation (a rough translation):

> "It is a conceptual and algorithmic theory which permits one to translate geometric and physical properties of a space into an analytic form which is independent of a particular selection of coordinates in the space in which it is intended to take place."

This quotation explicitly emphasizes that tensors, from their inception, were created to handle geometric and physical questions in a manner not restricted to the choice of a privileged set of coordinates used to *describe* the space under consideration. In other words, the tensor calculus is a mathematical tool, and not a new theory of space, but merely a new methodology, and its role is primarily a descriptive one. The generality inherent in it is admirable: any two coordinate systems x^r and $\overline{x}^r$

($r = 1, 2, \cdots, n$) that are related by *regular*, i.e. continuously differentiable (smooth) and invertible, *transformations* are admissible and equivalent from a purely mathematical point of view. In physical situations such generality is rarely needed. For example, the notion of an inertial frame (a Newtonian reference system) is not preserved under arbitrary regular transformations, i.e. in Newtonian dynamics the admissible transformations are restricted to the so-called Galilean transformations (with constant velocity). Likewise, in Special Relativity one employs Lorentz transformations (again involving a constant velocity) and these leave Maxwellian electrodynamics invariant although strictly speaking the theory is invariant under a larger group of transformations. General relativity permits the use of arbitrary regular transformations, but at the cost of the loss of the notion of an inertial frame. Indeed, in his final paper Einstein noted that general relativity could be logically characterized as a theory which avoided the introduction of the notion of an inertial system, see EINSTEIN and KAUFMANN (1955). While such a step may be aesthetically pleasing in that it eliminates unnecessary/extraneous notions it obviously complicates the physical interpretation of the theory and its ultimate reduction to Newtonian physics in limiting cases.

Such things are often overlooked by tensor enthusiasts who mistakenly believe that tensors and tensor equations automatically have a geometric/physical significance and that the generality available in the choice of a coordinate system is something other than a mathematical convenience. Hence, the ultimate weakness of the approach [i] is in the choice of a natural, or relevant, system of coordinates. In this sense, the theory is too general in that it rejects only those coordinate systems for which the functional (Jacobian) determinant

$$J := \det \left\| \frac{\partial x^r}{\partial \overline{x}^s} \right\| \tag{3.1}$$

vanishes or is undefined (i.e. infinite). It is immediately seen that if $0 < J < \infty$ then also $0 < J^{-1} < \infty$. The naturalness, or relevance, of the coordinate system depends on the geometry, or physical nature, of the situation, and the tensor calculus *per se* seldom provides any clue on how to select such systems. The difficulty is that — as suggested before — coordinates provide a method of describing space, and are not a fundamental defining property of space. This is clear since the notion of coordinates has no analogue in a synthetic approach to the geometry of space! The same situation holds in physics, but there the need for coordinates is more compelling. One can readily think of a swinging pendulum or a harmonic oscillator etc., but without the use of coordinates one is virtually helpless to *describe* its motion since there is no effective synthetic method for doing physical calculations.

In conclusion, the methodology of [i] works admirably and efficiently *if* one can make a natural choice of the coordinates. Coordinates, despite some purist opinions to the contrary, are not unsuitable or unwieldy quantities, and their use is both effective and elegant *provided* a suitable choice of them can be made. Otherwise, they may encumber the analysis, and yield results which for practical purposes are unintelligible and useless. In other words, tensor calculus is an efficient formalism in the presence of 'something more' and this 'something more' is the essence of the problem which determines its solvability. A final caveat must be noted: a desired set of coordinates need not always exist and possess the properties one hopes to have. Moreover, when such a system of coordinates has been found, the resulting coordinates need not be valid over an unlimited region of space. This is commonly illustrated in the transformation from rectangular Cartesian coordinates $y^r = (x, y, z)$ to spherical polar coordinates $x^r = (r, \vartheta, \varphi)$ in $\mathbf{E}_3$. The Cartesian system y^r is valid over all finite regions of $\mathbf{E}_3$ i.e. $|y^r| < \infty$ for $r = 1, 2, 3$, whereas upon computing the functional determinant J for the transformation $y^r \to x^r$

$$\begin{aligned} x &= r \sin\vartheta \cos\varphi \\ y &= r \sin\vartheta \sin\varphi \\ z &= r \cos\vartheta \end{aligned} \tag{3.2}$$

(where tentatively one takes $0 \leq r < \infty$, $0 \leq \vartheta \leq \pi$ and $0 \leq \varphi < 2\pi$) we find that

$$J = \frac{\partial(x, y, z)}{\partial(r, \vartheta, \varphi)} = r^2 \sin\vartheta \tag{3.3}$$

and hence

$$J^{-1} = \frac{1}{r^2} \csc\vartheta. \tag{3.4}$$

The value $r = 0$ must obviously be excluded, and by (3.2) it corresponds merely to the origin 0 of the y^r-system. On the other hand, for a finite $r \neq 0$ the values $\vartheta = 0$ and π are inadmissible. These correspond by (3.2) to the z-axis or more precisely the finite segment of it given by $|z| < r$. Thus, the change to spherical polar coordinates *precludes* the description of points lying on the z-axis from $-r < z < +r$, and the actual range of (r, ϑ, φ) permitted in (3.2) is given by: $0 < r < \infty$, $0 < \vartheta < \pi$ and $0 \leq \varphi < 2\pi$.

Ricci's second approach to 'tensors' appeared in his memoir RICCI (1895). This begins with the comment that he 'seeks to determine the significance of the notation' previously employed — an explicit admission of the formal character of his earlier work! Now motivated by the investigations of Darboux and his method of the motion of a mobile triad to describe curves and surfaces in $\mathbf{E}_3$, Ricci recast his theory into an elegant geometric form in n-dimensions. His fundamental idea was to resolve all components of vectors/tensors not along a set of arbitrary coordinate axes, but rather along the tangent vectors to a set of congruences of curves. By the latter, he meant that the region of the n-shape under consideration is assumed to have n orthogonal families of curves passing through each point. The resulting theory is commonly known as the *method of rotation coefficients* (which describe the geometric properties of curves) although it is more appropriate to call it the *Ricci congruence calculus*, or simply the *Ricci calculus*.

The resolutions of vectors/tensors along these congruences may well explicitly depend on *some* coordinate system; however the resolved components — provisionally we may regard them as 'congruence-components' — behave as scalars under a change of coordinates $x^r \to \overline{x}^r$. The admissible changes for such components are from one system of n congruences to another system of n congruences, and this is expressed by a non-singular transformation between these systems of orthonormal, i.e. pairwise orthogonal, unit tangent vectors. In other words, if we label these vectors by lower case Latin letters chosen from the beginning of the alphabet, i.e., a, b, c, d, ..., ranging over the values 1, 2, 3, ..., n then

$$\boldsymbol{\lambda}_a \mapsto \overline{\boldsymbol{\lambda}}_a = A_{ab}\boldsymbol{\lambda}_b \tag{3.5}$$

where

$$A := \det \|A_{ab}\| \neq 0. \tag{3.6}$$

Upon introducing the contravariant coordinate-dependent components of these vectors, (3.5) becomes

$$\lambda_a{}^r \mapsto \overline{\lambda}_a{}^r = A_{ab}\lambda_b{}^r \tag{3.7}$$

and likewise for the covariant components we have

$$\lambda_{ar} \mapsto \overline{\lambda}_{ar} = A_{ab}\lambda_{br}. \tag{3.8}$$

In these expressions the coordinate-dependent indices are denoted by Latin letters chosen from the latter part of the alphabet, i.e. typically r, s, p, q, m, n, ..., and it is convenient to always write the congruence label first, with the coordinate index being offset over a congruence index as in (3.7) or (3.8) since the former may be 'lowered' by using the components g_{rs} of the metric tensor. If there is a possibility of confusing numerical values of congruence and coordinate indices, then the former may be enclosed in parentheses, viz. $\boldsymbol{\lambda}_{(1)}, \boldsymbol{\lambda}_{(2)}, \cdots$ denote the various tangent vectors with $\lambda_{(1)}{}^r, \lambda_{(2)}{}^r, \cdots$, and $\lambda_{(1)r}, \lambda_{(2)r}, \cdots$ being their respective contravariant and covariant (tensor) components. The Einstein summation convention holds for both congruence and coordinate components, but for the former the repeated indices always occur as subscripts.

Differentiation with respect to a congruence label now becomes *directional differentiation*, i.e. for a smooth function F of the coordinates we have

$$\frac{\partial F}{\partial s^a} := \frac{dx^r}{ds^a}\frac{\partial F}{\partial x^r} = \lambda_a{}^r \frac{\partial F}{\partial x^r} \tag{3.9}$$

where s^a denotes the arc length along the congruence in the $\boldsymbol{\lambda}_a$ direction. In general, upon indicating the congruence (directional) differentiation by "/" and partial differentiation by ";" (3.10) may be written in the form

$$F_{/a} := \lambda_a{}^r F_{;r}. \tag{3.10}$$

The content of the Ricci calculus is very rich and it underlies much of the modern structure encountered in differential geometry. The n congruences of curves defined by the n tangent vectors at a point P of the space form an n-dimensional real vector space called the *tangent space* $\mathbf{T}_P$ to the space at P. Thus, a change of the congruences, (3.8), corresponds to a change of basis in $\mathbf{T}_P$. A familiar example is the two dimensional tangent plane to a surface $\mathbf{S}$ of P in $\mathbf{E}_3$. If we regard this plane as being spanned by a pair of vectors tangential to $\mathbf{S}$ at P, then the change (3.8) is merely to another pair of tangential congruences to $\mathbf{S}$ at P. Equation (3.9), or (3.10), also contains the important idea of vector fields — more precisely tangent vector fields — being identified with linear partial differential operators.

Suppose

$$\{\boldsymbol{\lambda}_a\}_{a=1}^n := \left\{\boldsymbol{\lambda}_{(1)}, \boldsymbol{\lambda}_{(2)}, \cdots, \boldsymbol{\lambda}_{(n)}\right\} \tag{3.11}$$

denotes a set of n tangent vectors. Then (3.10) associates the n directional derivatives with the operation of (ordinary) partial differentiation with respect to the coordinates x^r. The function F in (3.9) is superfluous, hence we may write

$$\frac{\partial}{\partial s^a} = \lambda_a{}^r \frac{\partial}{\partial x^r} \tag{3.12}$$

where the left hand side formally represents directional differentiation along $\boldsymbol{\lambda}_a$. A more suggestive form of this equation is then obtained by symbolically rewriting it as simply

$$\boldsymbol{\lambda}_a = \lambda_a{}^r \frac{\partial}{\partial x^r}. \tag{3.13}$$

In the case of three-dimensions, we may simplify the notation by simply naming the individual tangent vectors as follows

$$\{\boldsymbol{\lambda}_a\}_{a=1}^{3} = \{\boldsymbol{\lambda}, \boldsymbol{\mu}, \boldsymbol{\nu}\}, \tag{3.14}$$

or simply $\{\boldsymbol{\lambda}_a\}$, so (3.13) becomes

$$\boldsymbol{\lambda} = \lambda^r \frac{\partial}{\partial x^r}, \; \boldsymbol{\mu} = \mu^r \frac{\partial}{\partial x^r}, \; \boldsymbol{\nu} = \nu^r \frac{\partial}{\partial x^r}. \tag{3.15}$$

Under a change of coordinates $x^r \to \overline{x}^r$ these differential operators are independent of the coordinate description appearing on the right-hand side of these expressions. This is immediate by employing the familiar transformation rules for λ^r and $\dfrac{\partial}{\partial x^r}$ and upon 'barring': i.e.

$$\begin{aligned} \overline{\boldsymbol{\lambda}} := \overline{\lambda}^r \frac{\partial}{\partial \overline{x}^r} &= \frac{\partial \overline{x}^r}{\partial x^p} \lambda^p \frac{\partial x^q}{\partial \overline{x}^r} \frac{\partial}{\partial x^q} = \frac{\partial \overline{x}^r}{\partial x^p} \frac{\partial x^q}{\partial \overline{x}^r} \lambda^p \frac{\partial}{\partial x^q} \\ &= \delta^q_p \lambda^p \frac{\partial}{\partial x^q} = \lambda^p \frac{\partial}{\partial x^p} = \boldsymbol{\lambda}. \end{aligned} \tag{3.16}$$

Finally, we observe that unlike partial differentiation of a smooth function F, directional derivatives need not be permutable, viz. although

$$F_{;r;s} - F_{;s;r} = 0, \tag{3.17}$$

it does *not* follow that

$$F_{/a/b} - F_{/b/a} = 0. \tag{3.18}$$

In Chapter II these facts will be considered in greater detail.

The third approach to the tensor calculus is due to E. Cartan (1869-1951) and is the method of exterior differential forms, which we will call the *Cartan calculus.* By common assent he is regarded as one of the greatest mathematicians of this century and the founding father of modern differential geometry. Whereas the approaches [i] and [ii] were respectively analytical and geometrical, his approach [iii] is *structural.* By this we mean that although the Cartan calculus furnishes a very powerful symbolic method for doing calculations, the ideas behind it are deeply conceptual and have their roots in the abstract mathematical theories of Lie groups, Lie algebras, Pfaffian differential systems, and algebraic topology. Although Cartan enriched each of these disciplines with his genius, his methods were intuitive and guided by his profound algebraic and geometric insight. Indeed, much of modern mathematical work in these areas is still devoted to putting his work on a firm and rigorous basis. Fortunately, for practical purposes mastery of the purely computational aspects of the Cartan calculus requires only a few simple ideas, which we will indicate in the next section. Curiously enough Cartan's work on differential geometry began only in the later stage of his career, viz. in the 1920's, and was presented in his treatise CARTAN (1928). His starting point was the systematic use of differentials and their symbolic products. If (3.15) can be regarded as a method of introducing the contravariant components of $\{\boldsymbol{\lambda}_a\}$ in $\mathbf{T}_P$, then Cartan's approach involved introducing the covariant components in linear combinations of the differentials dx^r, i.e.

$$\theta_1 := \lambda_r dx^r,\ \theta_2 := \mu_r dx^r,\ \theta_3 := \nu_r dx^r; \tag{3.19}$$

or more generally by

$$\theta_a := \lambda_{ar} dx^r \tag{3.20}$$

where $a = 1, 2, \cdots, n$. These expressions are called *Pfaffians*, and strictly speaking they are defined in the dual vector space — the *cotangent space* — $\mathbf{T}_P^*$ of $\mathbf{T}_P$. Sometimes, to emphasize this fact, the expressions exhibited in (3.20) are called *covectors* to distinguish them from the *vectors* given in (3.15). In this sense, the Pfaffians can be regarded as the algebraic *duals* of the vectors $\boldsymbol{\lambda}$, $\boldsymbol{\mu}$, $\boldsymbol{\nu}$ or of $\{\boldsymbol{\lambda}_a\}$. Roughly speaking, this calculus is essentially concerned with giving a sensible meaning to expressions of the form (3.19) and (3.20), giving rules for forming products of such quantities, and defining an appropriate notion of differentiation for them. It is not surprising that the first step in formalizing Cartan's theory involves understanding the cotangent space $\mathbf{T}_P^*$. This is also the starting point of the abstract approach to tensors (iv) and will be considered in Section 4.

Before concluding this section it is natural to wonder to what extent Marussi and Hotine were aware of [ii] and [iii]. Apparently, Marussi was acquainted with the Ricci calculus and he made some mention of it in MARUSSI (1957). However, since it was largely unrelated to his beloved homographic calculus, he appears to have made no substantive use of it. Hotine employed the three tangent vectors (3.14) in his work, but he did not formalize his considerations into the framework of the Ricci calculus, or know of the identification given in (3.15). He must have been at least vaguely aware of it since two of the books quoted in HOTINE (1991), i.e. LEVI-CIVITÀ (1925) and WEATHERBURN (1938) contain rather detailed discussions of this material. In any case, regardless of his knowledge of the Ricci calculus he did implicitly employ some of the key ideas of it in his work. Even more remarkable are the following Hotine transformation formulas

$$\begin{aligned} \frac{\partial \overline{x}^r}{\partial x^s} &= \overline{\lambda}^r \lambda_s + \overline{\mu}^r \mu_s + \overline{\nu}^r \nu_s \\ \frac{\partial x^r}{\partial \overline{x}^s} &= \lambda^r \overline{\lambda}_s + \mu^r \overline{\mu}_s + \nu^r \overline{\nu}_s \end{aligned} \tag{3.21}$$

which he announced in [1.22] on page 7 of his treatise. He gave no proof of these formulas, but merely commented that upon multiplication of them respectively by λ^s, μ^s, ν^s and by λ_r, μ_r, ν_r they reproduce the classical-based tensor transformation laws for these components. These equations were not previously known and the possibility of exhibiting such expressions for the coordinate-based *transformation factors*, i.e. the left-hand sides of (3.21), is quite unexpected. Various properties of the transformation formulas are given in PROBLEM I.8 at the end of this chapter, and later in Chapter IV we will see that they are important in our exposition of the general leg calculus. It seems probable that since these equations were not reproduced in the exhaustive *Summary of Formulas* given at the end of his treatise, Hotine merely regarded them as 'nice equations' *without realizing* their importance, or their curious character.

The situation relative to the Cartan calculus is equally puzzling. While Marussi made no use of differential forms in his publications, the general formulation of tensor analysis in Chapter III (see in particular §3.5 pages 70-73) of MARUSSI (1988) is strikingly reminiscent of the discussion given in CARTAN (1928).) This could have been accidental, or common knowledge, but the evidence and general spirit of the presentation suggests that Marussi had at least some familiarity with Cartan's approach to the subject. In the case of Hotine, it is definitely known that in the final months of his life he was learning the Cartan theory, and was enthusiastic about its application in mathematical geodesy. This fact is explicitly stated by him in a letter to his colleague Bernard Chovitz, and this letter was reproduced in CHOVITZ (1982) as an appendix.

I.4 The Abstract Notion of a Tensor

The abstract approach leading to the notion of a tensor has been slow in its development and dissemination, and became well known to mathematicians only in the 1950s. A familiarity with this material among practitioners of tensor-theoretic methods in the physical sciences is still far from being common knowledge. This is both ironic and unfortunate since the theory is essentially a part of linear and multilinear algebra, and was first set out in explicit form by Hermann Weyl in his books on relativity and quantum theory, i.e. WEYL (1918) and WEYL (1928). Indeed, in his introduction to the latter, Weyl decried the slow acceptance of this branch of mathematics, which he said should be as well known as that of the calculus. Basically the delay was due to the time required to obtain a firm grasp of the ideas of a vector space $\mathbf{V}$ and its dual space $\mathbf{V}^*$, and to learn how to use these to construct 'bigger' vector spaces — the *tensor spaces* — by devising appropriate products of $\mathbf{V}$ with itself or with $\mathbf{V}^*$. Once this has been attained the general notion of a tensor is immediate and self-evident. Nevertheless, the first modern exposition of the subject seems to have been left aside until the appearance of BOURBAKI (1948).

Prior to this, a common complaint heard from many people — mainly mathematicians — had long been that essentially the notion of a tensor was enigmatic and had not been precisely defined. One knew how a tensor behaved under a coordinate transformation $x^r \rightarrow \overline{x}^r$, viz. the usual transformation law for its contravariant/covariant components, but not how it came to have such components. The transformation formulas are merely *properties of tensors*, and not a definition of what they are! Notwithstanding, the anarchist nature of such an assertion it was a valid complaint, and the fact that for several decades tensor-theorists were unable to meaningfully respond to it did not ameliorate the situation. In Section 1.3 we indicated three seemingly different approaches to the tensor calculus, and it is natural to wonder how these are related. Now, in this section we seek a general definition of a tensor which contains each of these realizations upon an appropriate specialization.

It is clear that such a task necessarily demands some mathematical preparation and sophistication, but the situation is considerably simplified by the happy circumstance that all our considerations take place in 3-dimensions and that for the requirements of differential geodesy we need only second order tensors. In the following we will attempt to sketch — with usually only a hint of the proofs — the major steps involved in setting out the theory. Our discussion makes no pretension of being general, indeed it is quite restricted, and admittedly a 'bare bones' approach. However, it should be accessible to anyone having a modicum of knowledge of linear algebra, i.e. the theory of finite-dimensional (real) vector spaces. In PROBLEMS I.1-4 we will give some examples, and indicate some details; the reader should try to fill in the gaps in our presentation.

We begin by recalling the definition of a vector space, and some of its elementary properties. A *vector space* $\mathbf{V}$ is a set of elements called *vectors* which is *closed* under the following two operations: multiplication of a vector $\boldsymbol{\xi}$ by a scalar (a real number) a, i.e. $a\boldsymbol{\xi}$, and addition of a pair of vectors $\boldsymbol{\xi}$ and $\boldsymbol{\eta}$ denoted by $\boldsymbol{\xi}+\boldsymbol{\eta}$. Closure means that the quantity $a\boldsymbol{\xi}+b\boldsymbol{\eta}$ is a well-defined vector in $\mathbf{V}$ for all scalars a and b and all vectors $\boldsymbol{\xi}$ and $\boldsymbol{\eta}$ in $\mathbf{V}$. The *dimension* of $\mathbf{V}$, i.e. $\dim \mathbf{V}$, is the maximum number of linearly-independent vectors in $\mathbf{V}$, and any such set of vectors is said to form a *basis* of $\mathbf{V}$. We denote a basis of $\mathbf{V}$ by $\{\mathbf{e}_i\}_{i=1}^n$ where $n = \dim \mathbf{V}$, or merely by $\{\mathbf{e}_i\}$, and in our considerations $n \leq 3$. The vectors $\mathbf{e}_1, \mathbf{e}_2, \cdots, \mathbf{e}_n$ of $\{\mathbf{e}_i\}$ are said to *generate*, or *span*, the space $\mathbf{V}$.

Two important kinds of mapping can be defined on $\mathbf{V}$. The first is a *linear transformation*, or *operator*, A, taking $\mathbf{V}$ into itself which we indicate by writing $A : \mathbf{V} \to \mathbf{V}$, and which satisfies the condition

$$A\,(a\boldsymbol{\xi} + b\boldsymbol{\eta}) = a\,(A\boldsymbol{\xi}) + b\,(A\boldsymbol{\eta}) \tag{4.1}$$

where a, b are scalars, $\boldsymbol{\xi}$, $\boldsymbol{\eta}$ are vectors of $\mathbf{V}$ and it is understood that $A\boldsymbol{\xi}$ and $A\boldsymbol{\eta}$ also belong to $\mathbf{V}$. The transformation A is non-singular, i.e. invertible, whenever it takes a basis $\{\mathbf{e}_i\}$ of $\mathbf{V}$ into another basis $\{\overline{\mathbf{e}}_i\}$ of $\mathbf{V}$. If A is represented by matrix $\|a_{ij}\|$ then the matrix is $n \times n$, and the requirement that A be non-singular is equivalent to the condition that the determinant of $\|a_{ij}\|$ be non-zero, i.e. $\det \|a_{ij}\| \neq 0$. The second kind of mapping is that of a *linear functional* L which takes vectors into scalars (the real numbers $\mathbf{R}$), i.e. $L : \mathbf{V} \to \mathbf{R}$, and which satisfies the condition

$$L\,(a\boldsymbol{\xi} + b\boldsymbol{\eta}) = aL\,(\boldsymbol{\xi}) + bL\,(\boldsymbol{\eta}) \tag{4.2}$$

as in (4.1) except that now $L\,(\boldsymbol{\xi})$ and $L\,(\boldsymbol{\eta})$ are real numbers. Linear functionals are also called linear forms or merely linear functions, since the distinction between (4.1) and (4.2) is simply that the former is a vector-valued function whereas the latter is a scalar-valued function.

The notion of a linear functional is crucial, and has several noteworthy consequences. First, it immediately leads to the idea of the dual space $\mathbf{V}^*$ of the vector space $\mathbf{V}$. By definition, $\mathbf{V}^* = \{L \mid L : \mathbf{V} \to \mathbf{R}\}$, i.e., it is the set of all linear functionals defined on $\mathbf{V}$. If $L \in \mathbf{V}^*$ and $L' \in \mathbf{V}^*$, then $\mathbf{V}^*$ may be made into a vector space with vector addition and scalar multiplication defined by:

$$(L + L')\,(\boldsymbol{\xi}) = L\,(\boldsymbol{\xi}) + L'\,(\boldsymbol{\xi})\,, \text{ and } (aL)\,(\boldsymbol{\xi}) = a\,[L\,(\boldsymbol{\xi})]\,,$$

respectively, or more succinctly, by

$$(aL + bL')\,(\boldsymbol{\xi}) = a\,[L\,(\boldsymbol{\xi})] + b\,[L'\,(\boldsymbol{\xi})]\,, \tag{4.3}$$

where a and b are arbitrary scalars. The reader is asked in PROBLEM I.4 to show that the dual vector space $\mathbf{V}^*$ has the same dimension as $\mathbf{V}$ i.e., that if any $\dim \mathbf{V} = n$, then $\dim \mathbf{V}^* = \dim \mathbf{V} = n$. Introducing n linear functionals $\{f^1, f^2, \cdots, f^n\}$ in $\mathbf{V}^*$, defined by the following action on elements of a basis $\{\mathbf{e}_j\}$ of $\mathbf{V}$:

$$f^i(\mathbf{e}_j) = \delta^i{}_j, \tag{4.4}$$

it can be shown that the set $\{f^i\}$ is a basis of $\mathbf{V}^*$. This particularly important basis of $\mathbf{V}^*$ is referred to as the basis *dual* to $\{\mathbf{e}_j\}$ or simply the *dual basis.* Since both $\mathbf{V}$ and $\mathbf{V}^*$ are real vector spaces of the same dimension, as such they are *isomorphic*, i.e. they have the *same* algebraic structure, and we express this by writing $\mathbf{V} \approx \mathbf{V}^*$. If one continues the above process and considers the set of linear functionals defined on $\mathbf{V}^*$, we obtain a vector space $\mathbf{V}^{**}$ which again has dimension n, and as before $\mathbf{V}^{**} \approx \mathbf{V}^* \approx \mathbf{V}$. The isomorphism $\mathbf{V} \approx \mathbf{V}^*$ as defined by (4.3) is clearly basis-dependent, however, it can be shown (the proof is non-trivial) that $\mathbf{V}^{**} \approx \mathbf{V}$ is basis-independent. Hence, we may regard $\mathbf{V}^{**}$ and $\mathbf{V}$ as being essentially identical. Thus, given a finite-dimensional vector space $\mathbf{V}$, the dual space $\mathbf{V}^*$ is the only other non-trivial vector space associated with it.

Relative to $\{\mathbf{e}_i\}$ a vector $\boldsymbol{\xi}$ of $\mathbf{V}$ admits the representation

$$\boldsymbol{\xi} = \xi^i \mathbf{e}_i \tag{4.5}$$

while the corresponding vector of $\mathbf{V}^*$, which we denote by ξ^* and call a *covector* for obvious reasons, has the representation

$$\xi^* = \xi_i f^i \tag{4.6}$$

relative to $\{f^i\}$. These equations exhibit a striking similarity with our coordinate-based expressions (3.15) and (3.19), and this suggests that vectors having contravariant components belong to $\mathbf{V}$, while those having covariant components belong to $\mathbf{V}^*$. This furnishes an algebraic *hint* of why vectors in $\mathbf{V}$ and $\mathbf{V}^*$ have *different* kinds of components associated with them. However, it must be kept in mind that the indices on the components of $\boldsymbol{\xi}$ and ξ^* appearing in (4.5) and (4.6) are *abstract* *indices* which also enumerate the basis elements appearing in $\{\mathbf{e}_i\}$ and $\{f^i\}$. These indices need not be identical with either the coordinate-based or the congruence indices discussed in Section 1.3. For the moment it is convenient to retain this degree of generality, and we will continue to do so in the remainder of this section.

By virtue of (4.3), (4.5) and (4.6) we may exhibit explicit functional-theoretic expressions for the abstract components ξ^i and ξ_i. These are obtained by considering the action of f^i on $\boldsymbol{\xi}$, i.e.

$$f^i(\boldsymbol{\xi}) = f^i\left(\xi^j \mathbf{e}_j\right) = \xi^j f^i(\mathbf{e}_j) = \xi^j \delta^i{}_j$$

which yields

$$f^i(\boldsymbol{\xi}) = \xi^i; \tag{4.7}$$

and similarly the action of ξ^* on $\mathbf{e}_i$, i.e.

$$\xi^*(\mathbf{e}_i) = \xi_j f^j(\mathbf{e}_i) = \xi_j \delta^j{}_i$$

which gives

$$\xi^*(\mathbf{e}_i) = \xi_i. \tag{4.8}$$

These equations conclusively show that there is no reason why ξ^i and ξ_i need to be the same quantities since they are defined by different expressions. It also indicates that our rather artificial choice of writing the indices on the basis vectors of $\mathbf{V}$ and $\mathbf{V}^*$ as subscripts and superscripts respectively (and possibly also attaching an asterisk to the ξ_i appearing in (4.6) if desired) would not have resulted in (4.7) and (4.8) giving the same expressions for the components of $\boldsymbol{\xi}$ and ξ^*. Hence, our choice was justified since the components in question are *different*!

Further insight into the nature of linear functionals and how they produce a *coupling* of the spaces $\mathbf{V}$ and $\mathbf{V}^*$ may be given by considering the action of ξ^* on a vector $\boldsymbol{\eta}$ of $\mathbf{V}$. By employing the usual properties

$$\begin{aligned} \xi^*(\boldsymbol{\eta}) &= \xi_i f^i(\eta^j \mathbf{e}_j) = \xi_i \eta^j f^i(\mathbf{e}_j) \\ &= \xi_i \eta^j \delta^i{}_j, \text{ viz.} \end{aligned}$$

$$\xi^*(\boldsymbol{\eta}) = \xi_i \eta^i. \tag{4.9}$$

This equation, when the abstract indices are taken to be the coordinate-based indices of the classical tensor calculus, is none other than the usual expression for the inner product $\langle \boldsymbol{\xi}, \boldsymbol{\eta} \rangle$ of a pair of vectors of $\mathbf{V}$. More precisely, in general upon introducing the abstract components $g_{ij} = g_{ji}$ of the metric tensor g, putting

$$\xi_i := g_{ij} \xi^j, \tag{4.10}$$

and (as in tensor calculus) defining

$$\langle \boldsymbol{\xi}, \boldsymbol{\eta} \rangle := g_{ij} \xi^i \eta^j, \tag{4.11}$$

we obtain the relationship

$$\xi^* (\boldsymbol{\eta}) = \langle \boldsymbol{\xi}, \boldsymbol{\eta} \rangle . \tag{4.12}$$

The inner product of a pair of vectors is symmetric, i.e. $\langle \boldsymbol{\xi}, \boldsymbol{\eta} \rangle = \langle \boldsymbol{\eta}, \boldsymbol{\xi} \rangle$, and by (4.12) this requires that

$$\xi^* (\boldsymbol{\eta}) = \boldsymbol{\eta} (\xi^*) \tag{4.13}$$

which may appear — at first glance — to be somewhat surprising. The right-hand side is curious, but readily explained by the isomorphism $\mathbf{V} \approx \mathbf{V}^{**}$ which allows $\boldsymbol{\eta}$ to be replaced by some η^{**}. Hence, $\eta^{**} (\xi^*)$ is a linear functional on $\mathbf{V}^*$ and up to an isomorphism involved in the above replacement of $\boldsymbol{\eta}$ by η^{**} the right-hand side of (4.13) is well-defined.

Equation (4.12) is an important result in that it shows that a linear functional may be identified with an inner product. But, the latter involves a pair of vectors in $\mathbf{V}$, while the former couples vectors in $\mathbf{V}^*$ and $\mathbf{V}$! We have already learned that $\mathbf{V} \approx \mathbf{V}^*$, and the above situation is resolved by (4.9) which suggests that the isomorphism between $\mathbf{V}$ and $\mathbf{V}^*$ is actually supplied by the abstract components g_{ij} of the metric tensor. Likewise the equation

$$\xi^i := g^{ij} \xi_j, \tag{4.14}$$

Which is a mate to (4.10), then suggests that the abstract components g^{ij} yield the inverse isomorphism between $\mathbf{V}^*$ and $\mathbf{V}$.

Both of these suggestions will turn out to be facts, however, they are premature until we can explain what a tensor *is* and how it comes to have components. In particular, why does the metric tensor g have the abstract components g_{ij} and g^{ij}? This will be our next task, and we will essentially restrict our considerations to second order tensors.

Before seeking an abstract definition of a tensor, it is useful to emphasize the corresponding abstract definition of a vector! This was mentioned at the beginning of our technical discussion in this section, but it bears repeating. A 'vector' is an element of a vector space $\mathbf{V}$, or $\mathbf{V}^*$, for which the operations involved in forming the expressions $a\boldsymbol{\xi}+b\boldsymbol{\eta}$, or $a\xi^* + b\eta^*$, where a, b are scalars, are *defined*! While these expressions are *conveniently* written in terms of components relative to *some* choice of basis, e.g. for *abstract components* we have

$$a\boldsymbol{\xi} + b\boldsymbol{\eta} \text{ in } \mathbf{V} \longleftrightarrow a\xi^i + b\eta^i \text{ relative to } \{\mathbf{e}_i\},$$

$$a\xi^* + b\eta^* \text{ in } \mathbf{V}^* \longleftrightarrow a\xi_i + b\eta_i \text{ relative to } \left\{f^i\right\};$$

these expressions *per se* are meaningful *independent* of $\{\mathbf{e}_i\}$ and $\{f^i\}$ respectively. Hence, the spaces $\mathbf{V}$ and $\mathbf{V}^*$ *per se* are *not basis-dependent.* This fact is often misunderstood due to the demoralizing influence of ordinary Gibbsian vector analysis in $\mathbf{E}_3$, and classical tensor calculus in more general spaces. Relative to the former one is *ab-initio* taught to think in terms of directed line segments (which fail to include bound vectors, or even — strictly speaking — vector fields), and to *automatically* resolve vectors into their components along the Cartesian coordinate axes, i.e. in terms of the very special constant Cartesian basis $\left\{\widehat{\imath}, \widehat{\jmath}, \widehat{k}\right\}$. The situation in classical tensor calculus is more general since more general coordinate systems are employed, but still vectors are automatically identified with their contravariant/covariant coordinate-based components and their corresponding coordinate-based bases are never mentioned. Actually, this is not quite true, the transformation laws:

$$\frac{\partial}{\partial x^r} \mapsto \frac{\partial}{\partial \overline{x}^r} = \frac{\partial x^p}{\partial \overline{x}^r}\frac{\partial}{\partial x^p} \tag{4.15}$$

$$dx^r \mapsto d\overline{x}^r = \frac{\partial \overline{x}^r}{\partial x^q} dx^q \tag{4.16}$$

under the change of coordinates $x^r \mapsto \overline{x}^r$ *are cited*, but it is not emphasized that

$$\left\{ \frac{\partial}{\partial x^r} \right\}_{r=1}^{3} \quad \text{is a } \textit{basis} \text{ of } \mathbf{T}_P,$$

and

$$\{dx^r\}_{r=1}^{3} \text{ is the } \textit{dual basis} \text{ of } \mathbf{T}_P^*$$

(see equations (3.15) and (3.19) with λ^*, μ^*, ν^* being written in place of θ_1, θ_2, θ_3; and of course with $\mathbf{T}_P$ and $\mathbf{T}_P^*$ being identified with $\mathbf{V}$ and $\mathbf{V}^*$). In other words, the abstract approach makes a sharp distinction between vectors and their components, which depend upon the choice of a basis in $\mathbf{V}$ and $\mathbf{V}^*$.

Let $\{\mathbf{e}_i\}$ be a basis of $\mathbf{V}$. Then any other basis $\{\overline{\mathbf{e}}_i\}$ of $\mathbf{V}$ is related to $\{\mathbf{e}_i\}$ by a non-singular linear transformation $A : \mathbf{V} \to \mathbf{V}$. Explicitly we write this change as

$$\mathbf{e}_i \mapsto \overline{\mathbf{e}}_i = A_i^k \mathbf{e}_k \tag{4.17}$$

where $\left\| A_i^k \right\|$ is a non-singular square matrix, i.e. $\det \left\| A_i^k \right\| \neq 0$. Denoting the inverse matrix of $\left\| A_i^k \right\|$ by $\left\| \breve{A}_i^k \right\|$ we have the analogous result

$$\overline{\mathbf{e}}_i \mapsto \mathbf{e}_i = \breve{A}_i^k \mathbf{e}_k \tag{4.18}$$

where

$$A_j^i \breve{A}_k^j = \breve{A}_j^i A_k^j = \delta_k^i. \tag{4.19}$$

Upon 'barring' (4.5) and remembering that since $\boldsymbol{\xi}$ itself is basis independent, i.e. $\overline{\boldsymbol{\xi}} = \boldsymbol{\xi}$, we have

$$\overline{\xi}^i \overline{\mathbf{e}}_i = \overline{\xi}^i A_i^k \mathbf{e}_k = \xi^i \mathbf{e}_i$$

which leads to the equation

$$\left(A_i^k \overline{\xi}^i - \delta_i^k \xi^i \right) \mathbf{e}_k = \mathbf{0}.$$

However, since the $\mathbf{e}_k$ are linearly independent, it follows that

$$\xi^k = A^k_i \overline{\xi}^i$$

so upon multiplying by $\breve{A}^j_k$ and using (4.19) we obtain, after renaming the unsummed index:

$$\overline{\xi}^i = \breve{A}^i_k \xi^k. \tag{4.20}$$

Thus, under a change of basis (4.18) the components ξ^i transform in a manner 'inverse to' the basis vectors.

By (4.3) the change (4.18) must certainly induce a change in the dual basis of $\mathbf{V}^*$. Writing this as

$$f^i \mapsto \overline{f}^i = B^i_k f^k,$$

the matrix $\|B^i{}_k\|$ must be non-singular, and we seek how the entries $B^i{}_k$ are related to $A^i{}_k$. Clearly,

$$\begin{aligned} \overline{f}^i(\overline{\mathbf{e}}_j) &= \overline{f}^i\left(A^k_j \mathbf{e}_k\right) = B^i_\ell f^\ell\left(A^k_j \mathbf{e}_k\right) \\ &= B^i_\ell A^k_j f^\ell(\mathbf{e}_k) = B^i_\ell A^k_j \delta^\ell_k \\ &= B^i_k A^k_j = \delta^i_j, \end{aligned}$$

and hence $B^i_k = \breve{A}^i_k$ by (4.19). Hence, under $A : \mathbf{V} \to \mathbf{V}$ we have the corresponding transformation laws in $\mathbf{V}^*$:

$$f^i \mapsto \overline{f}^i = \breve{A}^i_k f^k \tag{4.21}$$

and

$$\xi_i \mapsto \overline{\xi}_i = A^k_i \xi_k. \tag{4.22}$$

By (4.5), vectors belonging to $\mathbf{V}$ have abstract 'contravariant' components associated with them, while by (4.6) those of $\mathbf{V}^*$ have 'covariant' components. It is natural to expect that objects having pairs of such components, i.e. a pair of contravariant or covariant components, be obtained by forming an appropriate product of $\mathbf{V}$ with $\mathbf{V}$, or $\mathbf{V}^*$ with $\mathbf{V}^*$; or in the case of 'mixed' components, of a product of $\mathbf{V}$ with $\mathbf{V}^*$ or $\mathbf{V}^*$ with $\mathbf{V}$. This turns out to be precisely what happens. However, the rigorous procedure (there is more than one of them) for carrying out the construction of such a product is a delicate and intricate matter. For our purposes, we will regard the product space in question to be generated by forming a formal product, called the *tensor product of vectors* taken from the given space. The resulting space is called a *tensor space*, or the *tensor product* of the indicated spaces, and is sometimes said to be obtained by 'tensoring' the spaces.

For simplicity we will first consider the product of $\mathbf{V}$ with itself, and then the other cases involving $\mathbf{V}^*$ will be considered. Suppose it is known, i.e. it has been proven, that the resulting product space, denoted by $\mathbf{V} \otimes \mathbf{V}$ (or merely $\otimes^2\mathbf{V}$) is a vector space spanned by elements of the form $\boldsymbol{\xi}\otimes\boldsymbol{\eta}$ where $\boldsymbol{\xi}$ and $\boldsymbol{\eta}$ belong to $\mathbf{V}$ and this product is required to have the properties:

1^0 $\boldsymbol{\xi} \otimes (\boldsymbol{\eta} + \boldsymbol{\zeta}) := \boldsymbol{\xi} \otimes \boldsymbol{\eta} + \boldsymbol{\xi} \otimes \boldsymbol{\zeta},$
$\quad (\boldsymbol{\xi} + \boldsymbol{\eta}) \otimes \zeta := \boldsymbol{\xi} \otimes \boldsymbol{\zeta} + \boldsymbol{\eta} \otimes \boldsymbol{\zeta};$

2^0 $(a\boldsymbol{\xi}) \otimes \boldsymbol{\eta} := \boldsymbol{\xi} \otimes (a\boldsymbol{\eta}) = a(\boldsymbol{\xi} \otimes \boldsymbol{\eta})$
where a is a scalar; and

3^0 $\{\mathbf{e}_i \otimes \mathbf{e}_j\}$ is a basis for $\otimes^2\mathbf{V}$.

Upon examining the basis representation of $\boldsymbol{\xi}\otimes\boldsymbol{\eta}$, we immediately have that

$$\boldsymbol{\xi} \otimes \boldsymbol{\eta} = \xi^i \eta^j \mathbf{e}_i \otimes \mathbf{e}_j, \tag{4.23}$$

and if $\boldsymbol{\zeta}$ and $\boldsymbol{\tau}$ also belong to $\mathbf{V}$ we have

$$\boldsymbol{\zeta} \otimes \boldsymbol{\tau} = \zeta^i \tau^j \mathbf{e}_i \otimes \mathbf{e}_j. \tag{4.24}$$

Then by 1^0-3^0, upon making the basis-dependent definition

$$a(\boldsymbol{\xi} \otimes \boldsymbol{\eta}) + b(\boldsymbol{\zeta} \otimes \boldsymbol{\tau}) := \left(a\xi^i\eta^j + b\zeta^i\tau^j\right) \mathbf{e}_i \otimes \mathbf{e}_j,$$

the vector space structure of $\otimes^2\mathbf{V}$ is clear.

Moreover, if $\dim \mathbf{V} = n$, then it follows that

$$\dim\left(\otimes^2\mathbf{V}\right) = (\dim \mathbf{V})^2 = n^2.$$

In general, an element of $\otimes^2\mathbf{V}$ is not merely $\boldsymbol{\xi}\otimes\boldsymbol{\eta}$, but rather a finite sum of elements of this form. For example if $n = 3$ and $\boldsymbol{\xi}$, $\boldsymbol{\eta}$ and $\boldsymbol{\zeta}$ are linearly-independent,

$$\begin{aligned} &c_1\boldsymbol{\xi}\otimes\boldsymbol{\xi} + c_2\boldsymbol{\xi}\otimes\boldsymbol{\eta} + c_3\boldsymbol{\xi}\otimes\boldsymbol{\zeta} \\ &+c_4\boldsymbol{\eta}\otimes\boldsymbol{\xi} + c_5\boldsymbol{\eta}\otimes\boldsymbol{\eta} + c_6\boldsymbol{\eta}\otimes\boldsymbol{\zeta} \\ &+c_7\boldsymbol{\zeta}\otimes\boldsymbol{\xi} + c_8\boldsymbol{\zeta}\otimes\boldsymbol{\eta} + c_9\boldsymbol{\zeta}\otimes\boldsymbol{\zeta} \end{aligned} \tag{4.25}$$

where $c_1, \cdots, c_9$ are scalars, is the general form of an element T of $\otimes^2\mathbf{V}$. More concisely, we write simply

$$T = \tau^{ij}\mathbf{e}_i \otimes \mathbf{e}_j. \tag{4.26}$$

Likewise, we have the corresponding expressions:

$$T = \tau_{ij} f^i \otimes f^j, \tag{4.27}$$

$$T = \tau^i{}_j \mathbf{e}_i \otimes f^j, \tag{4.28}$$

$$T = \tau_i{}^j f^i \otimes \mathbf{e}_j, \tag{4.29}$$

when T belongs to $\otimes^2\mathbf{V}$, $\mathbf{V}\otimes\mathbf{V}^*$, and $\mathbf{V}^*\otimes\mathbf{V}$ respectively. It can be shown that $\mathbf{V}\otimes\mathbf{V}^* \approx \mathbf{V}^*\otimes\mathbf{V}$ so the latter two expressions of T may be regarded as being identical, and essentially depend on whether the contravariant or covariant index is written in the first or second position on τ. We choose the first alternate and take (4.28) as the mode of writing T in this mixed case.

In each of these representations T as an element of the tensor space does not *per se* depend on the choice of basis in $\mathbf{V}$, or $\mathbf{V}^*$. Hence, upon effecting the changes (4.17) and (4.21) in (4.27)-(4.29) and $T = \overline{T}$ this *forces* the following transformation laws on the 'τ-components':

$$\tau^{ij} \longrightarrow \overline{\tau}^{ij} = \breve{A}^i_k \breve{A}^j_\ell \tau^{k\ell} \tag{4.30}$$

$$\tau_{ij} \longrightarrow \overline{\tau}_{ij} = A^k_i A^\ell_j \tau_{k\ell} \tag{4.31}$$

$$\tau^i{}_j \longrightarrow \overline{\tau}^i{}_j = \breve{A}^i_k A^\ell_j \tau^k{}_\ell \tag{4.32}$$

which are completely analogous to those given in (4.20) and (4.22). Conversely, if we have these transformation formulas, and know (4.17) and (4.21), then in each of the cases (4.30)-(4.31) by virtue of (4.19) we have $\overline{T} = T$.

Finally, by employing the coordinated-based expressions for $\breve{A}^i_k$ and A^k_i, i.e.

$$\left\|\breve{A}^i_k\right\| \longleftrightarrow \left\|\frac{\partial \overline{x}^r}{\partial x^s}\right\|, \qquad \left\|A^k_i\right\| \longleftrightarrow \left\|\frac{\partial x^r}{\partial \overline{x}^s}\right\|;$$

we recover the classical transformation formulas

$$\overline{\xi}^r = \frac{\partial \overline{x}^r}{\partial x^p}\xi^p, \qquad \overline{\xi}_r = \frac{\partial x^q}{\partial \overline{x}^r}\xi_q;$$

and more generally

$$\overline{T}^{rs} = \frac{\partial \overline{x}^r}{\partial x^p}\frac{\partial \overline{x}^s}{\partial x^q}T^{pq} \text{ etc.},$$

which justifies our regarding (4.20), (4.22), (4.30)-(4.32) as being abstract tensor transformation laws for the respective components.

We conclude our excursion into the abstract approach to tensors with another approach which is important in that it reveals how the idea of a pair of vector spaces arises in the theory. This requires the notion of a bilinear functional on a pair of vector spaces. In effect, this is an immediate generalization of a linear functional to a pair of vector arguments. By definition, a *bilinear functional* B is a mapping

$$B : \mathbf{V} \times \mathbf{V} \to \mathbf{R}$$

such that

$$B(a\boldsymbol{\xi} + b\boldsymbol{\eta}, \boldsymbol{\zeta}) = aB(\boldsymbol{\xi}, \boldsymbol{\zeta}) + bB(\boldsymbol{\eta}, \boldsymbol{\zeta})$$

$$B(\boldsymbol{\xi}, a\boldsymbol{\eta} + b\boldsymbol{\zeta}) = aB(\boldsymbol{\xi}, \boldsymbol{\eta}) + bB(\boldsymbol{\xi}, \boldsymbol{\zeta})$$

where a, b are real scalars. Analogous definitions hold for $\mathbf{V}^* \times \mathbf{V}^*, \mathbf{V} \times \mathbf{V}^*$ and $\mathbf{V}^* \times \mathbf{V}$. The value of a bilinear functional is a scalar and when applied to a pair of basis vectors $(\mathbf{e}_i, \mathbf{e}_j)$ of $\mathbf{V} \times \mathbf{V}$ we obtain the scalars β_{ij}, i.e.

$$B(\mathbf{e}_i, \mathbf{e}_j) := \beta_{ij}. \tag{4.33}$$

Thus, in the case of a pair of arbitrary vectors $(\boldsymbol{\xi}, \boldsymbol{\eta})$ of $\mathbf{V} \times \mathbf{V}$ we have

$$\begin{aligned} B(\boldsymbol{\xi}, \boldsymbol{\eta}) &= B(\xi^i \mathbf{e}_i, \eta^j \mathbf{e}_j) \\ &= \xi^i \eta^j B(\mathbf{e}_i, \mathbf{e}_j) \\ &= \xi^i \eta^j \beta_{ij} \end{aligned} \tag{4.34}$$

which is clearly a scalar. Indeed, under the basis change (4.18), we have

$$\beta_{ij} \mapsto \overline{\beta}_{ij} = A_i^k A_j^\ell \beta_{k\ell} \tag{4.35}$$

so (4.34) is an invariant. This suggests taking a second order tensor in the abstract theory to be a bilinear functional, say T, with the components of the tensor being the values that T takes on the pair of basis vectors as in (4.33). This is our second abstract definition of a tensor and the components appearing on the right hand sides of (4.27)-(4.29) are then respectively given by

$$T\left(f^i, f^j\right) := \tau^{ij} \tag{4.36}$$

$$T(\mathbf{e}_i, \mathbf{e}_j) := \tau^{ij} \tag{4.37}$$

$$T\left(f^i, \mathbf{e}_j\right) := \tau^i{}_j. \tag{4.38}$$

Hence employing (4.34) we have

$$T(\boldsymbol{\xi}, \boldsymbol{\eta}) = T\left(\xi^i \mathbf{e}_i, \eta^j \mathbf{e}_j\right) = \xi^i \eta^j T(\mathbf{e}_i, \mathbf{e}_j) = \xi^i \eta^j \tau_{ij}. \tag{4.39}$$

Recall that a linear functional $L : \mathbf{V} \to \mathbf{R}$ yields $\mathbf{V}^*$, and $L : \mathbf{V}^* \to \mathbf{R}$ yields $\mathbf{V}^{**} \cong \mathbf{V}$; so explicitly in terms of the basis functionals $f^i : \mathbf{e}_j \to \delta^i{}_j$ via (4.3), and analogous to (4.13), we have $e_j^{**} = \mathbf{e}_j : f^i \to \delta^i{}_j$. Thus, for the bilinear functionals T defined by $\mathbf{V} \times \mathbf{V} \to \mathbf{R}$ and $\mathbf{V}^* \times \mathbf{V}^* \to \mathbf{R}$ we expect the results to be 'something' involving $\mathbf{V}^*$ and $\mathbf{V}^*$, and $\mathbf{V}$ and $\mathbf{V}$, respectively. This something turns out to be $\mathbf{V}^* \otimes \mathbf{V}^*$ and $\mathbf{V} \otimes \mathbf{V}$ and this identification is the key to defining the tensor products of these spaces. More precisely, for a bilinear functional T we take $\mathbf{V} \times \mathbf{V} \to \mathbf{R}$, $\mathbf{V}^* \times \mathbf{V}^* \to \mathbf{R}$, and $\mathbf{V}^* \times \mathbf{V} \to \mathbf{R}$ to be $\mathbf{V}^* \otimes \mathbf{V}^*$, $\mathbf{V} \otimes \mathbf{V}$ and $\mathbf{V} \otimes \mathbf{V}^*$ respectively. To justify the first of these we generalize (4.3) to

$$f^i \otimes f^j : (\mathbf{e}_k, \mathbf{e}_\ell) \mapsto \delta_k^i \delta_\ell^j, \tag{4.40}$$

and for an arbitrary pair of vectors $(\boldsymbol{\xi}, \boldsymbol{\eta})$ of $\mathbf{V} \times \mathbf{V}$ compute

$$\begin{aligned} T(\boldsymbol{\xi}, \boldsymbol{\eta}) &= \tau_{ij}\delta^i_k\delta^j_\ell\xi^k\eta^\ell \\ &= \tau_{ij}\xi^k\eta^\ell f^i \otimes f^j(\mathbf{e}_k, \mathbf{e}_\ell) \\ &= \tau_{ij} f^i \otimes f^j\left(\xi^k\mathbf{e}_k, \eta^\ell\mathbf{e}_\ell\right) \\ &= \tau_{ij} f^i \otimes f^i(\boldsymbol{\xi}, \boldsymbol{\eta}) \end{aligned}$$

which we may rewrite as

$$\left(T - \tau_{ij} f^i \otimes f^j\right)(\boldsymbol{\xi}, \boldsymbol{\eta}) = \mathbf{0}.$$

But since the pair of vectors in $\mathbf{V} \times \mathbf{V}$ was arbitrary, this expression yields (4.27).

Thus, in conclusion we have two ways of regarding a second order tensor, i.e. either as an element of the tensor product space admitting an expansion of forms (4.26)-(4.29), or as a bilinear functional acting on a product space as just defined. These two approaches are equivalent, and will turn out to be important in Chapter VII when we consider the theory of the Marussi tensor.

There are many elementary books on linear algebra (the theory of vector spaces), however many do not introduce the notion of a tensor. Among those that do include this material, we recommend GEL'FAND (1948), HALMOS (1958), and LICHNEROWICZ (1947). Likewise, the texts by LICHNEROWICZ (1950), WILLMORE (1959) and DEHEUVELS (1993) contain readable introductions to the abstract approach to tensors.

I.5 A Preview of the Leg Calculus

In the previous two sections we have considered three classical approaches to the tensor calculus and seen how the notion of a tensor can be formulated in terms of the theory of an arbitrary finite dimensional real vector space and various kinds of linear-functionals defined on it and its associated spaces. All of these things are usually subsumed — without specific reference to their origin or consideration of their computational convenience — into modern differential geometry. The result is one of the most imposing/impressive theories achieved by twentieth century mathematics and although it is rightfully acclaimed for its conceptual elegance, it is often ill suited for calculations. This is to be expected, since by design it was created to facilitate the interaction/interplay of geometry with other disciplines of mathematics. Indeed, modern mathematics is concerned with structural questions, and seldom turns its attention to the computational requirements of the physical sciences.

The formalism which we call the *leg calculus* is a unification of *all* of the tensor-theoretic methods which we have discussed in this chapter. As such — mathematically speaking — it is not a new discipline but rather a formalism which is designed specifically to provide a convenient methodology for the needs of differential geodesy and geophysics, viz. the Gaussian differential geometry of curves and surfaces in Euclidean 3-space $\mathbf{E}_3$. It is both a synthetic and an analytic calculus whose *raison d'être* is to provide an efficient method for doing calculations in a geopotential field. Both Marussi and Hotine perceived various aspects of it, but neither chose to formalize it into a coherent theory. Perhaps due to their predilection for coordinates, neither felt such a formalism was necessary. But now, with the benefit of hindsight and with the discovery of non-holonomic reference systems by GRAFAREND (1971), (1975) and GROSSMAN (1974), a rethinking of the utility and availability of coordinates as assumed by Marussi and Hotine is mandatory. Indeed, such an analysis is required if we are to understand the Marussi-Hotine approach to geodesy in contemporary language. This demands a mathematical formalism more general than that employed by them in their work, yet one which upon suitable specialization readily reproduces their work.

By definition, a 3-*leg* is a linearly independent system of 'vectors' in $\mathbf{E}_3$, and for convenience we take it to be an orthonormal system. Thus a 'vector' may be either a tangent vector to a congruence of curves (as in the Ricci calculus), or the dual notion of a covector which is a Pfaffian, i.e. an exterior differential 1-form (as in the Cartan calculus). Upon identifying the abstract vector spaces $\mathbf{V}$ and $\mathbf{V}^*$ with the tangent and cotangent spaces $\mathbf{T}_P$ and $\mathbf{T}_P^*$ respectively we obtain a theory which is specialized to $\mathbf{E}_3$. The spaces $\mathbf{T}_P$ and $\mathbf{T}_P^*$ are isomorphic and we may regard both to arise as special realizations of a general notion of a leg space. In other words, at each point of $\mathbf{E}_3$ we have a leg space which we may choose to be either $\mathbf{T}_P$ or $\mathbf{T}^*_P$. The former is a *vectorial* 3-*leg* of tangent vectors to curves in $\mathbf{E}_3$, while the latter is a *Pfaffian* 3-*leg* consisting of exterior differential 1-forms. The first is obviously geometrical while the second is especially well-suited to handling differential systems which naturally occur in differential geodesy. The results obtained in one approach may be readily carried over to the other approach, and thus the resulting unified methodology is arguably more powerful and flexible than either of the separate methodologies. This unified theory will be presented in Chapter IV, and subsequently adapted to Gaussian differential geometry in Chapter V. As preliminaries the Ricci calculus and Cartan calculus will be reviewed in Chapters II and III respectively. These reviews are not all inclusive, but are designed to exhibit those properties which will be needed in the leg calculus.

PROBLEMS FOR CHAPTER I

I.1 Show that the following are real vector spaces when addition and multiplication by a real scalar, denoted by c, are defined as follows:

i) vectors $\boldsymbol{\xi}$, $\boldsymbol{\eta}$ in an Euclidean 3-space $\mathbf{E}_3$ written as 3-tuples (ξ^1, ξ^2, ξ^3) etc., when

$$\begin{aligned}(\xi^1,\xi^2,\xi^3)+(\eta^1,\eta^2,\eta^3) &:= (\xi^1+\eta^1,\xi^2+\eta^2,\xi^3+\eta^3)\\ c(\xi^1,\xi^2,\xi^3) &:= (c\xi^1,c\xi^2,c\xi^3);\end{aligned}$$

ii) polynomials in the indeterminate t and of the same degree m, i.e.

$$p(t) = a_0 + a_1 t + \cdots + a_m t^m \text{ etc.,}$$

when

$$\begin{aligned}p(t)+q(t) &:= (a_0+b_0)+\cdots+(a_m+b_m)t^m\\ cp(t) &:= ca_0+\cdots+ca_m t^m;\end{aligned}$$

iii) $n \times n$ matrices $\|a_{ij}\|$ etc., when

$$\|a_{ij}\| + \|b_{ij}\| := \|a_{ij} + b_{ij}\|$$
$$c\,\|a_{ij}\| := \|c a_{ij}\|\,.$$

I.2 What is the dimension of each of the vector spaces in PROBLEM I.1?

I.3 Show that the following operations L define linear functionals on the indicated vector space $\mathbf{V}$:

i) $\mathbf{V}$ consists of continuous functions f on the closed interval $[0,1]$;

$$L(f) := \int_0^1 f(x)\,dx;$$

ii) $\mathbf{V}$ as in i) and

$$L(f) := f\left(x^0\right)$$

i.e. evaluation of the function f at an arbitrary point x^0 of $[0,1]$;

iii) $\mathbf{V}$ differentiable functions f on $[0,1]$ with x^0 an arbitrary point of $[0,1]$

$$L(f) := \left(\frac{df}{dx}\right)_{x^0}$$

i.e. evaluation of the derivative at x^0;

iv) $\mathbf{V}$ a 3-dimensional vector space having an inner, or "dot", product defined on it,

$$L(\mathbf{v}) := \mathbf{v} \cdot \mathbf{v}^0$$

where $\mathbf{v}^0$ is an arbitrary fixed vector of $\mathbf{V}$;

v) $\mathbf{V}$ the vector space of real $n \times n$ matrices $A := \|a_{ij}\|$,

$$L(A) := \sum_i a_{ii}$$

i.e. the trace of the matrix A.

I.4 Give a proof that if **V** is an n-dimensional vector space, then its dual (or conjugate) space $\mathbf{V}^*$ is also n-dimensional.

I.5 Show that if the bases $\{\mathbf{e}_i\}$ and $\{\overline{\mathbf{e}}_i\}$ of **V** are related by a non-singular linear transformation A as discussed in Section I.4, then if

$$\mathbf{e}_i \mapsto \overline{\mathbf{e}}_i = A^k_i \mathbf{e}_k \qquad (*)$$

and

$$\overline{\mathbf{e}}_i \mapsto \mathbf{e}_i = \breve{A}^k_i \overline{\mathbf{e}}_k \qquad (**)$$

where

$$A^i_j \breve{A}^j_k = \breve{A}^i_j A^j_k = \delta^i_k,$$

then ξ^i and $\overline{\xi}^i$ are related by

$$\xi^i \mapsto \overline{\xi}^i = \breve{A}^i_k \xi^k.$$

I.6 As in PROBLEM I.5 show that if the bases $\{f^i\}$ and $\left\{\overline{f}^i\right\}$ of $\mathbf{V}^*$ are related by

$$f^i \mapsto \overline{f}^i = B^i_k f^k$$

under a non-singular transformation A of **V**, then

$$B^i_k = \breve{A}^i_k,$$

and consequently

$$\xi_i \mapsto \overline{\xi}_i = A^k_i \xi_k.$$

I.7 Derive the Hotine-transformation formulas (3.21).

ZUND and WILKES (1988)

I.8 Observing that under a coordinate transformation $x^r \to \overline{x}^r$, $\lambda_r \to \overline{\lambda}_r = \dfrac{\partial x^p}{\partial \overline{x}^r} \lambda_p$ etc. for the tangential 3-leg vectors $\boldsymbol{\lambda}$,

$\boldsymbol{\mu}$, $\boldsymbol{\nu}$, show that in a 3-dimensional space if any vector $\boldsymbol{\xi}$ is expressed in the form

$$\boldsymbol{\xi} = a\boldsymbol{\lambda} + b\boldsymbol{\mu} + c\boldsymbol{\nu}$$

where a, b, and c are scalars, then one has the transformation laws

$$\overline{\xi}^r = \frac{\partial \overline{x}^r}{\partial x^p}\xi^p, \qquad \overline{\xi}_r = \frac{\partial x^q}{\partial \overline{x}^r}\xi_q.$$

I.9 Peruse DUHEM (1914), and, in particular, read Chapter IV of Part I. NB. Duhem's style is polemical and his arguments are somewhat outdated — essentially Chapter IV is an essay on Maxwellian electrodynamics — nevertheless, it is well-done and will challenge the reader to consider *how* a physical theory should be formulated.

I.10 Read HOTINE (1964) and MARUSSI (1959) noting the differences in style and viewpoint. Can one see a connection — in the mathematical sense — in the kinds of minds discussed by DUHEM (1914) given in PROBLEM I.9?

II

The Ricci Calculus

II.1 Introduction

We now begin our presentation of the Ricci calculus, or the calculus of congruences of curves, which will form one of the essential ingredients of the leg calculus. Our exposition closely follows the classical one given by Ricci which is recounted in the books of LEVI-CIVITÀ (1925), EISENHART (1926), and WEATHERBURN (1938), except for our use of language and some organizational changes. These changes are intended to ease the inclusion of the material into the leg calculus in CHAPTER IV. Essentially, relative to terminology, the usage in the classical literature is not uniform, and a system of orthonormal unit tangent vectors, which for us is a *vectorial* n-leg $\{\boldsymbol{\lambda}_a\}$, was called a *pyramid* (an n-dimensional generalization of a trihedron in 3-dimensions) by Levi-Cività, and an *orthogonal ennuple* by Eisenhart and Weatherburn. We regard such terminology as being obsolete and inferior to that of an n-leg which explicitly exhibits the dimensionality of the system. Likewise, these authors, following Ricci, set out the theory in an n-dimensional Riemannian space $\mathbf{V}_n$, while for our purposes we need only the case $n = 3$. Hence, we will consider only the case of a *curved* Riemannian $\mathbf{V}_3$ and ultimately specialize it to a *flat* Euclidean $\mathbf{E}_3$. Later, in CHAPTER V we will consider the case of a vectorial 2-leg, in which the curved Riemannian $\mathbf{V}_2$ is identified with a surface lying in an $\mathbf{E}_3$. The 'leg language' is especially convenient since it not only indicates the dimensionality of $\{\boldsymbol{\lambda}_a\}$, i.e. the range of the leg index a, but it readily allows one to specialize down from a *space-based* 3-leg to a *surface-based* 2-leg. Hotine in his treatise occasionally referred to his spatial system $\{\boldsymbol{\lambda}, \boldsymbol{\mu}, \boldsymbol{\nu}\}$ as a *triad*, which is an unhappy choice since then —

presumably — the analogous 2-dimensional system, i.e. $\{\boldsymbol{\lambda}, \boldsymbol{\mu}\}$, would be a *dyad* which is likely to be confused with the completely different notion of a dyad, or dyadic, in Gibbsian vector analysis. As the reader will note, our 'leg language' neatly avoids this dilemma!

II.2 3-Leg Algebra

We denote by $\{\boldsymbol{\lambda}_a\}_{a=1}^{3}$ an orthonormal set of vectors in $\mathbf{V}_3$ which respectively have the *contravariant* components $\lambda_a{}^r$ and the *covariant* components λ_{ar}. *Leg indices* will always be denoted by the initial Latin letters: a, b, c, d, e, f, g, h and *coordinate-based indices* by the latter Latin letters: m, n, p, q, r, s, t etc., and both kinds of indices range over the values 1, 2, 3. The leg indices always occur as subscripts and repeated indices obey the Einstein summation convention, which also holds (as in the tensor calculus) for repeated coordinate-based indices appearing as superscripts and subscripts. In the rare cases when repeated leg indices are not summed we write NS *before* the equation. To avoid the possible confusion of a numerical subscript on $\boldsymbol{\lambda}$ with a covariant index, it is convenient to write — following Hotine's practice of naming the individual vectors — the following *canonical identification*:

$$\{\lambda_a\} := \{\boldsymbol{\lambda}, \boldsymbol{\mu}, \boldsymbol{\nu}\}\,, \tag{2.1}$$

i.e.

$$\boldsymbol{\lambda}_1 := \boldsymbol{\lambda}, \quad \boldsymbol{\lambda}_2 := \boldsymbol{\mu}, \quad \boldsymbol{\lambda}_3 := \boldsymbol{\nu}. \tag{2.2}$$

The vectors $\{\boldsymbol{\lambda}_a\}$ are regarded as generating the tangent space $\mathbf{T}_P$ to $\mathbf{V}_3$ at a point P and, hence, can be written as linear combinations of the coordinate-related basis vectors as indicated in SECTION I-4, i.e.

$$\boldsymbol{\lambda}_a = \lambda_a{}^r \frac{\partial}{\partial x^r} \tag{2.3}$$

or upon using (2.2), we have

$$\boldsymbol{\lambda} = \lambda^r \frac{\partial}{\partial x^r}, \quad \boldsymbol{\mu} = \mu^r \frac{\partial}{\partial x^r}, \quad \boldsymbol{\nu} = \nu^r \frac{\partial}{\partial x^r}. \tag{2.4}$$

The covariant components of the leg vectors of $\{\boldsymbol{\lambda}_a\}$ are defined by use of the covariant components g_{rs} of the metric tensor, i.e.

$$\lambda_{ar} := g_{rs}\lambda_a{}^s. \tag{2.5}$$

In the Ricci calculus, the dual representation of (2.3) does not occur. However, it will be essentially the point of departure of the Cartan calculus in CHAPTER III. Hence, ultimately the coordinate differentials dx^r will play the same role as in the tensor calculus. Thus, strictly speaking, the Ricci calculus involves only $\mathbf{T}_P$ with *no mention* of its dual (cotangent) space $\mathbf{T}_P^*$.

The orthonormality conditions on the leg vectors of $\{\boldsymbol{\lambda}_a\}$ are summarized by the equations

$$\lambda_a{}^r\lambda_{br} = \delta_{ab} \tag{2.6}$$

which are of course equivalent to

$$g_{rs}\lambda_a{}^r\lambda_b{}^s = \delta_{ab}. \tag{2.7}$$

In these equations δ_{ab} is the familiar Kronecker delta, i.e. $\delta_{ab} = 1$ when $a = b$ (unsummed) and $\delta_{ab} = 0$ when $a \neq b$. The case $a \neq b$ yields the condition of orthogonality of a pair of vectors, while $a = b$ (unsummed) requires that each vector is a unit vector.

The equations corresponding to (2.6) and having a pair of summed leg indices are

$$\lambda_a{}^r\lambda_{as} = \delta^r_s. \tag{2.8}$$

Upon using the inverse of (2.5), i.e.,

$$\lambda_a{}^r := g^{rs}\lambda_{as}, \tag{2.9}$$

where the g^{rs} form the inverse matrix of metric tensor components, we have

$$\lambda_{ar}\lambda_{as} = g_{rs} \tag{2.10}$$

and

$$\lambda_a{}^r\lambda_a{}^s = g^{rs}. \tag{2.11}$$

Before commenting on the latter pair of equations, we note that (2.6) and (2.8) taken together are *analogous* to the algebraic expressions — written in terms of abstract indices —

$$a^i_k \breve{a}^k_j = \delta^i_j$$

$$a^k_i \breve{a}^j_k = \delta^j_i .$$

These state that if the quantities a^i_k are entries in a non-singular matrix, then $\breve{a}^i_k$ are the entries in the inverse matrix. Moreover, these equations are in 'correct' matrix multiplication form if we regard the superscripts as row labels and the subscripts as column labels. Clearly, (2.6) and (2.8) are not in 'correct' matrix multiplication form since their leg indices *always* occur at the same level. This turns out to be a minor inconvenience, one which can be remedied if we make the effort to exhibit the quantities $\lambda_a{}^r$ and λ_{ar} 'correctly' in 3×3 matrices and introduce a transpose operation into the matrix equations corresponding to (2.6) and (2.8). At the present time nothing is to be gained by doing this (we will do so in CHAPTER X) and we will not attempt to put such leg equations in matrix form. However, it is nevertheless convenient to think of (2.6) and (2.8) as essentially indicating the mutually inverse character of the quantities $\lambda_a{}^r$ and λ_{as}.

Equation (2.8), when written out using the identification (2.2), yields a leg expression for the Kronecker delta:

$$\lambda^r \lambda_s + \mu^r \mu_s + \nu^r \nu_s = \delta^r_s . \tag{2.12}$$

(see [2.07]). The corresponding equations (2.10) and (2.11) then provide leg representations for the components of the metric tensor

$$g_{rs} = \lambda_r \lambda_s + \mu_r \mu_s + \nu_r \nu_s , \tag{2.13}$$

$$g^{rs} = \lambda^r \lambda^s + \mu^r \mu^s + \nu^r \nu^s \tag{2.14}$$

(see [2.08] and [2.09]). It should be noted that the derivations of the indicated equations in Hotine's treatise are given for a Cartesian coordinate system in $\mathbf{E}_3$, whereas our approach is more general and holds for an arbitrary (orthonormal) 3-leg in $\mathbf{V}_3$. Equation (2.12) also follows from the Hotine transformation formulas I-(3.21) upon taking $\overline{x}^r = x^s$ or $x^r = \overline{x}^s$. We suspect that this is how he first came upon his remarkable formulas.

The above three equations represent the fundamental idea of the leg calculus, viz. to systematically resolve all the expressions occurring in tensor calculus in terms of (orthonormal) leg systems. In this regard (2.7) has a special significance. The left hand side of it produces quantities, say g_{ab}, which are the leg-components of the metric tensor. But the right-hand side shows that by virtue of the orthogonality of the leg vectors that

$$g_{ab} = \delta_{ab}. \tag{2.15}$$

Hence, in terms of the leg components — conveniently displayed in a 3×3 matrix — we have

$$\|g_{ab}\| = \begin{Vmatrix} 1 & 0 & 0 \\ 0 & 1 & 0 \\ 0 & 0 & 1 \end{Vmatrix} = \left\|g^{ab}\right\| = \|\delta^a_b\|\,. \tag{2.16}$$

This explains why in terms of our orthonormal vectorial 3-leg, we make no distinction between 'contravariant' and 'covariant' leg-indices. Hence, we may write — without loss of generality — all leg indices as subscripts!

Likewise the leg representation of any vector $\boldsymbol{\xi}$ in $\mathbf{T}_P$ is given by

$$\boldsymbol{\xi} = a\boldsymbol{\lambda} + b\boldsymbol{\mu} + c\boldsymbol{\nu} \tag{2.17}$$

where a, b, c are scalars, or more precisely as

$$\begin{aligned} \xi^r &= \xi_a \lambda_a^{\ r} \\ \xi_r &= \xi_a \lambda_{ar} \end{aligned} \tag{2.18}$$

where the ξ_a are the same leg components in both expressions given in (2.18). These expressions are obtained by using the components of the metric tensor to raise or lower the coordinate-based index as in (2.5) and (2.9). An immediate question arises as how to avoid confusing the numerical indices. In practice this problem does not arise since the context will explicitly indicate which type of index is being employed. However, if necessary one can underline a leg index, i.e. $\lambda_{\underline{1}}$, or put a 'hat' on it, viz. $\lambda_{\hat{1}}$. But this is seldom necessary. In writing out the summation in (2.18) only leg indices occur, whereas upon displaying the components in (2.12) *only* coordinate-based indices occur. Our use of the canonical identification (2.2) also tends to minimize the appearance of leg indices, and later in the leg calculus the use of such indices will be rare.

The equations inverse to (2.18) are given by

$$\xi_a := \lambda_{ar}\xi^r = \lambda_a{}^r\xi_r. \tag{2.19}$$

Finally, we note that for a general second order tensor T, in general, we have the leg representations:

$$T_{rs} = T_{ab}\lambda_{ar}\lambda_{bs}, \tag{2.20}$$

and

$$T^{rs} = T_{ab}\lambda_a{}^r\lambda_b{}^s, \tag{2.21}$$

and

$$T_{ab} := \lambda_a{}^r\lambda_b{}^sT_{rs} = \lambda_{ar}\lambda_{bs}T^{rs}. \tag{2.22}$$

In these expressions the leg components T_{ab} are called *leg coefficients* of T, and these — in contrast to those in (2.13) and (2.14) — need not be constants. An analogous remark of course holds for the ξ_a in (2.18) and (2.19).

II.3 Congruences

Following Ricci, the $\{\boldsymbol{\lambda}_a\}$ are regarded as unit tangent vectors to a set of congruences of curves in $\mathbf{V}_3$. These are denoted by $\boldsymbol{\Gamma}_a$ where $a = 1, 2, 3$ or more explicitly one also writes

$$\boldsymbol{\Gamma}_a := \boldsymbol{\Gamma}(\boldsymbol{\lambda}_a)$$

to indicate that $\boldsymbol{\lambda}_a$ is the unit tangent vector to $\boldsymbol{\Gamma}_a$. Geometrically one imagines $\mathbf{V}_3$, or at least a portion of it, to be filled with such congruences and for a given value of a there passes one and only one curve of $\boldsymbol{\Gamma}_a$ and that these curves do not intersect one another. Of course, for $a = 1,2,3$ the $\boldsymbol{\Gamma}_1$, $\boldsymbol{\Gamma}_2$ and $\boldsymbol{\Gamma}_3$ intersect each other orthogonally as in the case of the coordinate curves:

$$\begin{array}{llllll} \boldsymbol{\Gamma}_1: & x^1 & = & \text{variable}, \; x^2 & = \; x^3 & = \; \text{constant}; \\ \boldsymbol{\Gamma}_2: & x^2 & = & \text{variable}, \; x^3 & = \; x^1 & = \; \text{constant}; \\ \boldsymbol{\Gamma}_3: & x^3 & = & \text{variable}, \; x^2 & = \; x^2 & = \; \text{constant}; \end{array}$$

of a triply-orthogonal system of coordinates in $\mathbf{V}_3$. However, the use of such congruences does not imply the existence of a triply-orthogonal system of curves, or surfaces in $\mathbf{V}_3$, or that *all* three possible congruences are equally interesting. Physical examples of congruences include the lines of force, bundle of light rays, and the flow lines of a fluid. None of these need fill the entire space, which again emphasizes the local nature of our considerations. In differential geodesy we will generally consider an equipotential surface $\mathbf{S}$, or more precisely a portion of $\mathbf{S}$, having $\boldsymbol{\Gamma}_1 = \boldsymbol{\Gamma}(\boldsymbol{\lambda})$, $\boldsymbol{\Gamma}_2 = \boldsymbol{\Gamma}(\boldsymbol{\mu})$ as tangential congruences of curves to $\mathbf{S}$, and $\boldsymbol{\Gamma}_3 := \boldsymbol{\Gamma}(\boldsymbol{\nu})$ being a normal congruence to $\mathbf{S}$. Geodetically, then $\boldsymbol{\Gamma}_3$ corresponds to the plumb lines of $\mathbf{S}$. Many analytical difficulties, which are substantial and substantive, can be avoided by restricting our considerations to a single surface $\mathbf{S}$ and then taking one of the congruences, say $\boldsymbol{\Gamma}_3$ being normal to $\mathbf{S}$, written $\boldsymbol{\Gamma}_3 \perp \mathbf{S}$, and the tangential congruences $\boldsymbol{\Gamma}_1$ and $\boldsymbol{\Gamma}_2$ being orthogonal to $\boldsymbol{\Gamma}_3$, viz. $\boldsymbol{\Gamma}_3 \perp \boldsymbol{\Gamma}_1$ and $\boldsymbol{\Gamma}_3 \perp \boldsymbol{\Gamma}_2$ with $\boldsymbol{\Gamma}_1 \perp \boldsymbol{\Gamma}_2$. In this case, $\boldsymbol{\Gamma}_3$ is called a *primary congruence*, and $\boldsymbol{\Gamma}_1$ and $\boldsymbol{\Gamma}_2$ are known as *secondary congruences*. This will be the major case investigated in this book.

II.4 The Ricci Coefficients

Since we have now identified the leg vectors $\boldsymbol{\lambda}_a$ with the (unit) tangents to the congruences $\boldsymbol{\Gamma}_a$ it is clear that the geometry of these congruences will be defined by the derivatives of these vectors. These are conveniently described by a set of scalars which we call the *Ricci coefficients* which are defined by

$$\gamma_{abc} := \lambda_{ars} \lambda_b{}^r \lambda_c{}^s \tag{4.1}$$

where following Hotine we denote the covariant derivative by simply adding a covariant coordinate-based index, viz. we write

$$\lambda_{ars} := \lambda_{ar,s} \tag{4.2}$$

with the comma denoting covariant differentiation. Covariant differentiation of (2.6) obviously gives

$$\lambda_{ars}\lambda_b{}^r + \lambda_{ar}\lambda_b{}^r{}_s = 0,$$

viz.

$$\lambda_{ars}\lambda_b{}^r = -\lambda_{brs}\lambda_a{}^r,$$

so upon multiplication with $\lambda_c{}^s$ we obtain

$$\gamma_{abc} = -\gamma_{bac}. \tag{4.3}$$

Hence, in $\mathbf{V}_3$ there are *nine* independent Ricci coefficients. Inverting (4.1) and solving for the λ_{ars} yields the following leg representation

$$\lambda_{ars} = \gamma_{abc}\lambda_{br}\lambda_{cs} \tag{4.4}$$

and upon explicitly expanding the leg indices and using the identification (2.2) we obtain the following leg representations for λ_{rs}, μ_{rs}, and ν_{rs}:

$$\begin{aligned} \lambda_{rs} &= \gamma_{121}\mu_r\lambda_s + \gamma_{122}\mu_r\mu_s + \gamma_{123}\mu_r\nu_s \\ &\quad -\gamma_{311}\nu_r\lambda_s - \gamma_{312}\nu_r\mu_s - \gamma_{313}\nu_r\nu_s, \end{aligned} \tag{4.5}$$

$$\begin{aligned} \mu_{rs} &= -\gamma_{121}\lambda_r\lambda_s - \gamma_{122}\lambda_r\mu_s - \gamma_{123}\lambda_r\nu_s \\ &\quad +\gamma_{231}\nu_r\lambda_s + \gamma_{232}\nu_r\mu_s + \gamma_{233}\nu_r\nu_s, \end{aligned} \tag{4.6}$$

$$\begin{aligned} \nu_{rs} &= \gamma_{311}\lambda_r\lambda_s + \gamma_{312}\lambda_r\mu_s + \gamma_{313}\lambda_r\nu_s \\ &\quad -\gamma_{231}\mu_r\lambda_s - \gamma_{232}\mu_r\mu_s - \gamma_{233}\mu_r\nu_s. \end{aligned} \tag{4.7}$$

In these expressions we have employed the following *canonical cyclic pair enumeration scheme* for the pairs of skew-symmetric indices: 12, 31, 23, and so the nine γ_{abc} may *canonically displayed* as:

$$\begin{matrix} \gamma_{121} & \gamma_{122} & \gamma_{123} \\ \gamma_{311} & \gamma_{312} & \gamma_{313} \\ \gamma_{231} & \gamma_{232} & \gamma_{233}. \end{matrix} \tag{4.8}$$

Ricci interpreted these coefficients as describing rotations about the respective congruences and called them 'rotation coefficients'. However, this interpretation is of little use to us, and hence we prefer the less suggestive name of *Ricci coefficients* (see PROBLEM II.8). MARUSSI (1979), due to the manner in which they occurred in the homographic calculus in the rotational homography, chose to interpret them as the so-called *Pensa coefficients* π_{ab}. If ε_{abc} are the leg components of the permutation tensor, i.e.

$$\varepsilon_{abc} := +1 \text{ or } \varepsilon_{abc} = -1$$

according to whether a, b, c is an even or odd permutation of 1, 2, 3; and $\varepsilon_{abc} := 0$ otherwise, he wrote

$$\gamma_{abc} = \varepsilon_{abf}\pi_{cf} \tag{4.9}$$

or inversely

$$\pi_{ab} = \frac{1}{2}\gamma_{fga}\varepsilon_{fgb} \tag{4.10}$$

where

$$p_a := \pi_{fa}\lambda_f \tag{4.11}$$

is the *Pensa rotation vector.* Then (4.10) gives

$$2\pi_{ab} = \gamma_{12a}\varepsilon_{12b} + \gamma_{31a}\varepsilon_{31b} + \gamma_{23a}\varepsilon_{23b}$$

and consequently upon expansion we obtain

$$\|\pi_{ab}\| = \frac{1}{2}\begin{Vmatrix} \gamma_{231} & \gamma_{311} & \gamma_{121} \\ \gamma_{232} & \gamma_{312} & \gamma_{122} \\ \gamma_{233} & \gamma_{313} & \gamma_{123} \end{Vmatrix} \tag{4.12}$$

which, apart from the unessential numerical factor, is our canonical display (4.8) with an interchange of rows and columns. We will not employ the Pensa approach in our work.

II.5 The Geometry of Congruences

We now summarize the various conditions on the Ricci coefficients and how these successively specialize the congruences of curves $\boldsymbol{\Gamma}_a$ and surfaces $\mathbf{S}_a$ $(a = 1,2,3)$ in $\mathbf{V}_3$. In exhibiting these, it is convenient to employ the $\boldsymbol{\Gamma}_a \perp \mathbf{S}_a$ notation — no summation being implied — introduced at the end of Section 3. None of these conditions are difficult to prove; however, the proofs require some care and preparation which preclude their being included in our presentation. Complete and detailed proofs are given in Chapter VI of WEATHERBURN (1938).

1^0 The *curvatures* χ_a of the congruences $\boldsymbol{\Gamma}_a$ $(a = 1,2,3)$ are given by

$$\chi_1 = \sqrt{(\gamma_{121})^2 + (\gamma_{131})^2} = \sqrt{(\gamma_{121})^2 + (\gamma_{311})^2}, \tag{5.1}$$

$$\chi_2 = \sqrt{(\gamma_{212})^2 + (\gamma_{232})^2} = \sqrt{(\gamma_{122})^2 + (\gamma_{232})^2}, \tag{5.2}$$

$$\chi_3 = \sqrt{(\gamma_{313})^2 + (\gamma_{323})^2} = \sqrt{(\gamma_{313})^2 + (\gamma_{233})^2} \tag{5.3}$$

(the latter expressions are in terms of our canonical enumerations of their γ_{abc}, whereas the former indicate the general pattern);

2^0 $\boldsymbol{\Gamma}_a$ is a geodesic if and only if

$$\text{NS} \qquad \gamma_{aba} = 0 \qquad (a \neq b)\,, \tag{5.4}$$

i.e.

$$\text{for } \boldsymbol{\Gamma}_1 : \gamma_{121} = 0 \text{ and } \gamma_{131} = 0 \text{ (i.e. } \gamma_{311} = 0), \tag{5.5}$$

$$\text{for } \boldsymbol{\Gamma}_2 : \gamma_{212} = 0 \text{ (i.e. } \gamma_{122} = 0) \text{ and } \gamma_{232} = 0, \tag{5.6}$$

$$\text{for } \boldsymbol{\Gamma}_3 : \gamma_{313} = 0 \text{ and } \gamma_{323} = 0 \text{ (i.e. } \gamma_{233} = 0), \tag{5.7}$$

(N.B. these respectively require that $\chi_a = 0$ $(a = 1,2,3)$ and in an $\mathbf{E}_3$ such $\boldsymbol{\Gamma}_a$ $(a = 1,2,3)$ become straight lines);

3^0 the *tendency* of $\boldsymbol{\Gamma}_a$, or $\boldsymbol{\lambda}_a$, in the direction of $\boldsymbol{\Gamma}_b$, or $\boldsymbol{\lambda}_b$, is given by

$$\text{NS} \qquad \gamma_{abb} \qquad (a \neq b) \tag{5.8}$$

(N.B. Since we will be primarily concerned with $\boldsymbol{\Gamma}_3$ its tendencies in the directions of $\boldsymbol{\Gamma}_1$ and $\boldsymbol{\Gamma}_2$ are as follows:

$$\boldsymbol{\Gamma}_3 \textit{ towards } \boldsymbol{\Gamma}_1 : \gamma_{311} \tag{5.9}$$

$$\boldsymbol{\Gamma}_3 \textit{ towards } \boldsymbol{\Gamma}_2 : \gamma_{322} = -\gamma_{232}); \tag{5.10}$$

4⁰ $\boldsymbol{\Gamma}_a$ is *normal* to $\mathbf{S}_a$, i.e. $\boldsymbol{\Gamma}_a \perp \mathbf{S}_a$, if and only if

$$\gamma_{abc} = \gamma_{acb} \qquad (b \neq c)\,; \tag{5.11}$$

5⁰ $\boldsymbol{\Gamma}_b$ and $\boldsymbol{\Gamma}_c$ are *canonical with respect to* $\boldsymbol{\Gamma}_a$ $(b \neq c \neq a)$ if and only if

$$\gamma_{abc} = -\gamma_{acb} \qquad (a, b, c \text{ distinct}); \tag{5.12}$$

(N.B. In the case of $\boldsymbol{\Gamma}_3$, $\boldsymbol{\Gamma}_1$ and $\boldsymbol{\Gamma}_2$ are canonical whenever

$$\gamma_{312} = -\gamma_{321}$$

i.e.

$$\gamma_{312} = \gamma_{231}); \tag{5.13}$$

6⁰ $\boldsymbol{\Gamma}_a$, $\boldsymbol{\Gamma}_b$ and $\boldsymbol{\Gamma}_c$ $(a \neq b \neq c)$ are *normal* respectively to $\mathbf{S}_a$, $\mathbf{S}_b$ and $\mathbf{S}_c$, viz.

$$\boldsymbol{\Gamma}_a \perp \mathbf{S}_a, \quad \boldsymbol{\Gamma}_b \perp \mathbf{S}_b \text{ and } \boldsymbol{\Gamma}_c \perp \mathbf{S}_c,$$

if and only if

$$\gamma_{123} = 0 \tag{5.14}$$

(N.B. In this case $\mathbf{S}_a$, $\mathbf{S}_b$ and $\mathbf{S}_3$ are said to form a *triply-orthogonal system* of surfaces in $\mathbf{V}_3$).

Note that 4⁰ requires

$$\gamma_{abc} = +\gamma_{acb} \qquad (b \neq c) \tag{5.15}$$

while 5⁰ requires that

$$\gamma_{abc} = -\gamma_{acb} \qquad (b \neq c)\,. \tag{5.16}$$

Hence, 4⁰ and 5⁰, taken together demand that

$$\gamma_{abc} = 0 \qquad (\text{for } a \neq b \neq c) \tag{5.17}$$

which is (5.14).

II.6 Calculation of the Ricci Coefficients

In SECTION 5 we have seen that the Ricci coefficients are clearly of importance in describing the geometry of the congruence $\boldsymbol{\Gamma}_a$ ($a = 1, 2, 3$). Unfortunately (4.1) furnishes a very inefficient manner of computing them since explicitly it is

$$\gamma_{abc} = \left(\lambda_{ar;s} - \Gamma^p_{rs}\lambda_{ap}\right)\lambda_b{}^r\lambda_c{}^s \tag{6.1}$$

(where the semicolon denotes partial differentiation), which requires knowing the Christoffel symbols of the second kind, Γ^p_{rs}. Computation of these symbols is straightforward but frequently laborious and time-consuming. However, all of this can be avoided by the following trick involving skew-symmetrizing on the leg indices b and c. On the right-hand side of (4.1) this is clearly equivalent to skew-symmetrizing on the coordinate-based indices r and s. Hence, we have

$$\gamma_{abc} - \gamma_{acb} = \left(\lambda_{ar;s} - \lambda_{as;r}\right)\lambda_b{}^r\lambda_c{}^s \tag{6.2}$$

and *provisionally* we write

$$\gamma_{abc} - \gamma_{acb} := \#_{abc}, \tag{6.3}$$

where obviously

$$\#_{abc} = -\#_{acb}. \tag{6.4}$$

Then upon forming a cyclic permutation of the indices a, b, c in (6.3) we obtain three equations

$$\gamma_{abc} - \gamma_{acb} = \#_{abc},$$
$$\gamma_{bca} - \gamma_{bac} = \#_{bca},$$
$$\gamma_{cab} - \gamma_{cba} = \#_{cab}.$$

Upon adding the first two of these equations and subtracting the third, by virtue of (4.3) we obtain

$$2\gamma_{abc} = \#_{abc} + \#_{bca} - \#_{cab},$$

so that

$$\gamma_{abc} = \frac{1}{2}\left(\#_{abc} + \#_{bca} - \#_{cab}\right). \tag{6.5}$$

This shows that the Ricci coefficients may be computed by means of the nine $\#_{abc}$ *without* knowing the Γ^p_{rs}.

II.7 The Commutativity of the Leg Derivatives

In SECTION I-3 we observed that there was no reason to expect that the directional derivatives would be permutable, i.e. that I-(3.18) would be zero. We now re-examine the situation and obtain an explicit expression for the right-hand side of I-(3.18). By definition, viz. by I-(3.10), we recall that if F is an arbitrary smooth function of the coordinates x^r, then

$$F_{/a} := F_{;r}\lambda_a{}^r. \tag{7.1}$$

Upon computing the leg derivative of this expression we have

$$F_{/a/b} := (F_{,r}\lambda_a{}^r)_{,s}\lambda_b{}^s \tag{7.2}$$

since the partial derivative in (7.1) may be replaced by a covariant derivative. Then evaluation of (7.2) yields (upon omission of the commas à la Hotine)

$$\begin{aligned} F_{/a/b} &= (F_{rs}\lambda_a{}^r + F_r\lambda_a{}^r{}_s)\lambda_b{}^s \\ &= F_{rs}\lambda_a{}^r\lambda_b{}^s + F_r\gamma_{acd}\lambda_c{}^r\lambda_{ds}\lambda_b{}^s \\ &= F_{rs}\lambda_a{}^r\lambda_b{}^s + F_r\gamma_{acd}\lambda_c{}^r\delta_{bd} \\ &= F_{rs}\lambda_a{}^r\lambda_b{}^s + F_{/c}\gamma_{acb} \end{aligned}$$

hence

$$F_{/a/b} = F_{rs}\lambda_a{}^r\lambda_b{}^s - F_{/c}\gamma_{cab},$$

and similarly

$$F_{/b/a} = F_{rs}\lambda_b{}^r\lambda_a{}^s - F_{/c}\gamma_{cba}.$$

Thus, by subtraction we have

$$F_{/a/b} - F_{/b/a} = -F_{/c}\#_{cab}, \tag{7.3}$$

and again the $\#_{cab}$ appear. Both sides of this equation are skew-symmetric in a and b. However, since γ_{abc} is skew-symmetric in the first two indices it would be nicer if the coefficients appearing on the right-hand side had a similar character. This is readily achieved by replacing our provisional notation and writing

$$\alpha_{abc} := \#_{cab}. \tag{7.4}$$

These α-coefficients are called the *anholonomic coefficients* (they are essentially the objects of anholomity/non-holonomity employed by SCHOUTEN (1954) — in a different notation!). Then (7.3) becomes

$$F_{/a/b} - F_{/b/a} = -\alpha_{abc} F_{/c} \tag{7.5}$$

and likewise (6.5) assumes the form

$$\gamma_{abc} = \frac{1}{2}\left(\alpha_{bca} + \alpha_{cab} - \alpha_{abc}\right). \tag{7.6}$$

Henceforth, we may forget (7.3) and (6.5), and take (7.5) and (7.6) as the definitive versions of these equations in the Ricci calculus.

Explicitly, the relations between α_{abc} and γ_{abc} are given — using our canonical enumeration scheme — and

$$\alpha_{abc} = \gamma_{cab} - \gamma_{cba} \tag{7.7}$$

by the following expressions:

$$\begin{aligned}
\alpha_{121} &= -\gamma_{121}, \\
\alpha_{122} &= -\gamma_{122}, \\
\alpha_{123} &= \gamma_{312} + \gamma_{231}; \\
\\
\alpha_{311} &= -\gamma_{311}, \\
\alpha_{312} &= \gamma_{231} + \gamma_{123}, \\
\alpha_{313} &= -\gamma_{313}; \\
\\
\alpha_{231} &= \gamma_{123} + \gamma_{312}, \\
\alpha_{232} &= -\gamma_{232}, \\
\alpha_{233} &= -\gamma_{233};
\end{aligned} \tag{7.8}$$

or more conveniently:

$$
\begin{aligned}
\gamma_{121} &= -\alpha_{121}, \\
\gamma_{122} &= -\alpha_{122}, \\
\gamma_{123} &= \frac{1}{2}(\alpha_{231} - \alpha_{312} - \alpha_{123}); \\
\\
\gamma_{311} &= -\alpha_{311}, \\
\gamma_{312} &= \frac{1}{2}(\alpha_{123} - \alpha_{231} - \alpha_{312}), \\
\gamma_{313} &= -\alpha_{313}; \\
\\
\gamma_{231} &= \frac{1}{2}(\alpha_{312} - \alpha_{123} - \alpha_{231}), \\
\gamma_{232} &= -\alpha_{232}, \\
\gamma_{233} &= -\alpha_{233}.
\end{aligned}
\tag{7.9}
$$

In practice, (7.5) are required in exhibiting the integrability conditions of a leg differential equation. For example, the necessary and sufficient conditions for the existence of a function F satisfying the leg differential equation

$$F_{/a} = f_a \tag{7.10}$$

for a given f_a, is that one has

$$F_{/a/b} - F_{/b/a} = f_{a/b} - f_{b/a}, \tag{7.11}$$

and upon substitution of (7.5) we see that f_a must satisfy the equation

$$f_{a/b} - f_{b/a} = -\alpha_{abc} f_c. \tag{7.12}$$

Equation (7.10) is the leg version of the partial differential equation

$$F_{;r} = f_r \tag{7.13}$$

for a given f_r, and in this case the corresponding integrability equations are

$$F_{;r;s} - F_{;s;r} = f_{r;s} - f_{s;r}$$

and this reduces to merely

$$f_{r;s} - f_{s;r} = 0 \tag{7.14}$$

by I-(3.17), which is simpler than (7.12).

If (7.5) is explicitly written out using our canonical enumeration scheme we have $[F_{\mathcal{A}}]$, $\mathcal{A} = \mathrm{I, II, III}$.

$$
\begin{aligned}
&[F_{\mathrm{I}}] \quad F_{/1/2} - F_{/2/1} &=& \ \gamma_{121} F_{/1} + \gamma_{122} F_{/2} - (\gamma_{312} + \gamma_{231}) F_{/3}, \\
&[F_{\mathrm{II}}] \quad F_{/3/1} - F_{/1/3} &=& \ -\gamma_{131} F_{/1} - (\gamma_{123} + \gamma_{231}) F_{/2} + \gamma_{313} F_{/3}, \\
&[F_{\mathrm{III}}] \quad F_{/2/3} - F_{/3/2} &=& \ -(\gamma_{123} + \gamma_{312}) F_{/1} + \gamma_{232} F_{/2} + \gamma_{233} F_{/3}.
\end{aligned}
\tag{7.15}
$$

In practice, it is convenient to consider these expressions instead of (7.12). These equations are often called *commutators*, since they explicitly indicate the non-commutability of the leg derivatives (this terminology is due to LEVI-CIVITÀ (1925) who termed them *commutation formulae*). This notation is useful since it indicates the function F whose integrability conditions are being exhibited. In the Cartan theory of CHAPTER III it will be shown that these equations can be readily computed as they arise naturally in the analysis, however, the $[F_{\mathcal{A}}]$ notation is still useful.

II.8 The Lie Bracket

We now present another derivation of (7.5) which is based on the representation (2.3) of the leg vectors as differential operators. By definition, the *Lie bracket* of a pair of leg vectors is given by

$$[\boldsymbol{\lambda}_a, \boldsymbol{\lambda}_b] := \boldsymbol{\lambda}_a \boldsymbol{\lambda}_b - \boldsymbol{\lambda}_b \boldsymbol{\lambda}_a \tag{8.1}$$

where $a \neq b$ and each vector is represented by (2.3). In other words, when applied to a smooth function F of the coordinates, the right-hand side of this expression is

$$\boldsymbol{\lambda}_a (\boldsymbol{\lambda}_b F) - \boldsymbol{\lambda}_b (\boldsymbol{\lambda}_a F).$$

Hence, we have

$$[\boldsymbol{\lambda}_a, \boldsymbol{\lambda}_b] = -[\boldsymbol{\lambda}_b, \boldsymbol{\lambda}_a] \tag{8.2}$$

for $a \neq b$, and of course for an unsummed a

$$\text{NS} \qquad [\boldsymbol{\lambda}_a, \boldsymbol{\lambda}_a] = 0 \tag{8.3}$$

We now evaluate $[\boldsymbol{\lambda}_a, \boldsymbol{\lambda}_b]\, F$. Clearly by (2.3)

$$\begin{aligned}
[\boldsymbol{\lambda}_a, \boldsymbol{\lambda}_b]\, F &= \lambda_a{}^r \left(\lambda_b{}^s F_{;s}\right)_{;r} - \lambda_b{}^r \left(\lambda_a{}^s F_{;s}\right)_{;r} \\
&= \lambda_a{}^r \left(\lambda_b{}^s{}_{;r} F_{;s} + \lambda_b{}^s F_{;s;r}\right) - \lambda_b{}^r \left(\lambda_a{}^s{}_{;r} F_{;s} + \lambda_a{}^s F_{;s;r}\right) \\
&= \left(\lambda_a{}^s \lambda_b{}^r{}_{;s} - \lambda_b{}^s \lambda_a{}^r{}_{;s}\right) F_{;r}\,,
\end{aligned}$$

but

$$\lambda_b{}^r{}_{,s} = \lambda_b{}^r{}_{;s} - \Gamma^r_{ps} \lambda_b{}^p$$

so

$$\lambda_b{}^r{}_{;s} = \lambda_b{}^r{}_{,s} + \Gamma^r_{ps} \lambda_b{}^p.$$

Hence, we obtain

$$[\boldsymbol{\lambda}_a, \boldsymbol{\lambda}_b]\, F = \left(\lambda_a{}^s \lambda_b{}^r{}_{,s} - \lambda_b{}^s \lambda_a{}^r{}_{,s}\right) F_{;r} \tag{8.4}$$

which upon replacing the covariant derivatives by the Ricci coefficients ultimately yields

$$[\boldsymbol{\lambda}_a, \boldsymbol{\lambda}_b]\, F = \alpha_{abc} F_{/c}. \tag{8.5}$$

Thus, the commutators (7.5) are none other than the negative of the result obtained by applying the Lie bracket to a smooth function F.

II.9 The Jacobi and Schouten Identities

The derivation of the commutators given in the previous section might be taken to suggest that the Lie bracket is merely a notational device. This is not true since it overlooks the fact that it is an operator which describes the commutation properties of a pair of linear partial differential operators. In particular, upon omitting the function F from (8.4) we have the pure operator-theoretic equation

$$[\boldsymbol{\lambda}_a, \boldsymbol{\lambda}_b] = \alpha_{abc} \boldsymbol{\lambda}_c. \tag{9.1}$$

Such results form the starting point of the theory of Lie algebras and groups, and there the coefficients α_{abc} turn out to be constants C_{abc} which are known as *structural constants.* The values of these constants essentially determine the structure of the algebra/group generated by the bracket operation. In such a context the $\boldsymbol{\lambda}_a$ are regarded as abstract symbols which need not have a concrete geometric interpretation.

On the other hand, in the Ricci calculus the $\boldsymbol{\lambda}_a$ are unit tangent vectors to the congruence of curves $\boldsymbol{\Gamma}_a$, and there is no reason that the α_{abc} be constant. Indeed, in differential geodesy we will see that although some of the nine α_{abc} may be zero (recall the geometric results in SECTION 5, which by (7.9) could be restated in terms of these coefficients), in general they are functions of the coordinates x^r. Therefore, keeping this in mind it is worthwhile noting that there is a remarkable identity-known as the Jacobi identity — which concerns the iteration of the bracket operation. We now derive this identity.

By definition, for distinct a, b, c:

$$[[\boldsymbol{\lambda}_a, \boldsymbol{\lambda}_b], \boldsymbol{\lambda}_c] = [\boldsymbol{\lambda}_a, \boldsymbol{\lambda}_b]\, \boldsymbol{\lambda}_c - \boldsymbol{\lambda}_c\, [\boldsymbol{\lambda}_a, \boldsymbol{\lambda}_b] \tag{9.2}$$

and upon expansion gives

$$\begin{aligned}&(\boldsymbol{\lambda}_a\boldsymbol{\lambda}_b - \boldsymbol{\lambda}_b\boldsymbol{\lambda}_a)\, \boldsymbol{\lambda}_c - \boldsymbol{\lambda}_c\, (\boldsymbol{\lambda}_a\boldsymbol{\lambda}_b - \boldsymbol{\lambda}_b\boldsymbol{\lambda}_a)\\ &= \boldsymbol{\lambda}_a\boldsymbol{\lambda}_b\boldsymbol{\lambda}_c - \boldsymbol{\lambda}_b\boldsymbol{\lambda}_a\boldsymbol{\lambda}_c - \boldsymbol{\lambda}_c\boldsymbol{\lambda}_a\boldsymbol{\lambda}_b + \boldsymbol{\lambda}_c\boldsymbol{\lambda}_b\boldsymbol{\lambda}_a\end{aligned}$$

which is a complicated third order differential operator. However, upon performing a cyclic permutation of the distinct indices we may immediately exhibit the corresponding expressions for $[[\boldsymbol{\lambda}_b, \boldsymbol{\lambda}_c], \boldsymbol{\lambda}_a]$ and $[[\boldsymbol{\lambda}_c, \boldsymbol{\lambda}_a], \boldsymbol{\lambda}_b]$. Jacobi's observation was the rather surprising fact that

$$[[\boldsymbol{\lambda}_a, \boldsymbol{\lambda}_b], \boldsymbol{\lambda}_c] + [[\boldsymbol{\lambda}_b, \boldsymbol{\lambda}_c], \boldsymbol{\lambda}_a] + [[\boldsymbol{\lambda}_c, \boldsymbol{\lambda}_a], \boldsymbol{\lambda}_b] = 0 \tag{9.3}$$

and hence this is called the *Jacobi identity.* By virtue of (9.1) this imposes conditions on the α_{abc}, and although these are complicated they are easy to derive. Consider (9.2) when these coefficients are introduced by using (9.1). Then clearly,

$$[[\boldsymbol{\lambda}_a, \boldsymbol{\lambda}_b], \boldsymbol{\lambda}_c] = \alpha_{abf}\boldsymbol{\lambda}_f\boldsymbol{\lambda}_c - \boldsymbol{\lambda}_c\, (\alpha_{abd}\boldsymbol{\lambda}_d)$$

and since

$$\boldsymbol{\lambda}_c (F\boldsymbol{\lambda}_f) = F_{/c}\boldsymbol{\lambda}_f + F\boldsymbol{\lambda}_c\boldsymbol{\lambda}_f$$

for a smooth function F, we readily see that

$$\begin{aligned} [[\boldsymbol{\lambda}_a, \boldsymbol{\lambda}_b], \boldsymbol{\lambda}_c] &= \alpha_{abf} [\boldsymbol{\lambda}_f, \boldsymbol{\lambda}_c] - \alpha_{abd/c}\boldsymbol{\lambda}_d \\ &= \alpha_{abf}\alpha_{fcd}\boldsymbol{\lambda}_d - \alpha_{abd/c}\boldsymbol{\lambda}_d. \end{aligned}$$

Hence (9.3) yields the following set of *three* identities

$$\alpha_{abd/c} + \alpha_{bcd/a} + \alpha_{cad/b} - \alpha_{abf}\alpha_{fcd} - \alpha_{bcf}\alpha_{fad} - \alpha_{caf}\alpha_{fbd} = 0 \quad (9.4)$$

which are known as the *Schouten identities.* One could explicitly expand these, and, adopting the notation employed in SECTION 7 for enumerating the commutators, exhibit the set of Schouten identities $(\mathcal{S}_{\mathcal{A}})$ $(\mathcal{A} = \text{I, II, III})$. However, for the present, this is unnecessary and serves no useful purpose. Later in CHAPTER V when the language of leg coefficients has been introduced, we will exhibit the $(\mathcal{S}_{\mathcal{A}})$.

The important point to note is that the $(\mathcal{S}_{\mathcal{A}})$ furnish *three* differential identities on the α_{abc} — which hold gratuitously independent of any geometric specialization of the $\boldsymbol{\Gamma}_a$ — and that by (7.7), or (7.8), these may be restated in terms of the Ricci coefficients γ_{abc}. It is this form of the Schouten identities with which we will deal ultimately in CHAPTER V.

II.10 Interpretations of the Ricci Coefficients

In SECTION 5 we briefly mentioned a geometric interpretation of the Ricci coefficients, and this will be further explored in PROBLEM II.8. We now consider two purely algebraic interpretations of these coefficients.

The first, and probably the best one, is immediate from (4.4): the γ_{abc} are the leg coefficients of the covariant derivative of $\boldsymbol{\lambda}_a$ which occur in its leg representation. This interpretation is simple and has the virtue of indicating precisely how these coefficients will later be employed in the general leg calculus.

The second interpretation involves returning to the definition of the Ricci coefficients given in (4.1) and explicitly writing out the covariant derivative. This gives

$$\gamma_{abc} = \left(\lambda_{ar;s} - \Gamma^p_{rs}\lambda_{ap}\right)\lambda_b{}^r\lambda_c{}^s, \tag{10.1}$$

which may be solved for

$$\Gamma^p_{rs}\lambda_{ap}\lambda_b{}^r\lambda_c{}^s = \lambda_{ar;s}\lambda_b{}^r\lambda_c{}^s - \gamma_{abc}. \tag{10.2}$$

The left-hand side of this equation clearly is the resolution of Γ^p_{rs} along the leg vectors. One might be inclined to regard this as the leg representation of Γ^p_{rs}, but in SECTION 2 such terminology was employed *only* for tensors (recall equations (2.19)-(2.22)). It is not evident that it is meaningful for non-tensorial quantities and in any case (10.2) shows that such a representation is in general not γ_{abc}. However, upon multiplying (10.2) by three leg vectors and employing the orthogonality conditions we obtain

$$\Gamma^p_{rs} = \lambda_a{}^p\left(\lambda_{ar;s} - \gamma_{abc}\lambda_{br}\lambda_{cs}\right). \tag{10.3}$$

This equation exhibits the non-tensorial character of Γ^p_{rs} (the tensor character is spoiled by the appearance of $\lambda_{ar;s}$). Nevertheless, (10.3) provides our second interpretation: in it the γ_{abc} are the leg coefficients of the tensorial part of Γ^p_{rs}. This interpretation is decidedly less satisfying than our previous one, but it is valid and somewhat unexpected.

Moreover, by (4.1) we see that (10.3) may be rewritten in the form

$$\Gamma^p_{rs} = \lambda_a{}^p\left(\lambda_{ar;s} - \lambda_{ar,s}\right). \tag{10.4}$$

Upon expanding the leg index and using the canonical identification (2.2) we obtain the following equation

$$\begin{aligned}\Gamma^p_{rs} &= \lambda_{r;s}\lambda^p + \mu_{r;s}\mu^p + \nu_{r;s}\nu^p \\ &\quad -\lambda_{r,s}\lambda^p - \mu_{r,s}\mu^p - \nu_{r,s}\nu^p\end{aligned} \tag{10.5}$$

which is due to Hotine (see [12.125]). This result appeared in the midst of his discussion of (ω, ϕ, N) coordinate system, but properly speaking it has nothing to do with this coordinate system. Hotine did not employ Ricci coefficients in his work, and in fact his 'proof' of (10.5) is merely a verification that upon multiplication by each of the leg vectors λ_p, μ_p and ν_p respectively it yields correct results, e.g.

$$\Gamma^p_{rs}\lambda_p = \lambda_{r;s} - \lambda_{r,s} \text{ etc.}$$

Despite its cunning nature — and regardless of how it is derived — equation (10.5) is almost useless, since it requires knowing $\lambda_{r,s}$, $\mu_{r,s}$ and $\nu_{r,s}$ which involve the Christoffel symbols! Hence, using the language of logic one would say that (10.4) and (10.5) merely represent a *tautology*, i.e. they are tautological equations not unlike $1 = 1$, $2 = 2$, etc. On the other hand, (10.3) has a more substantive character.

II.11 Miscellaneous Leg Representations

It is natural to expect that scalar equations will have a meaningful leg representation in the Ricci calculus. For example, consider the Laplacian of a smooth function F, i.e.

$$\Delta F := g^{rs}F_{rs} \tag{11.1}$$

where

$$F_{rs} := F_{;r,s} = F_{,r,s}. \tag{11.2}$$

Clearly

$$F_r = F_{/a}\lambda_{ar}, \tag{11.3}$$

which is the inverse of permitting I-(3.12) to operate on F. Then by calculating

$$F_{rs} = F_{/a/b}\lambda_{ar}\lambda_{bs} + F_{/a}\lambda_{ars}$$

and using (5.4) we have

$$F_{rs} = F_{/a/b}\lambda_{ar}\lambda_{bs} + F_{/a}\gamma_{abc}\lambda_{br}\lambda_{cs}.$$

This expression, together with (11.1) and (2.11), immediately gives

$$\Delta F = F_{/a/b}\delta_{ab} + F_{/a}\gamma_{abc}\delta_{bc},$$

or more concisely, using the leg summation convention,

$$\Delta F = F_{/a/a} + \gamma_{abb} F_{/a}. \tag{11.4}$$

(See PROBLEM II.10 for an explicit expansion of (11.4).)

Likewise, the *Beltrami first order differential operators* (also known as the Beltrami differential parameters) are defined by

$$\nabla (F, G) := g^{rs} F_r G_s, \tag{11.5}$$

$$\nabla (F, F) := g^{rs} F_r F_s, \tag{11.6}$$

where F and G are smooth functions, are easily translated into their leg representations. One need only substitute (11.3) twice and use (2.11) to obtain

$$\nabla (F, G) = F_{/a} G_{/a} \tag{11.7}$$

$$\nabla (F, F) = F_{/a} F_{/a}. \tag{11.8}$$

In SECTION 5, we mentioned the leg components of the permutation symbol ε_{abc}. We may now use these to obtain a leg representation for the covariant/contravariant components of the coordinate-based components of this tensor. By our usual translation rule (2.20), now written for three components we have

$$\varepsilon_{rst} = \varepsilon_{abc} \lambda_{ar} \lambda_{bs} \lambda_{ct}. \tag{11.9}$$

Upon employing our canonical identification scheme (2.2) we then have

$$\varepsilon_{rst} = \lambda_r \left(\mu_s \nu_t - \nu_s \mu_t \right) + \mu_r \left(\nu_s \lambda_t - \lambda_s \nu_t \right) + \nu_r \left(\lambda_s \mu_t - \mu_s \lambda_t \right). \tag{11.10}$$

This expression indicates skew-symmetry in the r, t indices and the skew-symmetry in r, s follows upon re-arrangement, i.e.

$$\varepsilon_{rst} = \left(\mu_r \nu_s - \nu_r \mu_s \right) \lambda_t + \left(\nu_r \lambda_s - \lambda_r \nu_s \right) \mu_t + \left(\lambda_r \mu_s - \mu_r \lambda_s \right) \nu_t. \tag{11.11}$$

We leave it to the reader to verify the remaining (complete) skew-symmetries. The contravariant form of (11.10) follows upon raising the indices by multiplying with three metric tensors having contravariant components. The contravariant expression was given by Hotine (see his [2.26]) and employed by him to give

$$\varepsilon^{rst}\lambda_r\mu_s\nu_t = 1, \tag{11.12}$$

which exhibits the right-handedness of the vectorial 3-leg $\{\boldsymbol{\lambda}_a\}$.

Finally, we note that the Hotine transformation formulas I-(3.21) are equivalent to the following leg representation equations

$$\begin{aligned}\frac{\partial \overline{x}^r}{\partial x^s} &= \overline{\lambda}_a{}^r\lambda_{as},\\ \frac{\partial x^r}{\partial \overline{x}^s} &= \lambda_a{}^r\overline{\lambda}_{as},\end{aligned} \tag{11.13}$$

where the 'bar' refers to the transformed coordinate-based components (see ZUND and WILKES (1988)).

An advantage of such leg representations is not only their conciseness and elegance, but that they permit one to evaluate — in closed form — many expressions in the tensor calculus which would not admit such simple expression. This is due primarily to the use of a 3-leg and representation of quantities in terms of various combinations of the leg vectors, and the fact that the 3-leg in question is orthonormal.

II.12 The Lamé Equations

Up to now all the results in this chapter have been given for a curved 3-dimensional Riemannian $\mathbf{V}_3$ (and thus we have generalized versions of the corresponding equations which occurred in Hotine's treatise). We now consider the conditions which specialize $\mathbf{V}_3$ to a flat 3-dimensional Euclidean $\mathbf{E}_3$. It is well-known that these conditions require the vanishing of the Riemann-Christoffel curvature tensor R_{rsmn}, and the conditions

$$R_{rsmn} = 0 \tag{12.1}$$

are classically known as the *Lamé equations.* Our first task is to translate the curvature tensor into leg form, i.e. give a leg representation of R_{rsmn}, and then to examine the structure of the system of equations defined by the condition (12.1). In the tensor calculus the tensor $R^r{}_{smn}$ is expressed in terms of the Christoffel symbols by

$$R^r{}_{smn} = \Gamma^r{}_{sn;m} - \Gamma^r{}_{sm;n} + \Gamma^p_{sn}\Gamma^r{}_{pm} - \Gamma^p_{sm}\Gamma_{pn}, \tag{12.2}$$

however in obtaining the corresponding leg representation neither (10.3) nor (10.5) is a particularly appealing approach.

In the Ricci calculus the leg components of the curvature tensor are defined by

$$R_{abcd} := R_{rsmn}\lambda_a{}^r\lambda_b{}^s\lambda_c{}^m\lambda_d{}^n \tag{12.3}$$

and to obtain a workable form of R_{rsmn}, one uses its classical definition

$$\lambda_{ar,m,n} - \lambda_{as,n,m} = R_{rsmn}\lambda_a{}^r \tag{12.4}$$

which may be regarded as a coordinate-based commutator for λ_{as}, this indicates the non-commutability of the second covariant derivatives of a vector. Inverting (12.4) yields

$$R_{rsmn} = (\lambda_{as,m,n} - \lambda_{as,n,m})\,\lambda_{ar} \tag{12.5}$$

where the first covariant derivative of λ_{as} is expressed by (5.2). Upon substitution of (12.5) in (12.3) one finds that ultimately

$$R_{abcd} = \gamma_{abc/d} - \gamma_{abd/c} + \gamma_{fad}\gamma_{fbc} - \gamma_{fac}\gamma_{fbd} + \gamma_{abf}\left(\gamma_{fcd} - \gamma_{fdc}\right) \tag{12.6}$$

where — if desired — the terms in the parentheses could be replaced by α_{cdf} by using (7.7).

By virtue of (12.3) the usual symmetries of R_{rsmn} are translated into the leg equations

$$R_{abcd} = -R_{bacd} = -R_{abdc} = R_{cdab} \tag{12.7}$$

which mean that R_{abcd} possesses six independent components in $\mathbf{V}_3$. These are summarized in 3-dimensions by the leg analogue of the algebraic Bianchi identity

$$R_{abcd} + R_{acdb} + R_{adbc} = 0. \tag{12.8}$$

Thus, the flatness of $\mathbf{E}_3$, i.e. the conditions that $\mathbf{V}_3$ reduces to $\mathbf{E}_3$, is expressed by the leg version of the Lamé equations which is conveniently written in the form:

$$\gamma_{abc/d} - \gamma_{abd/c} = \gamma_{afc}\gamma_{fbd} - \gamma_{afd}\gamma_{bfc} - \gamma_{abf}\alpha_{cdf}. \tag{12.9}$$

As in the case of the Schouten identities in SECTION 9, we will not explicitly write out the system of Lamé equations which we denote by $(\mathcal{L}_{\mathcal{B}})$ where $(\mathcal{B} = \text{I, II, ..., IX})$ since three turn out to be alternate versions of three other equations. In the present case when the basic variables in the Ricci calculus are the coefficients γ_{abc} the repetition of these three equations is not entirely a trivial matter. Later in CHAPTER V when the general leg calculus has been set out this redundancy will be fully explained and resolved.

Finally, in the tensor calculus the components of the curvature tensor also satisfy another identity known as the *differential Bianchi identity*:

$$R_{rsmn,t} + R_{rsnt,m} + R_{rstm,n} = 0. \tag{12.10}$$

There is also a Ricci calculus analogue of this equation which we will not state. It is complicated, but in $\mathbf{E}_3$ it is not necessarily trivial since as is evident from (12.6) it involves second leg derivatives of γ_{abc} as well as first derivatives of γ_{abc}. These may be useful in obtaining further constraining equations of the Ricci coefficients which turn out to be the primary variables in the leg calculus. This possibility is further discussed in ZUND (1990b) in terms of the leg calculus.

PROBLEMS FOR CHAPTER II

II.1 Show that for the vectorial 3-leg $\{\boldsymbol{\lambda}, \boldsymbol{\mu}, \boldsymbol{\nu}\}$

$$\begin{aligned} \varepsilon_{rst}\lambda^t &= \mu_r\nu_s - \nu_r\mu_s, \\ \varepsilon_{rst}\mu^t &= \nu_r\lambda_s - \lambda_r\nu_s, \\ \varepsilon_{rst}\nu^t &= \lambda_r\mu_s - \mu_r\lambda_s. \end{aligned}$$

II.2 Use the results of PROBLEM II.1 to conclude that for any vector $\boldsymbol{\xi}$ of the form (2.17) one has

$$\varepsilon_{rst}\xi^t = a\,(\mu_r\nu_s - \nu_r\mu_s)+b\,(\nu_r\lambda_s - \lambda_r\nu_s)+c\,(\lambda_r\mu_s - \mu_r\lambda_s)\,.$$

Hence, the scalar coefficients a, b and c in (2.17) are given by

$$a = \varepsilon_{rst}\mu^r\nu^s\xi^t, \quad b = \varepsilon_{rst}\nu^r\lambda^s\xi^t, \quad c = \varepsilon_{rst}\lambda^r\mu^s\xi^t$$

i.e.

$$a = \xi^r\lambda_r, \quad b = \xi^r\mu_r, \quad c = \xi^r\nu_r.$$

II.3 Using $\varepsilon^{rst}\lambda_r\mu_s\nu_t = 1$, i.e. (11.12) or [2.26] of Hotine's treatise, derive his expression [2.27] for ε^{rst}, as well as his [2.28] for $g_{rm}\varepsilon^{rst}\varepsilon^{mnp}$.

II.4 By using the results of PROBLEMS II.1-II.3, or otherwise, evaluate the expressions

$$g_{rm}\varepsilon^{rst}\varepsilon^{mnp}, \quad g_{mr}\varepsilon^{rst},$$

$$g_{mr}g_{ns}\varepsilon^{rst}, \quad g_{mr}g_{ns}g_{pt}\varepsilon^{rst}.$$

II.5 Compute the Ricci coefficients for the line element in rectangular Cartesian coordinates $x^r := (x, y, z)$, i.e. $ds^2 = dx^2 + dy^2 + dz^2$ and show that the congruences $\boldsymbol{\Gamma}_a$ ($a = 1, 2, 3$), viz. $\boldsymbol{\Gamma}_1 = \boldsymbol{\Gamma}_x$ etc. have zero curvature.

II.6 Compute the Ricci coefficients for the line element in spherical polar coordinates $x^r := (\omega, \phi, r)$, i.e.

$$ds^2 = r^2 \cos^2 \phi d\omega^2 + r^2 d\phi^2 + dr^2.$$

(This is Hotine's way of writing spherical polar coordinates, see [page 5]).

II.7 Compute the Ricci coefficients for the line element in cylindrical polar coordinates $x^r := (\omega, z, \rho)$, i.e.

$$ds^2 = \rho^2 d\omega^2 + dz^2 + d\rho^2.$$

II.8 Derive the expression $\psi_{/c} = -\gamma_{abc}$ where ψ is the angle between the ath leg vector $\boldsymbol{\lambda}_a$ and a unit vector ϵ which is parallelly displaced along the cth congruence $\boldsymbol{\Gamma}_c$. (This shows that $-\gamma_{abc}$ is the arc-rate of rotation of $\boldsymbol{\lambda}_a$ about $\boldsymbol{\Gamma}_c$.)

WEATHERBURN (1938)

II.9 Let $\boldsymbol{\nu}$ be the tangent vector to a congruence of curves $\boldsymbol{\Gamma}_3$ which is orthogonal to a family of surfaces: $F(x, y, z) = C =$ constant. Then writing $F_r := F_{;r}$ for the partial derivative with respect to x^r, show that

$$F_{;1}/\nu_1 = F_{;2}/\nu_2 = F_{;3}/\nu_r.$$

Hence, the necessary and sufficient condition for the existence of such an F is

$$F_r (F_{s;t} - F_{t;s}) + F_s (F_{t;r} - F_{r;t}) + F_t (F_{r;s} - F_{s;r}) = 0,$$

or equivalently

$$F_r (F_{s,t} - F_{t,s}) + F_s (F_{t,r} - F_{r,t}) + F_t (F_{r,s} - F_{s,r}) = 0$$

where the comma denotes covariant differentiation.

II.10 Explicitly expand equation (11.4) to obtain the Ricci calculus version of the 3-dimensional Laplace operator.

— o —

In PROBLEMS II.11 and II.12 x^r is a curvilinear coordinate system in $\mathbf{E}_3$ and we label the individual coordinates, scale factors, and components of the Riemann tensor by the Latin indices i, j and k which do not obey the summation convention.

II.11 If

$$ds^2 = \sum_i \left(h_i dx^i\right)^2 ,$$

then show that

$$\lambda_{ai} = h_i \delta^i_a, \quad \lambda_a{}^i = h_i^{-1} \delta^i_a,$$

$$\gamma_{abb} = \left(h_i h_j\right)^{-1} h_{j;a} \delta^i_a \delta^j_b$$

with *no sum* on the leg indices.

II.12 For the line element of PROBLEM II.11 show that the components of R_{rsmn} are given by

$$R_{ijkl} = 0 \text{ for } i, j, k, l \text{ unequal;}$$

$$R_{ijjk} = h_i \left\{h_{j;i;k} - h_{j;i}(\log h_i)_{;k} - h_{j;k}(\log h_k)_{;i}\right\} \text{ for } i, j, k \text{ unequ}$$

$$R_{ijji} = h_i h_j \left\{(h_i^{-1} h_{j;i})_{;i} + (h_j^{-1} h_{i;j})_{;j} + \sum_k h_k^{-2} h_{i;k} h_{j;k}\right\}$$

where $\sum_k$ denotes a sum of 1, 2, 3 excluding the values i and j.

Note these values of R_{rsmn} explicitly exhibit the non-trivial dependence of the Lamé equations of SECTION 12 on the scale factors.

III

The Cartan Calculus

III.1 Introduction

As noted in the previous chapter (SECTION II-2), the Ricci calculus makes no use of the cotangent space $\mathbf{T}_P^*$, nor does it identify the basis of covectors in this space. For Cartan this omission is the starting point of his theory, i.e. the basis of $\mathbf{T}_P^*$ is taken to be $\{dx^r\}$ in the coordinate-based representation or more generally $\{\theta_a\}$ where, as in I-(3.19), these Pfaffians are linear combinations of the dx^r. Moreover, products of such elements are skew-symmetric and involve serious consideration of the exterior product space of $\mathbf{T}_P^*$ with itself; this is a subspace of the tensor product spaces $\otimes^2\mathbf{T}_P^*$, or more generally of $\otimes^3\mathbf{T}_P^*$. Hence, whereas the material in SECTION I-3 on the abstract notion of a tensor is somewhat of a luxury for the Ricci calculus, it now plays a central role in the Cartan calculus. Consequently, any attempt to understand the Cartan theory must be rooted in terms of an abstract setting. This has seriously impeded the dissemination of it among non-mathematicians and in elementary mathematical texts the theory is usually presented in an unpalatable and abbreviated form which conceals as much as it reveals to the reader. Unfortunately, the same criticism can probably be said of the presentation given in this chapter.

The Cartan calculus did not originate with him, although he is responsible for its formulation as a coherent theory and, in particular, he recognized its importance for differential geometry. The theory has its origin in differential equations, and more specifically in determining when an expression — called a *total differential equation* —

$$X_r dx^r = 0 \tag{1.1}$$

where X_r is a smooth function of the x^r, is *integrable*, or if it is *non-integrable* when does it admit an integrating factor? For $n = 2$, this is a standard topic in elementary differential equations and (1.1) is either an exact differential equation, or always admits an integrating factor $\mu(x^1, x^2)$ which can be used to make (1.1) into an exact equation. For the case $n \geq 3$ the situation is decidedly less simple, and even the great Euler had serious misconceptions about the possibilities of solving (1.1). In fact, he was dead wrong — one of the few cases in which this can be said about his mathematical work. The correction of his error leads to what is known as *Pfaff's Problem*, which in its most elementary case concerns the existence of an integral surface satisfying (1.1), e.g. when $n = 3$ and X_r is chosen to be the gradient of the geopotential N with $X_r = N_{;r}$, then the integral surfaces are equipotential surfaces. On the other hand, suppose $X_r \neq N_{;r}$ then what can be said about the construction of a solution of (1.1)?

III.2 Exterior Product

Our first step towards establishing the Cartan calculus is to explain the exterior product. This is denoted by the wedge, or 'hook' symbol $\wedge$ and should not be confused with the cross product $\times$ which is used in vector analysis, although in some textbooks — particularly British ones — the cross product is written with $\wedge$! In discussing the exterior product we revert to the abstract notation and terminology used in SECTION 1-4 where the vector spaces $\mathbf{V}$ and $\mathbf{V}^*$ replace $\mathbf{T}_P$ and $\mathbf{T}_P^*$, and we restrict our considerations to the case when

$$\dim \mathbf{V} = \dim \mathbf{V}^* = 3. \tag{2.1}$$

Then the exterior product of a pair of vectors $\boldsymbol{\xi}$ and $\boldsymbol{\eta}$ in $\mathbf{V}$ is defined by

$$\boldsymbol{\xi} \wedge \boldsymbol{\eta} := \boldsymbol{\xi} \otimes \boldsymbol{\eta} - \boldsymbol{\eta} \otimes \boldsymbol{\xi} \tag{2.2}$$

and likewise for ξ^* and η^* in $\mathbf{V}^*$ we have

$$\xi^* \wedge \eta^* := \xi^* \otimes \eta^* - \eta^* \otimes \xi^* \tag{2.3}$$

(where $\otimes$ is the tensor product). Both these expressions possess the familiar properties

$$\boldsymbol{\xi} \wedge \boldsymbol{\eta} = -\boldsymbol{\eta} \wedge \boldsymbol{\xi}, \text{ etc.} \tag{2.4}$$

and

$$\boldsymbol{\xi} \wedge \boldsymbol{\xi} = 0 \tag{2.5}$$

which are shared by the cross product $\times$ in vector analysis. However, the exterior product is associative, i.e.

$$(\boldsymbol{\xi} \wedge \boldsymbol{\eta}) \wedge \boldsymbol{\zeta} = \boldsymbol{\xi} \wedge (\boldsymbol{\eta} \wedge \boldsymbol{\zeta}) = \boldsymbol{\xi} \wedge \boldsymbol{\eta} \wedge \boldsymbol{\zeta} \tag{2.6}$$

which is not true of the cross product — recall that

$$\begin{aligned} (\mathbf{u} \times \mathbf{v}) \times \mathbf{w} &= (\mathbf{u} \cdot \mathbf{w})\, \mathbf{v} - (\mathbf{v} \cdot \mathbf{w})\, \mathbf{u} \\ \mathbf{u} \times (\mathbf{v} \times \mathbf{w}) &= (\mathbf{u} \cdot \mathbf{w})\, \mathbf{v} - (\mathbf{u} \cdot \mathbf{v})\, \mathbf{w} \end{aligned} \tag{2.7}$$

so

$$(\mathbf{u} \times \mathbf{v}) \times \mathbf{w} \neq \mathbf{u} \times (\mathbf{v} \times \mathbf{w}) \tag{2.8}$$

so in vector algebra $\mathbf{u} \times \mathbf{v} \times \mathbf{w}$ is not unambiguously defined. Another equally important distinction is that upon forming $\mathbf{u} \times \mathbf{v}$ the resulting products of the Cartesian basis vectors $\mathbf{i} \times \mathbf{j}$, $\mathbf{k} \times \mathbf{i}$, $\mathbf{j} \times \mathbf{k}$ are automatically identified with the basis vectors $\mathbf{k}$, $\mathbf{j}$, $\mathbf{i}$ respectively. Consequently, if one has chosen to regard $\mathbf{u} \times \mathbf{v}$ as being a vector — just like any other ordinary vector, — such an identification, purely mathematically speaking, is *forced.* Indeed one can regard the loss of the associative property for the cross product as the *price* of this forcing. On the other hand, in forming $\boldsymbol{\xi} \wedge \boldsymbol{\eta}$, or $\xi^* \wedge \eta^*$, one has by analogy with I-(4.23)

$$\boldsymbol{\xi} \wedge \boldsymbol{\eta} = \xi^i \eta^j \mathbf{e}_i \wedge \mathbf{e}_j \tag{2.9}$$

and the $\mathbf{e}_i \wedge \mathbf{e}_j$ is *not identified* with $\mathbf{e}_k$. By analogy with the notation $\otimes^2 \mathbf{V} := \mathbf{V} \otimes \mathbf{V}$ we may say that $\wedge^2 \mathbf{V}$ is the *second exterior power* of $\mathbf{V}$ and is the space generated by products of the form (2.2), or in the basis representation by (2.9). It is not difficult to show that $\wedge^2 \mathbf{V}$ is a vector space having dimension

$$\dim \wedge^2 \mathbf{V} = 3 \tag{2.10}$$

and so may be regarded as the subspace of $\otimes^2\mathbf{V}$ spanned by skew-symmetric tensors. Likewise there is a subspace of $\otimes^2\mathbf{V}$ spanned by symmetric tensors and it is 6-dimensional! All of this discussion holds equally well for $\mathbf{V}^*$, and indeed since the Cartan theory deals with T_P^*, we may focus our attention on $\wedge^2\mathbf{V}^*$ and forget about $\wedge^2\mathbf{V}$. In other words, we are primarily interested in expressions of the form:

$$\xi^* = \xi_i f^i, \tag{2.11}$$

$$\xi^* \wedge \eta^* = \xi_i \eta_j f^i \wedge f^j, \tag{2.12}$$

$$\xi^* \eta^* \wedge \zeta^* = \xi_i \eta_j \zeta_k f^i \wedge f^j \wedge f^k, \tag{2.13}$$

which respectively belong to $\mathbf{V}^*$, $\wedge^2\mathbf{V}^*$ and $\wedge^3\mathbf{V}^*$. Note that by (2.1) due to the complete skew-symmetry of the product $f^i \wedge f^j \wedge f^k$ (the reader should check this) the only basis element in $\wedge^3\mathbf{V}^*$ is $f^1 \wedge f^2 \wedge f^3$ so

$$\dim \wedge^3\mathbf{V}^* = 1. \tag{2.14}$$

We will now show that expressions of the form (2.11)-(2.13) naturally occur, and are in fact well-known in the several variable calculus. For this purpose let x^r be a Cartesian coordinate system

$$x^r = (x, y, z).$$

Then the coordinate-based expressions for the basis vectors of the spaces corresponding to $\mathbf{V}^*$, $\wedge^2\mathbf{V}^*$, and $\wedge^3\mathbf{V}^*$ are respectively:

$$\begin{gathered} dx,\ dy,\ dz; \\ dx \wedge dy,\ dz \wedge dx,\ dy \wedge dz; \\ dx \wedge dy \wedge dz. \end{gathered} \tag{2.15}$$

Then if F is a smooth function of x^r, the obvious analogue of (2.11) is

$$\begin{aligned} dF &= F_{;r} dx^r \\ &= F_x dx + F_y dy + F_z dz \end{aligned} \tag{2.16}$$

where in the latter expression the subscripts denote partial differentiation. In other words, the differential of a function is an exterior differential 1-form.

The corresponding analogue of (2.13) is equally simple but of a different character. Consider one of the products, say $dx \wedge dy$, in the second line of (2.15). Then it is well known that under a change of variable

$$(x,y) \mapsto (\overline{x},\overline{y})$$

in the xy Cartesian coordinate plane the product of differentials (appearing in double integrals) is

$$dxdy \mapsto d\overline{x}d\overline{y} = \frac{\partial(\overline{x},\overline{y})}{\partial(x,y)}dxdy \tag{2.17}$$

where the multiplying factor on the right-hand side is a *functional determinant* (also called a *Jacobian*). But being a determinant, it is skew-symmetric, viz.

$$\frac{\partial(\overline{x},\overline{y})}{\partial(x,y)} = -\frac{\partial(\overline{y},\overline{x})}{\partial(x,y)} = -\frac{\partial(\overline{x},\overline{y})}{\partial(y,x)}, \tag{2.18}$$

and this feature is not indicated in the symbolic product (2.17). Hence, property (2.17) should be written in a manner which indicates the skew-symmetries (2.18), and this is the exterior product version

$$d\overline{x} \wedge d\overline{y} = \frac{\partial(\overline{x},\overline{y})}{\partial(x,y)}dx \wedge dy. \tag{2.19}$$

Likewise, in changing variables in 3-dimensions, i.e.

$$(x,y,z) \mapsto (\overline{x},\overline{y},\overline{z})$$

the so called volume element in a triple integral has the transformation law

$$dxdydz \mapsto d\overline{x}d\overline{y}d\overline{z} = \frac{\partial(\overline{x},\overline{y},\overline{z})}{\partial(x,y,z)}dxdydz \tag{2.20}$$

where the multiplicative factor is again a functional determinant. Hence, the correct version of (2.20) should be

$$d\overline{x} \wedge d\overline{y} \wedge d\overline{z} = \frac{\partial(\overline{x},\overline{y},\overline{z})}{\partial(x,y,z)}dx \wedge dy \wedge dz. \tag{2.21}$$

Examples of the use of (2.19) and (2.21) include the transformations of (x, y) to polar coordinates, and (x, y, z) to spherical polar coordinates.

Thus, all three kinds of expressions exhibited in (2.15) occur in calculus; however, they are often not specifically written in forms which indicate their proper meaning. Actually (2.19) and (2.21) represent oriented area and volume elements which would require the exterior product in a rigorous formulation.

III.3 Exterior Differentiation

The notion of exterior differentiation is somewhat more subtle. We have already seen in SECTION 2 that the differential of a smooth function F, (2.16), is a linear combination of the coordinate differentials and hence may be regarded as a Pfaffian expression i.e. an *exterior differential* 1-*form*, with the function F being considered to be a 0-form. This is reasonable since one also makes the identification

$$\wedge^0 \mathbf{V}^* := \mathbf{R} \text{ and } \wedge^1 \mathbf{V}^* = \mathbf{V}^*. \tag{3.1}$$

Now consider the following 1-form

$$\alpha = Adx + Bdy + Cdz \tag{3.2}$$

where for simplicity we have chosen the variables x^r to be the Cartesian coordinates, and all our discussion holds for an arbitrary curvilinear coordinate system. We also suppose that A, B, and C as well as their derivatives and differentials are smooth functions. Then we define $d\alpha$ by

$$d\alpha := dA \wedge dx + dB \wedge dy + dC \wedge dz \tag{3.3}$$

which amounts to assuming that

$$d(dx) = d(dy) = d(dz) = 0. \tag{3.4}$$

The latter is a sensible assumption since the x^r are independent variables. Then applying (2.16) to A, B and C in (3.3) and using the anti-commutativity of the differentials, viz.

$$dx \wedge dy = -dy \wedge dx, \quad dz \wedge dx = -dx \wedge dz, \quad dy \wedge dz = -dz \wedge dy \tag{3.5}$$

and

$$dx \wedge dx = dy \wedge dy = dz \wedge dz = 0 \tag{3.6}$$

which are special cases of (2.4), we obtain

$$d\alpha = (B_x - A_y)\, dx \wedge dy + (A_z - C_x)\, dz \wedge dx + (C_x - B_z)\, dx \wedge dy. \tag{3.7}$$

Hence $d\alpha$ is a 2-form. Likewise, by considering the 2-form

$$\beta = A dy \wedge dz + B dz \wedge dx + C dx \wedge dy \tag{3.8}$$

we define

$$d\beta := dA \wedge dy \wedge dz + dB \wedge dz \wedge dx + dC \wedge dx \wedge dy \tag{3.9}$$

and obtain in an analogous manner

$$d\beta = (A_x + B_y + C_z)\, dx \wedge dy \wedge dz \tag{3.10}$$

where

$$dx \wedge dy \wedge dz = dy \wedge dz \wedge dx = dz \wedge dx \wedge dy \tag{3.11}$$

(this amounts to using (3.5) to write the exterior products in cyclic order). Hence $d\beta$ is a 3-form, and, since our space is 3-dimensional, we need not consider higher order forms.

Consider finally the quantity $d(dF)$ where dF is given by (2.16). Clearly

$$d(dF) = d(F_x) \wedge dx + d(F_y) \wedge dy + d(F_z) \wedge dz, \tag{3.12}$$

but

$$d(F_x) = F_{xx} dx + F_{xy} dy + F_{xz} dz \tag{3.13}$$

with similar expressions holding for $d(F_y)$ and $d(F_z)$. Upon substitution into (3.12) we obtain

$$\begin{aligned} d(dF) &= (F_{xy} dy + F_{xz} dz) \wedge dx \\ &\quad + (F_{yx} dx + F_{yz} dz) \wedge dy \\ &\quad + (F_{zx} dx + F_{zy} dy) \wedge dz \\ &= (F_{yx} - F_{xy})\, dx \wedge dy + (F_{xz} - F_{zx})\, dz \wedge dx \\ &\quad + (F_{zy} - F_{yz})\, dy \wedge dz. \end{aligned}$$

But since F and its partial derivatives are smooth each of these coefficients is identically zero. Hence, we have

$$d^2F := d(dF) = 0 \tag{3.14}$$

for any smooth function. By using (3.7) one can also immediately show that

$$d(d\alpha) := d^2\alpha = 0 \tag{3.15}$$

which suggests that d^2 applied to any p-form is equal to zero. Note that since all our considerations are 3-dimensional we need only the two cases $p = 1$ and $p = 2$ (see PROBLEM III.1).

The utility of this operation is seen by comparing the expressions (3.2), (3.8), and (2.16) with the corresponding formulas in ordinary vector analysis. Let

$$\mathbf{v} = A\mathbf{i} + B\mathbf{j} + C\mathbf{k}$$

and

$$\mathbf{r} = x\mathbf{i} + y\mathbf{j} + z\mathbf{k}.$$

Then obviously

$$\alpha = \mathbf{v} \cdot d\mathbf{r} \tag{3.16}$$

and the coefficients of the differentials appearing in (3.2) are analogous to the components of $\operatorname{curl}\mathbf{v}$ in the expression (3.7),

$$d\alpha \longleftrightarrow \operatorname{curl}\mathbf{v}. \tag{3.17}$$

Likewise by considering β in (3.8), the coefficients of the differentials given in (3.10) are reminiscent of

$$d\beta \longleftrightarrow \operatorname{div}\mathbf{v}. \tag{3.18}$$

Finally, in (2.16), the coefficients are analogous to those of grad F, and hence (3.14) corresponds to

$$d(dF) = 0 \longleftrightarrow \operatorname{curl}(\operatorname{grad} F) = \mathbf{0}. \tag{3.19}$$

The preceding calculations suggest the following interpretations of exterior differentiation:

i) when d is applied to a 0-form, i.e. a function, it produces a gradient-like expression i.e. (2.16);

ii) when d operators on 1-forms, i.e. (3.16) or dF; it yields a curl-like expression; i.e. (3.7) or (3.17) and (3.19); or

iii) when applied to a 2-form, i.e. (3.7) and (3.9) it gives a divergence-like expression

$$d\,(d\alpha) = 0 \longleftrightarrow \operatorname{div}\operatorname{curl}\mathbf{v} = 0. \tag{3.20}$$

III.4 Basic Properties of the Exterior Product and Differentiation

In the previous section we have indicated various properties of the exterior product and differentiation. We now summarize these — and other properties — within the framework of a 3-dimensional space which could be $\mathbf{V}_3$ or $\mathbf{E}_3$. Thus, while our discussion makes no pretense at any degree of generality, it deals with precisely the aspects of the Cartan calculus which we employ in differential geodesy.

For simplicity we will denote $\wedge^p\mathbf{V}^*$, or $\wedge^p\mathbf{T}^*_P$, by merely $\wedge^p$, and a p-form in this space is said to have degree p i.e. $\alpha \in \wedge^p \iff \deg\alpha = p$. Moreover, a smooth function F may be regarded as a 0-form, i.e. $\deg F = 0$.

Then the exterior product has the following properties which we state for a set $\{\vartheta_a\}$ of Pfaffians 1-forms (which are intended to suggest the properties enjoyed by the $\{\theta_a\}$):

$$1^0 \quad \vartheta_1 \wedge \vartheta_2 = -\vartheta_2 \wedge \vartheta_1 \text{ etc.} \tag{4.1}$$

$$2^0 \quad \vartheta_1 \wedge \vartheta_1 = 0 \text{ etc.} \tag{4.2}$$

$$3^0 \quad \vartheta_1 \wedge (\vartheta_2 \wedge \vartheta_3) = (\vartheta_1 \wedge \vartheta_2) \wedge \vartheta_3 = \vartheta_1 \wedge \vartheta_2 \wedge \vartheta_3 \tag{4.3}$$

$4^0 \quad \vartheta_1 \wedge \vartheta_2 \wedge \vartheta_3 \neq 0 \qquad (4.4)$
if and only if the ϑ_a are linearly independent

$5^0 \quad \vartheta_1 \wedge \vartheta_2 \wedge \vartheta_3 = 0 \qquad (4.5)$
if and only if the ϑ_a are linearly dependent

Exterior differentiation is a linear mapping

$$d : \Lambda^p \to \Lambda^{p+1} \qquad 0 \leq p \leq 2$$

such that

1^0 for $p = 0$, dF is identical with the differential of a smooth function, and

$$\deg(dF) = 1; \tag{4.6}$$

2^0 if F is a smooth function, and $\deg \omega = p$ ($p \leq 2$), then

$$d(F\omega) = (dF) \wedge \omega + F \cdot (d\omega) \tag{4.7}$$

with $\deg\{d(F\omega)\} = p + 1$;

3^0 if x^r are three independent variables then

$$d(dx^r) := d^2 x^r = 0, \tag{4.8}$$

and if F is a smooth function of x^r then

$$d(dF) := d^2 F = 0, \tag{4.9}$$

and more generally for any smooth 1-form ϑ

$$d(d\vartheta) := d^2\vartheta = 0; \tag{4.10}$$

4^0 for any smooth ϑ_s

$$d(\vartheta_1 \wedge \vartheta_2) = d\vartheta_1 \wedge \vartheta_2 - \vartheta_1 \wedge d\vartheta_2, \text{ etc.} \tag{4.11}$$

Note that it is tempting to summarize (4.8)-(4.10) into the requirement

$$d(d\omega) := d^2\omega = 0 \tag{4.12}$$

for any p-form ω. This is true, however it is usually meant for $p > 0$. The case $p = 0$ may be included but still does not necessarily include (4.8). Indeed (4.9), (4.10) and (4.12) assume this property, and although the conditions (4.8)-(4.10) seem tedious in comparison to (4.12) they are the *full* set of conditions.

III.5 Cartan's Viewpoint

By the Marussi Condition (i) of SECTION I-2 the theoretical framework of differential geometry is a 3-dimensional Euclidean space $\mathbf{E}_3$. The Cartan calculus in $\mathbf{E}_3$ may be developed in two manners: either by considering the geometry of a 3-dimensional Riemannian space $\mathbf{V}_3$ and then imposing the Lamé equations II-(12.9) which reduce $\mathbf{V}_3$ to $\mathbf{E}_3$, or by *ab initio* working in $\mathbf{E}_3$ by imposing certain very restrictive assumptions (see PROBLEM III.9). As in our discussion of the Ricci calculus in CHAPTER II, we will follow the first approach, although the second one has already enjoyed some success — at the hands of Grafarend and Grossman — in their beautiful contributions to theoretical geodesy. Our choice is essentially one of methodology, and *not* that one approach is right and the other wrong, but rather that the first approach seems more convenient for our purposes in making contact with the work of Marussi and Hotine. This motivation is especially convincing with regard to recasting Hotine's analysis in terms of the general leg calculus. It is less so in the case of Marussi since his original conception of intrinsic geometry was rather restrictive, and subsequently generalized by Hotine in his treatise.

The starting point of our approach is that the line element

$$ds^2 = g_{rs}dx^r dx^s \tag{5.1}$$

in a coordinate neighborhood Ω of $\mathbf{V}_3$ can be rewritten in the Cartan theory as

$$ds^2 = (\theta_1)^2 + (\theta_2)^2 + (\theta_3)^2 \tag{5.2}$$

where, as in I-(3.20), we write

$$\theta_a := \lambda_{ar}dx^r, \tag{5.3}$$

with $\{\theta_a\}$ being the Pfaffian basis in $\mathbf{T}_P^*$ that is dual to the vectorial basis $\{\boldsymbol{\lambda}_a\}$ in $\mathbf{T}_P$. We call this the *orthonormality hypothesis.* To understand the meaning of this, we revert to the discussion and notation employed in Section I-4, with the abstract vector spaces $\mathbf{V}$ and $\mathbf{V}^*$ replacing the tangent and cotangent spaces $\mathbf{T}_P$ and $\mathbf{T}_P^*$ respectively. Then the metric tensor g, regarded as a symmetric bilinear functional, has the following abstract components on the basis vectors of $\mathbf{V}$ and $\mathbf{V}^*$

$$g(\mathbf{e}_i, \mathbf{e}_j) := g_{ij} \tag{5.4}$$

$$g\left(f^i, f^j\right) := g^{ij} \tag{5.5}$$

which are 'inverses' of each other, viz.

$$g^{ik} g_{jk} = \delta^i_j. \tag{5.6}$$

In terms of leg indices in $\mathbf{T}_P$ and $\mathbf{T}_P^*$ (where both congruence and Pfaffian indices are denoted by the Latin indices a, b, c, ...), equations (5.4)-(5.6) become

$$g(\boldsymbol{\lambda}_a, \boldsymbol{\lambda}_b) := g_{ab} \tag{5.7}$$

$$g\left(\theta^a, \theta^b\right) := g^{ab} \tag{5.8}$$

$$g^{ac} g_{bc} = \delta^a_b. \tag{5.9}$$

But since the $\{\boldsymbol{\lambda}_a\}$ are orthonormal $g_{ab} = \delta_{ab}$, hence $g^{ab} = \delta^{ab}$ as displayed in II-(2.16). Thus, (5.8) expresses the orthonormality of the Pfaffian basis $\{\theta^a\}$ and the θ^a may be identified with the θ_a.

The orthonormality hypothesis is a mathematical convenience, and not a restrictive assumption which involves any loss of generality in our analysis. This follows immediately from the observation that in a 3-dimensional vector space $\mathbf{V}$ any arbitrary set of three linearly independent non-unit vectors may be replaced by a set of three orthonormal (unit) vectors. This is well known in elementary linear algebra where such a replacement process is usually called the *Gram-Schmidt orthogonalization process*, e.g. see HALMOS (1958). Our previous argument was designed to indicate that *mutatis mutandi* the *same* argument holds in $\mathbf{V}^*$, or more specifically in $\mathbf{T}_P^*$!

It is important to note that the orthonormality of the systems $\{\boldsymbol{\lambda}_a\}$ and $\{\theta_a\}$ *does not imply* that the underlying coordinate system x^r is an orthogonal system. This can occur for simple choices of these basis systems but it is by no means obligatory. For example in $\mathbf{E}_3$, if one chooses x^r to be spherical polar coordinates one obtains a very simple form for ds^2. To illustrate this, let us choose Hotine's way of writing spherical polar coordinates (see his discussion [page 5]) $x^r = (\omega, \phi, r)$ where these curvilinear coordinates are respectively the longitude, latitude and radial coordinate. Then

$$ds^2 = \left(r^2 \cos^2 \phi\right) d\omega^2 + r^2 d\phi^2 + dr^2, \tag{5.10}$$

and obviously

$$\begin{aligned} \theta_1 &:= r \cos \phi d\omega, \\ \theta_2 &:= r d\phi, \\ \theta_3 &:= dr; \end{aligned} \tag{5.11}$$

i.e.

$$\begin{aligned} \lambda_{1r} &:= \lambda_r = (r \cos \phi)\, \delta^1_r, \\ \lambda_{2r} &:= \mu_r = (r)\, \delta^2_r, \\ \lambda_{3r} &:= \nu_r = \delta^3_r. \end{aligned} \tag{5.12}$$

However, in his more general (ω, ϕ, N) coordinate system — which is also a curvilinear coordinate system in $\mathbf{E}_3$ — *all six covariant components of the metric tensor are non-zero.* Thus, this ds^2 has the form

$$\begin{aligned} ds^2 = {} & g_{11} d\omega^2 + g_{22} d\phi^2 + g_{33} dN^2 \\ & + 2g_{12} d\omega d\phi + 2g_{13} d\omega dN \\ & + 2g_{23} d\phi dN, \end{aligned} \tag{5.13}$$

where each $g_{rs} \neq 0$. In this case the Pfaffian basis forms turn out to be

$$\begin{aligned} \theta_1 &:= A_1 d\omega + B_1 d\phi + C_1 dN, \\ \theta_2 &:= A_2 d\omega + B_2 d\phi + C_2 dN, \\ \theta_3 &:= C_3 dN; \end{aligned} \tag{5.14}$$

and

$$\begin{aligned} \lambda_{1r} &:= \lambda_r = A_1 \delta^1_r + B_1 \delta^2_r + C_1 \delta^3_r, \\ \lambda_{2r} &:= \mu_r = A_2 \delta^1_r + B_2 \delta^2_r + C_2 \delta^3_r, \\ \lambda_{3r} &:= \nu_r = C_3 \delta^3_r. \end{aligned} \tag{5.15}$$

The explicit values of the coefficients $A_1, B_1, \cdots, C_3$ are given in his [12.041]–[12.043] with the values of g_{rs} being displayed by Hotine in [12.069].

Hence, an orthogonal coordinate system x^r occurs when the θ_a are each multiples of a single coordinate differential as in (5.11), however a non-orthogonal coordinate system x^r arises when the θ_a are linear combinations of the coordinate differentials, e.g. in (5.14).

III.6 Structural Equations of Cartan

The essence of Cartan's approach to Riemannian geometry is, starting from the line element ds^2, i.e. (5.2) on which we have imposed the orthonormality hypothesis, the basic geometric structure of $\mathbf{V}_3$ is derivable from a pair of structural equations: 'l'equations de structure'. These equations are meaningful in the specialization which reduces $\mathbf{V}_3$ to $\mathbf{E}_3$, and in a geometric sense they are the central feature of the Cartan calculus. Their role is somewhat curious in that they are not derived, but rather represent a new approach to Riemannian geometry. Essentially what Cartan did was to identify the Pfaffian basis $\{\theta_a\}$ as a new structural element, re-write ds^2 in the form (5.2), and then to say that from this viewpoint the remainder of Riemannian geometry could be deduced from this pair of equations involving various combinations of the exterior derivatives of the θ's and their exterior products. Needless to say, such a radical approach — radical compared with that employed in the tensor calculus for several decades — was not understood by Cartan's contemporaries, and was felt to be both unnecessary and confusing. No doubt some of this was due to Cartan's mode of presentation which although ingenious, was quite concise and often appeared to be based on heuristic and intuitive grounds. An enlightening and interesting account relating how it was received is given in SCHOUTEN (1954) (see his discussions in Chapter II §12 and III §§9, 10). Schouten, to his credit, recognized the importance of Cartan's approach but still termed it 'Cartan's symbolical method' which implied it was largely a new notation. However, he did emphasize that

> "Both methods have their advantages (Cartan himself performed miracles with his method!) and both should be used by every author, each in the place where it suits best."

We completely agree with this assessment, and in effect the general leg calculus that we will employ in this book is yet another attempt to merge Cartan's ideas with those of Ricci, and convince people that the resulting synthesis is worth learning and using in differential geodesy.

The two structural equations in $\mathbf{V}_3$, subject to the orthonormality hypothesis, are as follows:

$$\Sigma_a := d\theta_a + \omega_{ab} \wedge \theta_b = 0 \tag{6.1}$$

$$\Omega_{ab} := d\omega_{ab} + \omega_{ac} \wedge \omega_{cb} \tag{6.2}$$

where Σ_a is the *torsion 2-form* (which vanishes for a Riemannian geometry), $\Omega_{ab} = -\Omega_{ba}$ is the *curvature 2-form*, and $\omega_{ab} = -\omega_{ba}$ is a 1-form known as the *connection 1-form*. We will determine explicit expressions for Ω_{ab} and ω_{ab} as our discussion unfolds, and the three values of the skew-symmetric index pair will always be enumerated by our canonical labelling scheme, viz. $ab = 12, 31$ and 23 which was employed previously in CHAPTER II (recall II-(4.8)).

In effect Cartan's methodology involved solving (6.1) for the ω_{ab}, i.e.

$$\omega_{ab} = \beta_{abc}\theta_c \ , \tag{6.3}$$

where in our notation and terminology the β's are *some* leg coefficients:

$$\beta_{abc} = -\beta_{bac} \tag{6.4}$$

which can be exhibited for a choice of the $\{\theta_a\}$ used in expressing ds^2; then, by (6.2), these β_{abc} are used to determine the leg coefficients which occur in Ω_{ab}. By virtue of (6.4) the β_{abc} turn out to be the Ricci coefficients — up to a sign difference — while the coefficients appearing in Ω_{ab} will be none other than the R_{abcd} which are the leg coefficients given in II-(12.6). Once we have established these facts, then the charm and power of the Cartan calculus is that it allows one to calculate these quantities *without* recourse to the more complicated formulas which occur in the Ricci calculus. Then the Lamé equations ultimately reduce to the three 2-form expressions

$$\Omega_{ab} = 0 \tag{6.5}$$

which reduce $\mathbf{V}_3$ to an $\mathbf{E}_3$.

We now proceed to justify these conclusions and establish the required identifications.

Exterior differentiation of (5.3) immediately gives

$$\begin{aligned} d\theta_a &= d\lambda_{ar} \wedge dx^r \\ &= \lambda_{ar;s} dx^s \wedge dx^r \\ &= -\lambda_{ar,s} dx^r \wedge dx^s \end{aligned} \tag{6.6}$$

by virtue of the skew-symmetry of the coordinate-based indices, and

$$\lambda_{ar;s} - \lambda_{as;r} = \lambda_{ar,s} - \lambda_{as,r}. \tag{6.7}$$

But the inverse of (5.3) is

$$dx^r = \lambda_a{}^r \theta_a \tag{6.8}$$

so

$$\begin{aligned} d\theta_a &= -\lambda_{ar,s} \lambda_b{}^r \lambda_c{}^s \theta_b \wedge \theta_c \\ &= -\gamma_{abc} \theta_b \wedge \theta_c \\ &= \gamma_{abc} \theta_c \wedge \theta_b \end{aligned} \tag{6.9}$$

by recalling II-(4.1). Then, (6.1) and (6.3) show that

$$\beta_{abc} = -\gamma_{abc}, \tag{6.10}$$

i.e. the Cartan coefficients β_{abc} are the negative Ricci coefficients γ_{abc}! The second expression in (6.9) involves skew-symmetrization over the last pair of indices in γ_{abc}, i.e. $-\frac{1}{2}(\gamma_{abc} - \gamma_{acb})$, which is equal to the provisional quantity $-\#_{abc}$ of SECTION II-6, which ultimately leads to $-\alpha_{bca}$, and these are the negative anholonomic coefficients of SECTION II-7. Thus, an alternate version of (6.9) is given by

$$d\theta_a = \frac{1}{2}\alpha_{bca}\theta_b \wedge \theta_c. \tag{6.11}$$

The identification (6.10) is important, but to use it in (6.3) immediately leads to the appearance of a number of minus signs in our analysis. These can be avoided by retaining the β-coefficients in many of the Cartan equations and then employing (6.10) in the final result. This will be our procedure.

Now we consider the second structural equation (6.2). Exterior differentiation of (6.3) yields

$$\begin{aligned} d\omega_{ab} &= \beta_{abc/d}\theta_d \wedge \theta_c + \beta_{abc}d\theta_c \\ &= -\beta_{abc/d}\theta_c \wedge \theta_d + \beta_{abc}\left(-\omega_{cd} \wedge \theta_d\right) \\ &= -\left(\beta_{abc/d} + \beta_{abf}\beta_{fdc}\right)\theta_c \wedge \theta_d \end{aligned} \tag{6.12}$$

while

$$\begin{aligned} \omega_{af} \wedge \omega_{fb} &= \beta_{afc}\beta_{fbd}\theta_c \wedge \theta_d \\ &= -\beta_{afc}\beta_{bfd}\theta_c \wedge \theta_d. \end{aligned} \tag{6.13}$$

Hence, Ω_{ab} by definition is a linear combination of the exterior products $\theta_c \wedge \theta_d$ which provisionally we will write as

$$\Omega_{ab} = \frac{1}{2}\#_{abcd}\theta_c \wedge \theta_d \tag{6.14}$$

where

$$\begin{aligned} \#_{abcd} &= -\beta_{abc/d} + \beta_{abd/c} - \beta_{abf}\beta_{fdc} + \beta_{abf}\beta_{fcd} \\ &\quad -\beta_{afc}\beta_{bfd} + \beta_{afd}\beta_{bfc}. \end{aligned} \tag{6.15}$$

Substitution of (6.10) in (6.15) establishes that $\#_{abcd} = R_{abcd}$, i.e., (6.15) is identical with the expression II-(12.6) in the Ricci calculus.

Note that (6.1) not only gives (6.3), but it is essential to have it to replace the $d\theta$ term which inevitably arises upon exterior differentiation of a p-form expression (see the second line in (6.12)). The 'correct' version of (6.14) is then

$$\Omega_{ab} := \frac{1}{2}R_{abcd}\theta_c \wedge \theta_d \tag{6.16}$$

and hence the Cartan version of the algebraic Bianchi identity II-(12.8) is given by

$$\Omega_{ab} \wedge \theta_b = 0. \tag{6.17}$$

The Lamé equations, which are equivalent to the vanishing of *all* of the leg coefficients R_{abcd}, then obviously correspond to (6.5) as claimed.

Now as an example of the methodology of the Cartan calculus, we calculate ω_{ab}, Ω_{ab} for the line element (5.10). The $\{\theta_a\}$ are given in (5.11) and the corresponding differentials are obviously

$$\begin{aligned} d\omega &= (r^{-1}\sec\phi)\,\theta_1, \\ d\phi &= (r^{-1})\,\theta_2, \\ dr &= \theta_3. \end{aligned} \tag{6.18}$$

Exterior differentiation of (5.10) yields

$$\begin{aligned} d\theta_1 &= \cos\phi dr \wedge d\omega - r\sin\phi d\phi \wedge d\omega, \\ d\theta_2 &= dr \wedge d\phi, \\ d\theta_3 &= 0; \end{aligned} \tag{6.19}$$

and, upon using (6.18) to reintroduce the Pfaffian basis 1-forms, we obtain

$$\begin{aligned} d\theta_1 &= (r^{-1}\tan\phi)\,\theta_1 \wedge \theta_2 + (r^{-1})\,\theta_3 \wedge \theta_1 \\ d\theta_2 &= (-r^{-1})\,\theta_2 \wedge \theta_3 \\ d\theta_3 &= 0. \end{aligned} \tag{6.20}$$

hence by (6.11) we see that

$$\begin{aligned} \alpha_{121} &= -r^{-1}\tan\phi, \\ \alpha_{311} &= -r^{-1}, \\ \alpha_{231} &= 0; \\ \\ \alpha_{122} &= 0, \\ \alpha_{312} &= 0, \\ \alpha_{232} &= r^{-1}; \\ \\ \alpha_{123} &= 0, \\ \alpha_{313} &= 0, \\ \alpha_{233} &= 0. \end{aligned} \tag{6.21}$$

Expansion of (6.1) gives

$$\begin{aligned}
d\theta_1 &= -\omega_{12} \wedge \theta_2 - \omega_{13} \wedge \theta_3 \\
&= -\omega_{12} \wedge \theta_2 + \omega_{31} \wedge \theta_3, \\
d\theta_2 &= -\omega_{21} \wedge \theta_1 - \omega_{23} \wedge \theta_3 \\
&= \omega_{12} \wedge \theta_1 - \omega_{23} \wedge \theta_3, \\
d\theta_3 &= -\omega_{31} \wedge \theta_1 - \omega_{32} \wedge \theta_2 \\
&= -\omega_{31} \wedge \theta_1 + \omega_{23} \wedge \theta_2,
\end{aligned} \tag{6.22}$$

and comparison with (6.20) then yields

$$\begin{aligned}
\omega_{12} &= (-r^{-1} \tan \phi)\, \theta_1, \\
\omega_{31} &= (-r^{-1})\, \theta_1, \\
\omega_{23} &= (r^{-1})\, \theta_2.
\end{aligned} \tag{6.23}$$

Exterior differentiation of these expressions gives

$$\begin{aligned}
d\omega_{12} &= (r^{-2})\, \theta_1 \wedge \theta_2, \\
d\omega_{31} &= (-r^{-2} \tan \phi)\, \theta_1 \wedge \theta_2, \\
d\omega_{23} &= 0;
\end{aligned} \tag{6.24}$$

and we also have from (6.23) the following exterior products:

$$\begin{aligned}
\omega_{31} \wedge \omega_{23} &= (-r^{-2})\, \theta_1 \wedge \theta_2, \\
\omega_{12} \wedge \omega_{23} &= (-r^{-2} \tan \phi)\, \theta_1 \wedge \theta_2, \\
\omega_{12} \wedge \omega_{31} &= 0.
\end{aligned} \tag{6.25}$$

Upon expanding (6.2) and rewriting them in our 'canonical' labelling, we obtain, using (6.24) and (6.25),

$$\begin{aligned}
\Omega_{12} &= d\omega_{12} + \omega_{31} \wedge \omega_{23} = 0, \\
\Omega_{31} &= d\omega_{31} - \omega_{12} \wedge \omega_{23} = 0, \\
\Omega_{23} &= d\omega_{23} + \omega_{12} \wedge \omega_{31} = 0;
\end{aligned} \tag{6.26}$$

i.e., the Lamé equations (6.5) are identically satisfied, as expected.

In the leg calculus, each of the Ricci coefficients will be replaced by a specific function — a leg coefficient — associated with the congruences Γ_a $(a = 1,2,3)$. The α_{abc} as indicated in SECTION II-6, will not be used, but given explicit expressions for the connection 1-forms, the values of the α_{abc} can be read off from the first structural equation (6.11) as the coefficients occurring on the right-hand sides of the $d\theta$'s. Later, the explicit values, as they arise in the Grafarend anholonomity matrices of CHAPTER IV, are not *per se* of any significance — the important fact is whether the anholonomic coefficients are non-zero or not. Likewise, relative to (6.16) the R_{abcd} will be equated to zero by virtue of the Lamé equations II-(12.9).

III.7 Commutators and Schouten Identities

The first structural equation upon exterior differentiation immediately yields the commutators and, more generally, the Schouten identities.

Let F be an arbitrary smooth function and consider the Pfaffian expression of dF, i.e.

$$dF = F_{/a}\theta_a. \tag{7.1}$$

Then upon exterior differentiation, since $d\,(dF) := d^2F = 0$, we have

$$\begin{aligned} &dF_{/a} \wedge \theta_a + F_{/a} d\theta_a = 0 \\ &F_{/a/b}\theta_b \wedge \theta_a + F_{/a}\,(-\omega_{ab} \wedge \theta_b) = 0. \end{aligned} \tag{7.2}$$

From (7.2) it follows that

$$\begin{aligned} F_{/a/b}\theta_a \wedge \theta_b &= F_{/a}\,(-\omega_{ab} \wedge \theta_b) \\ &= -F_{/c}\omega_{cb} \wedge \theta_b \\ &= -F_{/c}\beta_{cba}\theta_a \wedge \theta_b \\ &= F_{/c}\gamma_{cba}\theta_a \wedge \theta_b \\ &= -F_{/c}\gamma_{cab}\theta_a \wedge \theta_b \end{aligned} \tag{7.3}$$

with skew-symmetrization occurring on both sides of this equation. Thus

$$F_{/a/b} - F_{/b/a} = -\alpha_{abc}F_{/c} \tag{7.4}$$

gives the F-commutators $[F_{\mathcal{A}}]$ $(\mathcal{A} = \text{I, II, III})$. This shows that the $[F_{\mathcal{A}}]$ $(\mathcal{A} = \text{I, II, III})$ of II-(7.15) need not be remembered explicitly, since they may be immediately obtained by exterior differentiation of (7.1), which amounts to forming and exhibiting the integrability conditions of this leg equation.

For the Schouten identities, consider exterior differentiation of (6.11). Then $d(d\theta_a) = d^2\theta_a = 0$ viz.

$$(d\alpha_{bca}) \wedge \theta_b \wedge \theta_c + \alpha_{bca} d(\theta_b \wedge \theta_c) = 0$$

or upon re-labelling

$$(d\alpha_{abf}) \wedge \theta_a \wedge \theta_b + \alpha_{abf} d(\theta_a \wedge \theta_b) = 0, \tag{7.5}$$

i.e.

$$\alpha_{abf/c}\theta_c \wedge \theta_a \wedge \theta_b + \alpha_{abf} d\theta_a \wedge \theta_b - \alpha_{abf}\theta_a \wedge d\theta_b = 0$$

$$\alpha_{abf/c}\theta_a \wedge \theta_b \wedge \theta_c + \alpha_{abf}\left(-\frac{1}{2}\alpha_{ega}\theta_e \wedge \theta_g \wedge \theta_b\right)$$

$$-\alpha_{abf}\left(-\frac{1}{2}\alpha_{egb}\theta_a \wedge \theta_e \wedge \theta_g\right) = 0$$

$$\left(\alpha_{abf/c} + \frac{1}{2}\alpha_{hbf}\alpha_{ach} + \frac{1}{2}\alpha_{ahf}\alpha_{bch}\right)\theta_a \wedge \theta_b \wedge \theta_c = 0. \tag{7.6}$$

Hence

$$\begin{aligned}\alpha_{abd/c} \; &+ \alpha_{bcd/a} + \alpha_{cad/b} + \frac{1}{2}(\alpha_{fbd}\alpha_{acf} + \alpha_{fcd}\alpha_{baf} + \alpha_{fad}\alpha_{cbf} \\ &+ \alpha_{afd}\alpha_{bcf} + \alpha_{bfd}\alpha_{caf} + \alpha_{cfd}\alpha_{abf}) = 0\end{aligned} \tag{7.7}$$

which is equivalent to II-(9.4).

III.8 Cartan Calculus and the Classical Tensor Calculus

We now show how the Cartan structural equations include as a special case the familiar expressions which occur in the coordinate-based tensor calculus. This not only demonstrates the power of his approach, but also the scope of the generalization implicit in the methods previously given in this chapter.

Our first step is the obvious one that the two bases $\{\theta_a\}$ and $\{dx^r\}$ are analogous and in the classical theory the former reduces to the latter. These bases are related by (5.3) and (6.8) which, if they are identical must require that

$$\lambda_a{}^r = \delta^r_a \tag{8.1}$$

i.e.

$$dx^r = \delta^r_a \theta_a \tag{8.2}$$

i.e.

$$dx^1 = \theta_1, \qquad dx^2 = \theta_2, \quad dx^3 = \theta_3. \tag{8.3}$$

These equations require some comment: clearly they mix leg and coordinate-based indices, viz. the indices on the left-hand sides of (8.3) are coordinate-based while those on the right-hand sides are Pfaffian indices (the latter occur as subscripts by virtue of the orthonormality hypothesis). We then replace (6.3) and (6.16) respectively, by

$$\omega^r{}_s := \Gamma^r_{st} dx^t \tag{8.4}$$

and

$$\Omega^r{}_s := \frac{1}{2} R^r_{smn} dx^m \wedge dx^n \tag{8.5}$$

where we regard the Γ^r_{st} and $R^v{}_{smn}$ as being unknown quantities. We will show that the structural equations (which we formally write in terms of coordinate-based indices):

$$\Sigma^r := d\theta^r + \omega^r{}_s \wedge \theta^s = 0 \tag{8.6}$$

and

$$\Omega^r{}_s := d\omega^r{}_s + \omega^r{}_t \wedge \omega^t{}_s \tag{8.7}$$

will yield an identification of Γ^r_{st} with the Christoffel symbols of the second kind, and $R^r{}_{smn}$ with the Riemann-Christoffel curvature tensor. Note that in these equations the distinction between contravariant and covariant indices has forced us to express (8.6) and (8.7) in this 'mixed' form.

Before considering the meaning of these equations it is useful to recall what is known as the *Fundamental Theorem of Riemannian Geometry* which asserts that if we are given:

(1⁰) a symmetric second order tensor with covariant components g_{rs}

$$\det \|g_{rs}\| \neq 0; \tag{8.8}$$

(2⁰) a symmetric set of connection coefficients

$$\Gamma^p_{rs} = \Gamma^p_{sr}; \tag{8.9}$$

(3⁰) a covariant derivative, denoted by a comma, with respect to the connection coefficients such that

$$g_{rs,t} := g_{rs;t} - \Gamma^p_{rt} g_{ps} - \Gamma^p_{st} g_{rp} = 0; \tag{8.10}$$

then the g_{rs} may be identified with the components of a metric tensor, and the Γ^p_{rs} become the Christoffel symbols of the second kind.

All three of these conditions are familiar in the tensor calculus in an N-dimensional Riemannian space $\mathbf{V}_N$, but the point of the theorem is that the conditions are *structural features* of $\mathbf{V}_N$, i.e. all three are required to particularize $\mathbf{V}_N$ from some more general N-dimensional *non-Riemannian* space $\mathbf{X}_N$. If any one of these conditions are omitted, one does not have a $\mathbf{V}_N$.

For example, (1⁰) yields the existence of a symmetric inverse tensor g^{rs} which satisfies the coordinate-based analogue of (5.6) or (5.9). The familiar expression for Γ^p_{rs}, i.e.

$$\Gamma^p_{rs} := \frac{1}{2} g^{pt} \left(g_{st;r} + g_{tr;s} - g_{rs;t} \right) \tag{8.11}$$

clearly satisfies (2⁰), and for such Γ^p_{rs} we also have (3⁰) — the so-called *Ricci identity/lemma.* It is interesting to reproduce the part of the fundamental theorem which leads to (8.11).

To do this, rewrite (8.10) as

$$g_{rs;t} = \Gamma^p_{rt} g_{ps} + \Gamma^p_{st} g_{rp} \tag{8.12}$$

and multiply by dx^t to obtain

$$dg_{rs} = g_{ps} \omega^p{}_r + g_{rp} \omega^p{}_s = \omega_{sr} + \omega_{rs}. \tag{8.13}$$

Then if g_{rs} are constants we have $\omega_{rs} + \omega_{sr} = 0$, which is the analogue of $\omega_{ab} + \omega_{ba} = 0$, expressing the skew-symmetry of the connection 1-forms, i.e. ultimately the skew-symmetry of the Ricci or Cartan coefficients.

Upon forming a cyclic permutation of the indices r, s and t in (8.12) we obtain

$$g_{st;r} = \Gamma^p_{sr} g_{pt} + \Gamma^p_{tr} g_{sp} \tag{8.14}$$

$$g_{tr;s} = \Gamma^p_{ts} g_{pr} + \Gamma^r_{rs} g_{tp}. \tag{8.15}$$

Adding (8.12), (8.14) and (8.15), remembering (8.9), then gives

$$g_{st;r} + g_{tr;s} - g_{rs}; t = 2\Gamma^p_{rs} g_{pt} \tag{8.16}$$

and hence (8.11) follows by solving for Γ^p_{rs} (i.e. by multiplying by g^{qt}). This procedure is classically known as the *Christoffel elimination scheme*, and we recall that an analogous method was employed in SECTION II-6 for obtaining an expression for γ_{abc} in terms of the anholonomic coefficients α_{abc}.

Now we consider the first structural equation (8.6). By virtue of (8.3), we have $d\theta^r = d\,(dx^r) = d^2x^r = 0$ so that (8.6) reduces to

$$\omega^r{}_s \wedge \theta^s = 0. \tag{8.17}$$

By (8.4) this reduces to the conditions

$$\Gamma^r_{st} - \Gamma^r_{ts} = 0$$

viz. (8.9). If this were not satisfied, i.e. condition (2^0) of the fundamental theorem was omitted then we would have

$$\Sigma^r = \frac{1}{2}\,(\Gamma^r_{st} - \Gamma^r_{tr})\, dx^s \wedge dx^t \neq 0. \tag{8.18}$$

In this case Σ^r is known as the *torsion* 2-form and the resulting space $\mathbf{X}_N$ is said to have non-zero torsion. Condition (2^0), i.e. (8.9), excludes such spaces from our consideration, (although such spaces have been studied in theoretical geodesy).

Finally, we show that the second structural equation (8.7) leads us to the classical components of the curvature tensor. Clearly,

$$\begin{aligned} d\omega^r{}_s &= d\Gamma^r_{st} \wedge dx^t \\ &= \Gamma^r_{st;n} dx^n \wedge dx^t \\ &= -\Gamma^r_{sm;n} dx^m \wedge dx^n, \end{aligned} \tag{8.19}$$

and

$$\begin{aligned} \omega^r{}_t \wedge \omega^t{}_s &= \Gamma^r_{tp} dx^p \wedge \Gamma^t_{qs} dx^q \\ &= \Gamma^r_{tp} \Gamma^t_{qs} dx^p \wedge dx^q \\ &= \Gamma^r_{tm} \Gamma^t_{ns} dx^m \wedge dx^n. \end{aligned} \tag{8.20}$$

Each of the final expressions in (8.19) and (8.20) is skew-symmetric, and hence by (8.5) we obtain

$$R^r{}_{smn} = \Gamma^r_{sn;m} - \Gamma^r_{sm;n} + \Gamma^t_{sn} \Gamma^r_{tm} - \Gamma^t_{sm} \Gamma^r_{tn} \tag{8.21}$$

which is the classical tensor calculus equation. The algebraic Bianchi identity then follows from

$$\Omega^r{}_s \wedge dx^s = 0, \tag{8.22}$$

i.e.,

$$R^r{}_{smn} dx^m \wedge dx^n \wedge dx^s = 0,$$

which yields

$$R^r{}_{smn} + R^r{}_{mns} + R^r{}_{nsm} = 0. \tag{8.23}$$

III.9 Coordinates in the Cartan and Ricci Calculus

In our discussion of the Cartan and Ricci calculus the use of local coordinates x^r defined on a coordinate neighborhood Ω of $\mathbf{V}_3$ or $\mathbf{E}_3$ has been minimal, and our analysis has primarily focused on the leg indices, i.e. Pfaffian and congruence indices. This has been intentional and the passive role of the coordinates x^r has not been accidental. Indeed, the coordinate-based indices have been 'place-holders,' and they were necessary to make expressions like (5.3), i.e.,

$$\theta_a = \lambda_{ar} dx^r, \tag{9.1}$$

and II-(2.3), i.e.

$$\lambda_a = \lambda_a{}^r \frac{\partial}{\partial x^r}, \tag{9.2}$$

meaningful. In other words, such expressions were required to relate θ_a and $\boldsymbol{\lambda}_a$ to their 'familiar' coordinate-based covariant and contravariant components relative to *some* local coordinate system $\{x^r\}$ on Ω in $\mathbf{V}_3$ or $\mathbf{E}_3$. The set $\{x^r\}$ is known as an ***ambient*** (local) ***coordinate system*** on Ω and, strictly speaking, all that is demanded of this system is that it be ***admissible,*** i.e. on Ω one can make the changes

$$\begin{aligned} x^r \mapsto \overline{x}^r &= f^r\left(x^1, x^2, x^3\right) \\ \overline{x}^r \mapsto x^r &= g^r\left(\overline{x}^1, \overline{x}^2, \overline{x}^3\right) \end{aligned} \tag{9.3}$$

where f^r and g^r are smooth functions. Apart from this regularity hypothesis on Ω, nothing is assumed about the x^r, and they need not have any geodetic/geometric significance.

The motivation is that leg-based components of such differential-geometric quantities such as vectors, directional derivatives, or tensors may have a more immediate interpretation since, in addition to the orthonormality conditions, one may choose the $\{\theta_a\}$ and $\{\boldsymbol{\lambda}_a\}$ as desired. Indeed, not all of these directions need be of equal importance, however, as noted in SECTION II-3 we will often be interested in $\boldsymbol{\Gamma}_3$, which is normal to a surface $\mathbf{S}$. If, moreover, $\mathbf{S}$ is defined by the equation

$$F\left(x^1, x^2, x^3\right) = \text{constant},$$

then $F_{;r}$ will be collinear with $\boldsymbol{\lambda}_3 := \boldsymbol{\nu}$ and hence

$$\theta_3 := \nu_r dx^r \tag{9.4}$$

will be proportional to dF. Thus, the intrinsic geometry of $\mathbf{S}$ is characterized by the condition

$$\theta_3 = 0. \tag{9.5}$$

While $\boldsymbol{\Gamma}_3$ has an immediate interpretation, the vectors $\boldsymbol{\lambda}_1 := \boldsymbol{\lambda}$ and $\boldsymbol{\lambda}_2 := \boldsymbol{\mu}$ have *some* importance since they are directions orthogonal to $\boldsymbol{\nu}$.

III.10 Conclusions and Summary

The previous sections have outlined the principal features of the Cartan calculus, and we have seen several aspects of it which are pleasing. First, it works exclusively in $\mathbf{T}_P^*$ which complements the Ricci calculus in $\mathbf{T}_P$. By SECTION I-4 we know that both these spaces are required for a complete abstract picture of tensors, and now we have formalisms which deal with each of them. No geometrical interpretation of the Cartan coefficients was presented, but since up to a sign these coefficients are the Ricci coefficients we know the meaning of them. Strictly speaking, this non-geometrical aspect of the Cartan theory is illusory: geometrical interpretations can be given within the context of the theory, but they are not as immediate as in the Ricci theory. Such a situation is to be expected since the interpretations given in SECTION II-5 are in terms of the congruences (only 6^0 translates readily into families of surfaces) while the Ricci theory *ab initio* works in terms of (unit) tangent vectors to the $\boldsymbol{\Gamma}_a$, whereas the Cartan theory deals with the duals of these tangent vectors, i.e. the $\{\theta_a\}$. The discussion in SECTION 1 explicitly indicated this, and if (1.1) was identified with one of the Pfaffians, say θ, then $\theta = 0$ would define an integral surface in $\mathbf{V}_3$ or $\mathbf{E}_3$.

For our purposes, apart from the conciseness — largely a consequence of the two structural equations — the real advantage of the Cartan calculus is the facility and naturalness with which it handles the integrability conditions associated with a system of partial differential equations. This is immediately evident in the treatment of the commutators given in SECTION II-7. Much of differential geometry essentially reduces to the study of systems of partial differential equations, and the same can be said of differential geodesy. Obviously, it is advantageous to employ a formalism which permits one to immediately exhibit these integrability conditions in a natural and efficient manner without recourse to the commutators $[F_{\mathcal{A}}]$ $\mathcal{A} = \mathrm{I, II, III}$ for each function F which occurs in a leg differential equation. This is the great virtue of the Cartan theory, and it is no accident — the entire apparatus (apart from the structural equations) is derived from the theory of systems of differential equations.

Finally, we note that although it may appear that certain aspects of the theory are somewhat contrived, the whole theory has applications in continuous, or Lie, groups — including Lie algebras — and algebraic topology, which suggest that it has a deeper conceptual content and is the 'right' way to do things. These include the seminal notions of the exterior product and exterior differentiation, both of which are missing from the Ricci calculus, which is essentially a calculus of congruences. The above features of both theories suggest that for our needs — which are admittedly very special — these two theories are complementary, and that a unification of them will furnish a formalism which is more robust than either of them considered separately. Thus, in the next chapter this unification will be accomplished with special consideration to the needs of the Marussi-Hotine approach to differential geodesy.

PROBLEMS FOR CHAPTER III

III.1 Verify that $d^2\alpha = 0$ for any differential p-form in $\mathbf{E}_3$, and more generally that this property holds for any $p < n$ form in n-dimensions. The property $d^2 = 0$ is known as the *Poincaré property.*

III.2 Let F, G and H be smooth functions of rectangular Cartesian coordinates (x, y, z) on some region $\mathcal{R}$ of $\mathbf{E}_3$. Then evaluate the exterior product

$$dF \wedge dG \wedge dH.$$

III.3 Using exterior differentiation, find the necessary and sufficient condition that the total differential equation $\omega = 0$ i.e.

$$\omega = M dx + N dy$$

in two independent variables (x, y) where M and N are smooth functions, be an *exact differential equation.*

III.4 Show that any differential equation $\omega = 0$ in two independent variables (as in PROBLEM III.3) is either an exact differential equation or can be made exact upon multiplication by a suitable integrating factor $\mu = \mu(x, y)$.

III.5 Compute the connection 1-forms ω_{ab} and the curvature 2-forms Ω_{ab} for the line element given in PROBLEM II.5.

III.6 Compute the connection 1-forms ω_{ab} and the curvature 2-forms Ω_{ab} for the line element given in PROBLEM II.7.

III.7 Upon writing $\omega := X_r dx^r$, so that $\omega = 0$ expresses the total differential equation (1.1), show that the condition for the *complete integrability* of this equation is

$$\omega \wedge d\omega = 0 \Longleftrightarrow \varepsilon^{rst} X_r X_{s;t} = 0$$

(compare with PROBLEM II.9).

III.8 For the total differential equation $\omega = 0$ discussed in PROBLEM III.7, one may also consider the condition for *ordinary integrability*

$$d\omega = 0 \Longleftrightarrow \varepsilon^{rst} X_{s;t} = 0.$$

These notions are not equivalent — check! — and show that *one* of them is invariant under multiplication of $\omega = 0$ by an *integrating factor* μ, while the other is not invariant under the change $\omega \mapsto \mu\omega$.

III.9 Cartan's approach to differential geometry in $\mathbf{E}_3$, see CARTAN (1928), was based on the following intuitive arguments. Denote by $\mathbf{P}$, the position vector of a point in $\mathbf{E}_3$ relative to some fixed origin 0, and let x^r be an arbitrary system of curvilinear coordinates. Then he wrote

$$\mathbf{e}_r = \mathbf{P}_{;r} \tag{*}$$

where the semi-colon denotes partial differentiation (NB: $\mathbf{P}_{;r}$ are *not* the covariant components of a vector!). Hence, in terms of differentials one can write

$$d\mathbf{P} = \mathbf{e}_r dx^r \qquad [*]$$

and partial differentiation of (*) yields

$$\mathbf{e}_{r;s} = \mathbf{P}_{;r;s}$$

which is obviously symmetric in r and s. Again in terms of differentials one can write

$$d\mathbf{e}_r = \mathbf{e}_{r;s} dx^s$$

and by the symmetry previously noted, he made the identification

$$\mathbf{e}_{r;s} = \Gamma^m_{rs} \mathbf{e}_m \tag{**}$$

where Γ^m_{rs} are the Christoffel symbols of the second kind. Thus putting the

$$\omega^m_{\ n} := \Gamma^m_{ns} dx^s$$

we have

$$d\mathbf{e}_r = \omega^m_{\ r} \mathbf{e}_m. \qquad [**]$$

By partial differentiation of (**), form and identify the difference

$$\mathbf{e}_{r;s;t} - \mathbf{e}_{r;t;s}.$$

(The alert reader will note the apparently *strange usage* of a 'coordinate index' on the **e** vector fields in (*). For example, on the left-hand side of (*) the index obviously labels the vector fields, whereas on the right-hand side of (*) it is clearly a coordinate index viz. partial differentiation with respect to x^r. This cunning mixture of indices and their interpretation, bedevils Cartan's original theory and makes it very easy to become confused. Moreover, (*) and (**), and their differential versions [*] and [**], are valid *only* in a Euclidean space. Our preference, as given in the text, is to avoid this methodology and follow a Riemannian approach and then specialize it to Euclidean geometry by imposing the Lamé equations.)

III.10 For Hotine's (ω, ϕ, N) coordinate system (see Chapter 12 for a discussion, with the components of the metric tensor g_{rs} being given by [12.069]), show that the three Pfaffian forms (5.14) are given by

$$\begin{aligned}\theta_1 &= \{-k_2 \cos\phi d\omega + t_1 d\phi + (k_2\gamma_1 - t_1\gamma_2)\, dN/n\}/K,\\ \theta_2 &= \{t_1 \cos\phi d\omega - k_1 d\phi + (k_1\gamma_2 - t_1\gamma_1)\, dN/n\}/K,\\ \theta_3 &= dN/n.\end{aligned}$$

ZUND (1993)

— o —

In PROBLEMS III.11-14, the individual components and components of the differential forms are labelled by the Latin indices i, j, k which do not obey the summation convention in $\mathbf{E}_3$.

III.11 Evaluate dF when

(i) $F = \left(\sum_i x^i\right)^2$,

(ii) $F = \sum_{i\neq j} x^i x^j$.

III.12 If ω is the 1-form

$$\omega = \sum_i f_i dx^i$$

then show that

$$d\omega = \sum_{i,j} f_{i;j} dx^j \wedge dx^i = \sum_{i<j} (f_{i;j} - f_{j;i})\, dx^j \wedge dx^i.$$

What is the value of $d\omega$ when f_i depends only on x^i?

III.13 If ω is the 2-form

$$\omega = \sum_{i<j} f_{ij} dx^i \wedge dx^j,$$

then show that

$$d\omega = \sum_{i<j<k} (f_{ij;k} + f_{ki;j} + f_{jk;i})\, dx^i \wedge dx^j \wedge dx^k.$$

III.14 Let

$$\omega = \sum_i f_i dx^i,$$

then ω is said to admit a *primitive* F, or be an *exact* 1-form whenever $\omega = dF$. If one has the conditions:

(1⁰) ω admits a primitive;

(2⁰) $d\omega = 0$;

(3⁰) $f_{i;j} = f_{j;i}$ for all i, j

then show that $(1^0) \Rightarrow (2^0) \Leftrightarrow (3^0)$.

— o —

In PROBLEMS III.15-.17, $x^r = (x, y, z)$ are Cartesian coordinates in $\mathbf{E}_3$, and $\mathbf{E}_2$ denotes the xy plane.

III.15 If ω is the 2-form

$$\omega = f dy \wedge dz + g dz \wedge dx + h dx \wedge dy$$

where f, g, h are smooth functions of (x, y, z), then show that

$$*\omega = f dx + g dy + h dz,$$

$$**\omega = \omega,$$

$$d\omega = \left(\frac{\partial f}{\partial x} + \frac{\partial g}{\partial y} + \frac{\partial h}{\partial z}\right) dx \wedge dy \wedge dz.$$

III.16 If $\omega := y dx - x dy$ show that ω does not admit a primitive in $\mathbf{E}_2$.

III.17 If $\omega = (3x^2 + 2xy + y^2)\, dx + (x^2 + 2xy + 3y^2)\, dy$ show that

$$F = x^3 + x^2 y + x y^2 + y^3 + C,$$

where C is a constant, is a primitive of ω in $\mathbf{E}_2$.

IV

The General Leg Calculus

IV.1 Introduction

We now proceed to merge the approaches of the Ricci and Cartan calculus given in the two previous chapters into a single unified theory. Our point of departure is the observation that given an arbitrary vector $\boldsymbol{\xi}$ in $\mathbf{E}_3$, we may regard it *either* as an element $\boldsymbol{\xi}$ of $\mathbf{T}_P$ having contravariant components ξ^r, *or* as an element ξ^* of $\mathbf{T}_P^*$ having covariant components ξ_r. The choice — if one is indicated — may be dictated by geometrical or physical considerations. However, since the spaces $\mathbf{T}_P$ and $\mathbf{T}_P^*$ are isomorphic, the choices are algebraically equivalent, and the transitions between them are given by the rules

$$\mathbf{T}_P \to \mathbf{T}_P^* : \xi^r \mapsto \xi_r = g_{rs}\xi^s \tag{1.1}$$

$$\mathbf{T}_P^* \to \mathbf{T}_P : \xi_r \mapsto \xi^r = g^{rs}\xi_s \tag{1.2}$$

where the components of the metric tensors g_{rs} and g^{rs} satisfy

$$g^{rt}g_{st} = \delta^r_s. \tag{1.3}$$

Our viewpoint is that one may simply step back and regard $\mathbf{T}_P$ and $\mathbf{T}_P^*$ as special *realizations* of a more general notion of a *leg space* $\mathbf{L}_P$ attached to $\mathbf{E}_3$ at a point P. We need not formalize this more general notion, but it is there and in terms of synthetic geometry, one would say that both $\mathbf{T}_P$ and $\mathbf{T}_P^*$ are *models* of $\mathbf{L}_P$ which is merely a 3-dimensional (real) vector space associated with $\mathbf{E}_3$ at P. In this sense one might think of $\mathbf{L}_P$ as $\mathbf{V}$ or $\mathbf{V}^*$ without the reference to P. Since *both* realizations of $\mathbf{L}_P$ are taken to have orthonormal bases, we may regard $\mathbf{L}_P$ as having such a basis, and this explains why leg indices may be written as subscripts, i.e. there is *no distinction* between contravariance and covariance for leg indices. Later in SECTION 2, we will give a nice expression for the coordinate-based contravariant and covariant components $\lambda_a{}^r$ and λ_{ar} respectively.

IV.2 Fundamental Idea of the Leg Calculus

The fundamental idea of the leg calculus is that all vectorial/tensorial quantities should be resolved along a set of leg vectors and, without loss of generality, we may take this 3-leg system to be an orthonormal system (recall the orthonormality hypothesis of SECTION III-5). This 3-leg may be chosen to be either a vectorial 3-leg $\{\boldsymbol{\lambda}_a\}$ consisting of tangent vectors to a set of congruences of curves $\boldsymbol{\Gamma}_a$, or a Pfaffian 3-leg $\{\theta_a\}$ consisting of exterior differential 1-forms which are dual to the $\{\boldsymbol{\lambda}_a\}$. The choice between these leg systems is merely one of convenience and depends on the nature of the situation under consideration.

For example, let F be a smooth function of the ambient coordinates in some region Ω of $\mathbf{E}_3$. Then the *leg derivative* of F, i.e. the *directional derivative of* F in the direction $\boldsymbol{\lambda}_a$ is given by

$$F_{/a} := \lambda_a{}^r F_r \tag{2.1}$$

where $F_r := F_{;r}$, with the inverse expression being given by

$$F_{;r} = \lambda_{ar} F_{/a}. \tag{2.2}$$

These are equivalent to each other and lead to the same expression for the differential dF of F. Upon multiplication of (2.1) by θ_a, we obtain

$$\begin{aligned} F_{/a}\theta_a &= \lambda_a{}^r F_r \lambda_{as} dx^s \\ &= \left(\lambda_a{}^r \lambda_{as}\right) F_r dx^s \\ &= \delta^r_s F_r dx^s \\ &= F_{;r} dx^r = dF \end{aligned} \tag{2.3}$$

using II-(2.8) and III-(5.3). Likewise upon multiplying (2.2) by dx^r we have

$$F_r dx^r = \lambda_{ar} F_{/a} dx^r = F_{/a} \lambda_{ar} dx^r = F_{/a}\theta_a. \tag{2.4}$$

Thus we have two different representations for dF i.e.

$$dF = F_{;r}dx^r = F_{/a}\theta_a, \tag{2.5}$$

which means that dF has the components $F_{;r}$ with respect to the coordinate basis $\{dx^r\}$ of $\mathbf{T}^*_P$ while it has the components $F_{/a}$ relative to the Pfaffian basis $\{\theta_a\}$ of $\mathbf{T}^*_P$.

The coordinate bases $\{dx^r\}$ of $\mathbf{T}^*_P$ and $\left\{\frac{\partial}{\partial x^s}\right\}$ of $\mathbf{T}_P$ are related by

$$dx^r\left(\frac{\partial}{\partial x^s}\right) = \delta^r{}_s \tag{2.6}$$

while the arbitrary bases $\{\theta_a\}$ of $\mathbf{T}^*_P$ and $\{\boldsymbol{\lambda}_a\}$ of $\mathbf{T}_P$ are related by

$$\theta_a(\boldsymbol{\lambda}_b) = \delta_{ab}. \tag{2.7}$$

But both the $\{dx^r\}$ and $\{\theta_a\}$ are linear functionals on $\mathbf{T}_P$, hence

$$\begin{aligned} dx^r(\boldsymbol{\lambda}_a) &= dx^r\left(\lambda_a{}^s\frac{\partial}{\partial x^s}\right) \\ &= \lambda_a{}^s dx^r\left(\frac{\partial}{\partial x^s}\right) \\ &= \lambda_a{}^r \end{aligned} \tag{2.8}$$

and

$$\begin{aligned} \theta_a\left(\frac{\partial}{\partial x^r}\right) &= \lambda_{as}dx^s\left(\frac{\partial}{\partial x^r}\right) \\ &= \lambda_{as}\delta^s_r \\ &= \lambda_{ar}. \end{aligned} \tag{2.9}$$

Equations (2.8) and (2.9) give abstract definitions of $\lambda_a{}^r$ and λ_{ar} which classically were regarded merely as the contravariant and covariant components respectively of the leg vectors $\boldsymbol{\lambda}_a$. In this sense, one can regard these quantities as being representatives, (matrices of the non-singular linear transformations) relating the different bases of $\mathbf{T}_P$ and $\mathbf{T}^*_P$ respectively. Thus II-(2.3) and III-(5.3) give

$$\frac{\partial}{\partial x^r} \mapsto \boldsymbol{\lambda}_a = \lambda_a{}^r\frac{\partial}{\partial x^r} \text{ of } \mathbf{T}_P \to \mathbf{T}_P \tag{2.10}$$

$$dx^r \mapsto \theta_a = \lambda_{ar} dx^r \text{ of } \mathbf{T}_P^* \to \mathbf{T}_P^*. \tag{2.11}$$

Note that both these expressions are invariant under a change of the local ambient coordinates $x^r \to \overline{x}^r$ on Ω, i.e.

$$\overline{\boldsymbol{\lambda}}_a = \boldsymbol{\lambda}_a \text{ and } \overline{\theta}_a = \theta_a. \tag{2.12}$$

The inverses of (2.10) and (2.11) are given respectively by

$$\lambda_{ar} \boldsymbol{\lambda}_a = \frac{\partial}{\partial x^r} \tag{2.13}$$

and

$$dx^r = \lambda_a{}^r \theta_a \tag{2.14}$$

and can be obtained by multiplying the original equations by λ_{as} and $\lambda_a{}^s$ respectively and using II-(2.8). It is interesting to note that both (2.13) and (2.14) can also be derived by using the Hotine transformation formula II-(11.13).

To obtain (2.13), multiply the first equation in II-(11.13) by $\dfrac{\partial}{\partial \overline{x}^r}$:

$$\frac{\partial \overline{x}^r}{\partial x^s} \frac{\partial}{\partial \overline{x}^r} = \overline{\lambda}_a{}^r \lambda_{as} \frac{\partial}{\partial \overline{x}^r} = \lambda_{as} \overline{\lambda}_a{}^r \frac{\partial}{\partial \overline{x}^r},$$

so that

$$\frac{\partial}{\partial x^s} = \lambda_{as} \overline{\boldsymbol{\lambda}}_a = \lambda_{as} \boldsymbol{\lambda}_a;$$

similarly for (2.14), we multiply the second equation in II-(11.13) by $d\overline{x}^s$:

$$\frac{\partial x^r}{\partial \overline{x}^s} d\overline{x}^s = \lambda_a{}^r \overline{\lambda}_{as} d\overline{x}^s$$

so that

$$dx^r = \lambda_a{}^r \overline{\theta}_a = \lambda_a{}^r \theta_a,$$

where, in both these derivatives, we have used the invariance properties (2.12).

Note that since, dF is an invariant quantity,

$$\begin{aligned} dF(\boldsymbol{\lambda}_a) &= F_{/b} \theta_b (\boldsymbol{\lambda}_a) \\ &= F_{/b} \delta_{ba} \\ &= F_{/a}, \end{aligned} \tag{2.15}$$

hence in abstract terms the leg (directional) derivative $F_{/a}$ may be regarded as the value of the linear functional dF on the leg basis $\boldsymbol{\lambda}_a$ of $\mathbf{T}_P$, and also

$$\begin{aligned} dF\left(\frac{\partial}{\partial x^r}\right) &= F_{;s}dx^s\left(\frac{\partial}{\partial x^r}\right) \\ &= F_{;s}\delta^s{}_r \\ &= F_{;r}\ , \end{aligned} \tag{2.16}$$

hence the partial derivative $F_{;r}$ is the value of the linear functional dF on the coordinate basis $\dfrac{\partial}{\partial x^r}$ of $\mathbf{T}_P$.

Given a symmetric second order tensor T, its components are determined by its values as a bilinear (symmetric) functional on the basis vectors. Suppose $\{\boldsymbol{\lambda}_a\}$ is a basis of $\mathbf{T}_P$, then

$$\begin{aligned} T_{ab} := T\left(\boldsymbol{\lambda}_a, \boldsymbol{\lambda}_b\right) &= T\left(\lambda_a{}^r\frac{\partial}{\partial x^r}, \lambda_b{}^s\frac{\partial}{\partial x^s}\right) \\ &= \lambda_a{}^r\lambda_b{}^s T\left(\frac{\partial}{\partial x^r}, \frac{\partial}{\partial x^s}\right) \end{aligned}$$

i.e.,

$$T_{ab} = \lambda_a{}^r\lambda_b{}^s T_{rs}; \tag{2.17}$$

or,

$$\begin{aligned} T_{rs} := T\left(\frac{\partial}{\partial x^r}, \frac{\partial}{\partial x^s}\right) &= T\left(\lambda_{ar}\boldsymbol{\lambda}_a, \lambda_{bs}\boldsymbol{\lambda}_b\right) \\ &= \lambda_{ar}\lambda_{bs}T\left(\boldsymbol{\lambda}_a, \boldsymbol{\lambda}_b\right) \end{aligned}$$

i.e.,

$$T_{rs} = T_{ab}\lambda_{ar}\lambda_{bs}. \tag{2.18}$$

Equation (2.17) defines the leg components T_{ab}, whereas (2.18) is a leg representation of the covariant components of the tensor in terms of the 3-leg. Likewise, the contravariant version of (2.18) is given by

$$T^{rs} = T_{ab}\lambda_a{}^r\lambda_b{}^s \tag{2.19}$$

where the T_{ab} coefficients in (2.18) and (2.19) are identical.

A particularly important case is that of the metric tensor $\mathbf{g}$, for which

$$\begin{aligned} g_{rs} = g\left(\frac{\partial}{\partial x^r}, \frac{\partial}{\partial x^s}\right) &= g\left(\lambda_{ar}\boldsymbol{\lambda}_a, \lambda_{bs}\boldsymbol{\lambda}_b\right) \\ &= \lambda_{ar}\lambda_{bs} g\left(\boldsymbol{\lambda}_a, \boldsymbol{\lambda}_b\right) \\ &= \lambda_{ar}\lambda_{bs}\delta_{ab}, \end{aligned}$$

so

$$g_{rs} = \lambda_{ar}\lambda_{as}. \tag{2.20}$$

Finally, the connection 1-form ω_{ab} may also be regarded as a linear functional on $\mathbf{T}_P$, hence

$$\begin{aligned} \omega_{ab}\left(\boldsymbol{\lambda}_c\right) &= \beta_{abd}\theta_d\left(\boldsymbol{\lambda}_c\right) \\ &= \beta_{abd}\delta_{cd} \\ &= \beta_{abc} \end{aligned}$$

and

$$\omega_{ab}\left(\boldsymbol{\lambda}_c\right) = -\gamma_{abc}, \tag{2.21}$$

so that the Ricci coefficients may be regarded as the negative values of the linear functional ω_{ab} on $\boldsymbol{\lambda}_c$.

IV.3 Identification of the Ricci Coefficients

We now denote the *nine* Ricci coefficients γ_{abc} by the following scheme:

$$\left\| \begin{matrix} \gamma_{121}, \gamma_{122}, \gamma_{123} \\ \gamma_{311}, \gamma_{312}, \gamma_{313} \\ \gamma_{231}, \gamma_{232}, \gamma_{233} \end{matrix} \right\| = \left\| \begin{matrix} \sigma_1, & \sigma_2, & \varepsilon_3 \\ -k_1, & t_2, & \gamma_1 \\ t_1, & k_2, & -\gamma_2 \end{matrix} \right\| \tag{3.1}$$

where

$$k_1, k_2, t_1, t_2, \gamma_1, \gamma_2, \sigma_1, \sigma_2, \varepsilon_3 \tag{3.2}$$

are names, whose geometric meaning will be established shortly. Indeed, the first *eight* of these *leg coefficients* correspond to Hotine's usage in his treatise (see [pages 40-45, 74-75]) except his notation was slightly different:

$$k_1 \sim k,\ k_2 \sim k^*,\ t_1 \sim t,\ t_2 = -t_1 \sim -t,\ \sigma_1 \sim \sigma,\ \sigma_2 \sim \sigma^* \tag{3.3}$$

with γ_1 and γ_2 unchanged. Hotine never introduced a notation for ε_3, but as we will see — under a suitable specialization — it corresponds to the Cayley-Darboux equation [see Chapter 16]. Indeed, the *five* leg coefficients

$$k_1, k_2, t_1, \gamma_1, \gamma_2 \tag{3.4}$$

will turn out in CHAPTER VI to be Hotine's *curvature parameters.*

Analogous to (3.1) we also have the corresponding scheme for the anholonomity coefficients:

$$\left\| \begin{matrix} \alpha_{121}, \alpha_{122}, \alpha_{123} \\ \alpha_{311}, \alpha_{312}, \alpha_{313} \\ \alpha_{231}, \alpha_{232}, \alpha_{233} \end{matrix} \right\| = \left\| \begin{matrix} -\sigma_1, & -\sigma_2, & t_1 + t_2 \\ k_1, & t_1 + \varepsilon_3, & -\gamma_1 \\ t_2 + \varepsilon_3, & -k_2, & \gamma_2 \end{matrix} \right\| \tag{3.5}$$

IV.4 The Permutation Algorithm

By virtue of the identifications established in (3.1), we have

$$k_1 = \gamma_{131} = \lambda_{r,s}\nu^r\lambda^s = -\nu_{r,s}\lambda^r\lambda^s, \tag{4.1}$$

$$k_2 = \gamma_{232} = \mu_{r,s}\nu^r\mu^s = -\nu_{r,s}\mu^r\mu^s, \tag{4.2}$$

$$t_1 = \gamma_{231} = \mu_{r,s}\nu^r\lambda^s = -\nu_{r,s}\mu^r\lambda^s, \tag{4.3}$$

$$t_2 = \gamma_{312} = \nu_{r,s}\lambda^r\mu^s = -\lambda_{r,s}\nu^r\mu^s, \tag{4.4}$$

$$\varepsilon_3 = \gamma_{123} = \lambda_{r,s}\mu^r\nu^s = -\mu_{r,s}\lambda^r\nu^s, \tag{4.5}$$

$$\gamma_1 = \gamma_{313} = \nu_{r,s}\lambda^r\nu^s = -\lambda_{r,s}\nu^r\nu^s, \tag{4.6}$$

$$\gamma_2 = \gamma_{323} = \nu_{r,s}\mu^r\nu^s = -\mu_{r,s}\nu^r\nu^s, \tag{4.7}$$

$$\sigma_1 = \gamma_{121} = \lambda_{r,s}\mu^r\lambda^s = -\mu_{r,s}\lambda^r\lambda^s, \tag{4.8}$$

$$\sigma_2 = \gamma_{122} = \lambda_{r,s}\mu^r\mu^s = -\mu_{r,s}\lambda^r\mu^s. \tag{4.9}$$

Note that the first *seven* of these leg coefficients are *extrinsic*, i.e. they involve $\boldsymbol{\nu}$; while the last *two* are *intrinsic*, i.e. they are independent of $\boldsymbol{\nu}$. This terminology is borrowed from Gaussian differential geometry which speaks of extrinsic and intrinsic properties of a surface according to whether they depend on the normal $\boldsymbol{\nu}$ of the surface $\mathbf{S}$ or not.

Relative to these coefficients, under an interchange

$$\begin{aligned} &\boldsymbol{\lambda} \to \boldsymbol{\mu} \text{ (an interchange: } 1 \to 2), \\ &\boldsymbol{\mu} \to \boldsymbol{\lambda} \text{ (an interchange: } 2 \to 1), \\ &\boldsymbol{\nu} \to \boldsymbol{\nu} \text{ (no interchange: } 3 \to 3), \end{aligned} \tag{4.10}$$

we have

$$\begin{gathered} k_1 \to k_2,\ k_2 \to k_1,\ \gamma_1 \to \gamma_2,\ \gamma_2 \to \gamma_1, \\ \sigma_1 \to \sigma_2,\ \sigma_2 \to -\sigma_1,\ t_1 \to -t_2,\ t_2 \to -t_1, \\ \varepsilon_3 \to -\varepsilon_3. \end{gathered} \tag{4.11}$$

The list of interchanges summarized in (4.11) defines the *permutation algorithm* relative to the indicated changes. This can be used to check and detect errors that might occur in the leg equations which play a key role in our analysis.

For example, consider the commutators II-(7.15) rewritten in terms of the leg coefficients as

$$\begin{aligned} &[\mathrm{F_I}]\colon F_{/1/2} - F_{/2/1} = \sigma_1 F_{/1} + \sigma_2 F_{/2} - (t_1 + t_2)\, F_{/3}, \\ &[\mathrm{F_{II}}]\colon F_{/3/1} - F_{/1/3} = -k_1 F_1 - (t_1 + \varepsilon_3)\, F_{/2} + \gamma_1 F_{/3}, \\ &[\mathrm{F_{III}}]\colon F_{/2/3} - F_{/3/2} = -\,(t_2 + \varepsilon_3)\, F_{/1} + k_2 F_2 - \gamma_2 F_{/3}. \end{aligned} \tag{4.12}$$

Then under the interchange:

$$1 \to 2,\ 2 \to 1 \text{ with } 3 \to 3, \tag{4.13}$$

and (4.10), we have

$$\begin{aligned} &[\mathrm{F_I}] \;\to -[\mathrm{F_I}], \\ &[\mathrm{F_{II}}] \;\to -[\mathrm{F_{III}}], \\ &[\mathrm{F_{III}}] \;\to -[\mathrm{F_{II}}], \end{aligned} \tag{4.14}$$

which checks the validity of (4.12). More importantly, $[\mathrm{F_{II}}]$ implies the form of $[\mathrm{F_{III}}]$ and likewise $[\mathrm{F_{III}}]$ dictates the form of $[\mathrm{F_{II}}]$. In other words, *if we knew only* $[\mathrm{F_I}]$ and $[\mathrm{F_{II}}]$, the permutation algorithm would show us that our analysis was incomplete, and that $[\mathrm{F_{III}}]$ *must also appear* by virtue of the *symmetry* implicit in (4.11)! All of our analysis will explicitly involve the symmetries exhibited in (4.10) and (4.11) and the corresponding expressions indicated in (4.12) are said to be *conjugate* to each other, viz. $[F_{II}]$ and $[F_{III}]$ are conjugate, while $[F_I]$ is *self-conjugate.* Geometrically (4.10) expresses the fact that the 1 and 2-directions are tangential to the surface **S**, while the 3-direction is normal to **S**. This interpretation will be further justified by the surface leg calculus discussed in CHAPTER V, when $\boldsymbol{\nu}$ has a privileged direction normal to **S**. Nevertheless, even before this requirement is imposed, the permutation algorithm is useful as a check on the internal consistency of our leg equations.

IV.5 Interpretation of the Leg Coefficients

We may now restate the results of SECTION II-5 in terms of the various leg coefficients as follows.

1[0] The *curvatures* χ_a of the congruences $\boldsymbol{\Gamma}_a$ are given by

$$\begin{aligned} \chi_1 &:= \sqrt{k_1^2 + \sigma_1^2} \text{ for } \boldsymbol{\Gamma}_1, \\ \chi_2 &:= \sqrt{k_2^2 + \sigma_2^2} \text{ for } \boldsymbol{\Gamma}_2, \\ \chi_3 &:= \sqrt{\gamma_1^2 + \gamma_2^2} \text{ for } \boldsymbol{\Gamma}_3. \end{aligned} \tag{5.1}$$

2[0] The $\boldsymbol{\Gamma}_a$ are *geodesic congruences*, i.e. straight lines in $\mathbf{E}_3$, if and only if

$$\begin{aligned} \boldsymbol{\Gamma}_1 &: k_1 = 0,\ \sigma_1 = 0,\ \text{viz. } \chi_1 = 0, \\ \boldsymbol{\Gamma}_2 &: k_2 = 0,\ \sigma_2 = 0,\ \text{viz. } \chi_2 = 0, \\ \boldsymbol{\Gamma}_3 &: \gamma_1 = 0,\ \gamma_2 = 0,\ \text{viz. } \chi_3 = 0. \end{aligned} \tag{5.2}$$

3[0] The *tendencies* of $\boldsymbol{\Gamma}_3$ (or $\boldsymbol{\nu}$) in the directions of $\boldsymbol{\Gamma}_1$ (or $\boldsymbol{\lambda}$) and $\boldsymbol{\Gamma}_2$ (or $\boldsymbol{\mu}$) are given by $-k_1$ and $-k_2$ respectively.

4[0] $\boldsymbol{\Gamma}_a$ is *normal* to a surface $\mathbf{S}_a$, written $\boldsymbol{\Gamma}_a \perp \mathbf{S}_a$, whenever

$$\begin{aligned} t_1 &= -\varepsilon_3 \text{ for } \boldsymbol{\Gamma}_1 \perp \mathbf{S}_1, \\ t_2 &= -\varepsilon_3 \text{ for } \boldsymbol{\Gamma}_2 \perp \mathbf{S}_2, \\ t_2 &= -t_1 \text{ for } \boldsymbol{\Gamma}_3 \perp \mathbf{S}_3. \end{aligned} \tag{5.3}$$

The last of these conditions will be very important in our analysis and will be called *property* (T) for $\boldsymbol{\Gamma}_3$.

5^0 The congruences $\boldsymbol{\Gamma}_1$ and $\boldsymbol{\Gamma}_2$ are *canonical with respect to* $\boldsymbol{\Gamma}_3$ whenever

$$t_2 = t_1. \tag{5.4}$$

6^0 $\boldsymbol{\Gamma}_1, \boldsymbol{\Gamma}_2, \boldsymbol{\Gamma}_3$ are normal to $\mathbf{S}_1$, $\mathbf{S}_2$, $\mathbf{S}_3$ respectively, i.e. $\mathbf{S}_1$, $\mathbf{S}_2$, and $\mathbf{S}_3$ is a *triply-orthogonal system of surfaces* if and only if

$$\varepsilon_3 = 0. \tag{5.5}$$

Note that 6^0 requires both 4^0 and 5^0, so it is implicit in this case that we have not only (5.5), but *also*

$$t_1 = t_2 = 0. \tag{5.6}$$

Later we will see how (5.6) arises as the condition for the complete integrability of a differential 1-form. Hence, the suitable specializations which lead to the identification of (5.5) with the Cayley-Darboux equation are given by (5.6) and require *more* than the complete integrability of a differential 1-form.

IV.6 Basic Equations of the Leg Calculus

The identification scheme (3.1) allows us to express many of the equations of the Ricci and Cartan calculus in terms of the nine leg coefficients. First, we have the following analogues of II-(4.4)-(4.6):

$$\begin{aligned}\lambda_{rs} &= \sigma_1\mu_r\lambda_s + \sigma_2\mu_r\mu_s + \varepsilon_3\mu_r\nu_s \\ &\quad +k_1\nu_r\lambda_s - t_2\nu_r\mu_s - \gamma_1\nu_r\nu_s,\end{aligned} \tag{6.1}$$

$$\begin{aligned}\mu_{rs} &= -\sigma_1\lambda_r\lambda_s - \sigma_2\lambda_r\mu_s - \varepsilon_3\lambda_r\nu_s \\ &\quad +t_1\nu_r\lambda_s + k_2\nu_r\mu_s - \gamma_2\nu_r\nu_s,\end{aligned} \tag{6.2}$$

$$\begin{aligned}\nu_{rs} &= -k_1\lambda_r\lambda_s + t_2\lambda_r\mu_s + \gamma_1\lambda_r\nu_s \\ &\quad -t_1\mu_r\lambda_s - k_2\mu_r\mu_s + \gamma_2\mu_r\nu_s.\end{aligned} \tag{6.3}$$

In terms of the Cartan calculus, we have III-(6.3) and III-(6.10) which yield

$$\omega_{12} = -\sigma_1\theta_1 - \sigma_2\theta_2 - \varepsilon_3\theta_3 \tag{6.4}$$

$$\omega_{31} = k_1\theta_1 - t_2\theta_2 - \gamma_1\theta_3 \tag{6.5}$$

$$\omega_{23} = -t_1\theta_1 - k_2\theta_2 + \gamma_2\theta_3. \tag{6.6}$$

Then the first structural equation III-(6.1) gives

$$d\theta_1 = \sigma_1\theta_1 \wedge \theta_2 - k_1\theta_3 \wedge \theta_1 - (t_2 + \varepsilon_3)\,\theta_2 \wedge \theta_3, \tag{6.7}$$

$$d\theta_2 = \sigma_2\theta_1 \wedge \theta_2 - (t_1 + \varepsilon_3)\,\theta_3 \wedge \theta_1 + k_2\theta_2 \wedge \theta_3, \tag{6.8}$$

$$d\theta_3 = -(t_1 + t_2)\,\theta_1 \wedge \theta_2 + \gamma_1\theta_3 \wedge \theta_1 - \gamma_2\theta_2 \wedge \theta_3. \tag{6.9}$$

Each of the above sets of equations, qualifies as a *basic equation* of the leg calculus and we will use them frequently. Equations (6.7)-(6.9) also lead to the useful expressions

$$d(\theta_1 \wedge \theta_2) = -(k_1 + k_2)\,\theta_1 \wedge \theta_2 \wedge \theta_3, \tag{6.10}$$

$$d(\theta_3 \wedge \theta_1) = -(\gamma_2 + \sigma_1)\,\theta_1 \wedge \theta_2 \wedge \theta_3, \tag{6.11}$$

$$d(\theta_2 \wedge \theta_3) = -(\gamma_1 - \sigma_2)\,\theta_1 \wedge \theta_2 \wedge \theta_3; \tag{6.12}$$

which follow from using III-(4.11) with $\vartheta_a = \theta_a$ $(a = 1,2,3)$. It is interesting to note that upon contracting (6.1)-(6.3) we have

$$\lambda^r{}_r = \sigma_2 - \gamma_1, \tag{6.13}$$

$$\mu^r{}_r = -\sigma_1 - \gamma_2, \tag{6.14}$$

$$\nu^r{}_r = -k_1 - k_2. \tag{6.15}$$

Hence, an alternate version of (6.10)-(6.12) is given by

$$d(\theta_1 \wedge \theta_2) = \nu^r{}_{,r}\theta_1 \wedge \theta_2 \wedge \theta_3, \tag{6.16}$$

$$d(\theta_3 \wedge \theta_1) = \mu^r{}_{,r}\theta_1 \wedge \theta_2 \wedge \theta_3, \tag{6.17}$$

$$d(\theta_2 \wedge \theta_3) = \lambda^r{}_{,r}\theta_1 \wedge \theta_2 \wedge \theta_3, \tag{6.18}$$

and these equations indicate an unexpected aspect of our canonical enumeration scheme (recall SECTION II-4).

For purposes of reference we also note that

$$\begin{aligned}\lambda_{r,s} - \lambda_{s,r} &= -\sigma_1 (\lambda_r \mu_s - \mu_r \lambda_s) + \varepsilon_3 (\mu_r \nu_s - \nu_r \mu_s) \\ &\quad + k_1 (\nu_r \lambda_s - \lambda_r \nu_s) + t_2 (\mu_r \nu_s - \nu_r \mu_s),\end{aligned} \tag{6.19}$$

$$\begin{aligned}\mu_{r,s} - \mu_{s,r} &= -\sigma_2 (\lambda_r \mu_s - \mu_r \lambda_s) + \varepsilon_3 (\nu_r \lambda_s - \lambda_r \nu_s) \\ &\quad + t_1 (\nu_r \lambda_s - \lambda_r \nu_s) - k_2 (\mu_r \nu_s - \nu_r \mu_s),\end{aligned} \tag{6.20}$$

$$\begin{aligned}\nu_{r,s} - \nu_{s,r} &= t_2 (\lambda_r \mu_s - \mu_r \lambda_s) - \gamma_1 (\nu_r \lambda_s - \lambda_r \nu_s) \\ &\quad + t_1 (\lambda_r \mu_s - \mu_r \lambda_s) + \gamma_2 (\mu_r \nu_s - \nu_r \mu_s);\end{aligned} \tag{6.21}$$

and

$$\begin{aligned}\lambda_{r,s}\lambda^s &= \sigma_1 \mu_r + k_1 \nu_r, \\ \lambda_{r,s}\mu^s &= \sigma_2 \mu_r - t_2 \nu_r, \\ \lambda_{r,s}\nu^s &= \varepsilon_3 \mu_r - \gamma_1 \nu_r;\end{aligned} \tag{6.22}$$

$$\begin{aligned}\mu_{r,s}\lambda^s &= -\sigma_1 \lambda_r + t_1 \nu_r, \\ \mu_{r,s}\mu^s &= -\sigma_2 \lambda_r + k_2 \nu_r, \\ \mu_{r,s}\nu^s &= -\varepsilon_3 \lambda_r - \gamma_2 \nu_r;\end{aligned} \tag{6.23}$$

$$\begin{aligned}\nu_{r,s}\lambda^s &= -k_1 \lambda_r - t_1 \mu_r, \\ \nu_{r,s}\mu^s &= t_2 \lambda_r - k_2 \mu_r, \\ \nu_{r,s}\nu^s &= \gamma_1 \lambda_1 + \gamma_2 \mu_r;\end{aligned} \tag{6.24}$$

and that each of these sets of equations are linear combinations of the pairs of leg vectors $\{\boldsymbol{\mu}, \boldsymbol{\nu}\}$, $\{\boldsymbol{\lambda}, \boldsymbol{\nu}\}$ and $\{\boldsymbol{\lambda}, \boldsymbol{\mu}\}$ respectively.

IV.7 Commutators and Lie Brackets

We may illustrate the use of the basic equations of SECTION 6 by obtaining the integrability conditions of the set of leg differential equations:

$$F_{/a} = f_a, \tag{7.1}$$

which are equivalent to the Pfaffian expression

$$dF = f_a \theta_a. \tag{7.2}$$

Expansion of (7.2) yields

$$dF = f_1 \theta_1 + f_2 \theta_2 + f_3 \theta_3 \tag{7.3}$$

and hence exterior differentiation yields

$$df_1 \wedge \theta_1 + df_2 \wedge \theta_2 + df_3 \wedge \theta_3 + f_1 d\theta_1 + f_2 d\theta_2 + f_3 d\theta_3 = 0 \quad (7.4)$$

since $d\,(dF) := d^2 F = 0$ by III-(4.9). But we have

$$\begin{aligned} df_1 &= f_{1/1}\theta_1 + f_{1/2}\theta_2 + f_{1/3}\theta_3, \\ df_2 &= f_{2/1}\theta_1 + f_{2/2}\theta_2 + f_{2/3}\theta_3, \\ df_3 &= f_{3/1}\theta_1 + f_{3/2}\theta_2 + f_{3/3}\theta_3; \end{aligned} \quad (7.5)$$

so substitution into (7.4) gives

$$\begin{aligned} &\left(f_{1/1}\theta_1 + f_{1/2}\theta_2 + f_{1/3}\theta_3\right) \wedge \theta_1 \\ &\quad + \left(f_{2/1}\theta_1 + f_{2/2}\theta_2 + f_{2/3}\theta_3\right) \wedge \theta_2 \\ &\quad + \left(f_{3/1}\theta_1 + f_{3/2}\theta_2 + f_{3/3}\theta_3\right) \wedge \theta_3 \\ &\quad + f_1 d\theta_1 + f_2 d\theta_2 + f_3 d\theta_3 = 0, \end{aligned} \quad (7.6)$$

viz.

$$\begin{aligned} &\left(f_{2/1} - f_{1/2}\right)\theta_1 \wedge \theta_2 + \left(f_{1/3} - f_{3/1}\right)\theta_3 \wedge \theta_1 \\ &\left(f_{3/2} - f_{2/3}\right)\theta_2 \wedge \theta_3 + f_1 d\theta_1 + f_2 d\theta_2 + f_3 d\theta_3 = 0. \end{aligned} \quad (7.8)$$

Finally, recalling (6.7)-(6.9), we have

$$\begin{aligned} &\left\{f_{2/1} + f_{1/2} + f_1\sigma_1 + f_2\sigma_2 - f_3\left(t_1 + t_2\right)\right\}\theta_1 \wedge \theta_2 \\ &\quad + \left\{f_{1/3} - f_{3/1} - f_1 k_1 - f_2\left(t_1 + \varepsilon_3\right) + f_3\gamma_1\right\}\theta_3 \wedge \theta_1 \\ &\quad + \left\{f_{3/2} - f_{2/3} - f_1\left(t_1 + \varepsilon_3\right) + f_2 k_2 - f_3\gamma_2\right\}\theta_2 \wedge \theta_3 = 0. \end{aligned} \quad (7.9)$$

However, since the products $\theta_1 \wedge \theta_2$, $\theta_3 \wedge \theta_1$ and $\theta_2 \wedge \theta_3$ are linearly independent in $\wedge^2 \mathbf{T}_P^*$, each of the curly brackets must be identically zero and this yields the commutators $[\mathrm{F}_{\mathcal{A}}]$ $\mathcal{A} = \mathrm{I}$, II, III which we have previously displayed in (4.12) using (3.1). The essence of the argument and the advantage of employing the leg calculus is that equations (4.12) are *not needed*! They are *immediate consequences* of exterior differentiation of (7.2) upon using (7.5) and the expressions (6.7)-(6.9). All that need be remembered are the rules of exterior differentiation and the basic equations (6.7)-(6.9).

As a second example we compute the Lie bracket of the vectors $\boldsymbol{\lambda}$ and $\boldsymbol{\mu}$. This may be done as in SECTION II-8 by recalling II-(8.4) which is now

$$[\boldsymbol{\lambda}, \boldsymbol{\mu}]\, F = \left(\lambda^s \mu^r{}_{,s} - \mu^s \lambda^r{}_{,s}\right) F_r. \tag{7.10}$$

The values of the covariant derivatives of the leg vectors are given in (6.2) and (6.1), hence we have — upon displaying only the non-zero contributions —

$$\begin{aligned} [\boldsymbol{\lambda}, \boldsymbol{\mu}]\, F \quad &= \{\lambda^r [\cdots + (-\sigma_1 \lambda^r + t_1 \mu^r)\, \lambda_s + \cdots] \\ &\quad -\mu^s [\cdots + (\sigma_2 \lambda^r - t_2 \mu^r)\, \mu_s + \cdots]\}\, F_r \\ &= \{-\sigma_1 \lambda^r - \sigma_2 \mu^r + (t_1 + t_2)\, \nu^r\}\, F_r. \end{aligned}$$

Thus, omitting the function F we obtain the operator equation

$$[\boldsymbol{\lambda}, \boldsymbol{\mu}] = -\sigma_1 \boldsymbol{\lambda} - \sigma_2 \boldsymbol{\mu} + (t_1 + t_2)\, \boldsymbol{\nu}.$$

The same result may also be derived by exterior differentiation of (2.14) which upon expansion is

$$dx^r = \lambda^r \theta_1 + \mu^r \theta_2 + \nu^r \theta_3. \tag{7.11}$$

Then we have

$$d\lambda^r \wedge \theta_1 + d\mu^r \wedge \theta_2 + d\nu^r \wedge \theta_3 + \lambda^r d\theta_1 + \mu^r d\theta_2 + \nu^r d\theta_3 = 0. \tag{7.12}$$

But

$$\begin{aligned} d\lambda^r &= \lambda^r{}_{/1} \theta_1 + \lambda^r{}_{/2} \theta_2 + \lambda^r{}_{/3} \theta_3, \\ d\mu^r &= \mu^r{}_{/1} \theta_1 + \mu^r{}_{/2} \theta_2 + \mu^r{}_{/3} \theta_3, \\ d\nu^r &= \nu^r{}_{/1} \theta_1 + \nu^r{}_{/2} \theta_2 + \nu^r{}_{/3} \theta_3, \end{aligned} \tag{7.13}$$

by analogy with (7.5), so as in the previous derivation we have

$$\begin{aligned} &\left(\mu^r{}_{/1} - \lambda^r{}_{/2}\right) \theta_1 \wedge \theta_2 + \left(\lambda^r{}_{/3} - \nu^r{}_{/1}\right) \theta_3 \wedge \theta_1 \\ &\quad + \left(\nu^r{}_{/2} - \mu^r{}_{/3}\right) \theta_2 \wedge \theta_3 + \lambda^r d\theta_1 + \mu^r d\theta_2 + \nu^r d\theta_3 = 0. \end{aligned} \tag{7.14}$$

Note that

$$\begin{aligned} \mu^r{}_{/1} - \lambda^r{}_{/2} &= \lambda^s \mu^r{}_{;s} - \mu^s \lambda^r{}_{;s} \\ &= \lambda^s \mu^r{}_{,s} - \mu^s \lambda^r{}_{,s} \\ &= [\boldsymbol{\lambda}, \boldsymbol{\mu}]^r, \end{aligned} \tag{7.15}$$

so we need only consider the contributions of $\theta_1 \wedge \theta_2$ (note that such terms also occur in $d\theta_1$, $d\theta_2$ and $d\theta_3$) to obtain the required result. By (7.10) and (7.12) we then have

$$\mu^r{}_{/1} - \lambda^r{}_{/2} + \lambda^r \sigma_1 + \mu^r \sigma_2 - \nu^r (t_1 + t_2) = 0,$$

and our result is clear. For purposes of reference we once again list the Lie brackets:

$$\begin{aligned} [\boldsymbol{\lambda}, \boldsymbol{\mu}] &= -\sigma_1 \boldsymbol{\lambda} - \sigma_2 \boldsymbol{\mu} + (t_1 + t_2)\, \boldsymbol{\nu}, \\ [\boldsymbol{\nu}, \boldsymbol{\lambda}] &= k_1 \boldsymbol{\lambda} + (t_1 + \varepsilon_3)\, \boldsymbol{\mu} - \gamma_1 \boldsymbol{\nu}, \\ [\boldsymbol{\mu}, \boldsymbol{\nu}] &= (t_2 + \varepsilon_3)\, \boldsymbol{\lambda} - k_2 \mu + \gamma_2 \boldsymbol{\nu}. \end{aligned} \tag{7.16}$$

Finally, one can obtain the Schouten identities by using (7.16) in the Jacobi identity

$$[[\boldsymbol{\lambda}, \boldsymbol{\mu}], \boldsymbol{\nu}] + [[\boldsymbol{\mu}, \boldsymbol{\nu}], \boldsymbol{\lambda}] + [[\boldsymbol{\nu}, \boldsymbol{\lambda}], \boldsymbol{\mu}] = \mathbf{0}, \tag{7.17}$$

as in SECTION II-9. An even simpler approach is to use the Cartan calculus (recall SECTION III-3) and compute $d\,(d\theta_a) = d^2\theta_a = 0$ $(a = 1, 2, 3)$. We illustrate this for the Pfaffian basis form θ_1. Then $d\,(d\theta_1) := d^2\theta_1 = 0$ yields, by exterior differentiation of (6.7),

$$\begin{aligned} &d\sigma_1 \wedge \theta_1 \wedge \theta_2 - dk_1 \wedge \theta_3 \wedge \theta_1 - d\,(t_2 + \varepsilon_3) \wedge \theta_2 \wedge \theta_3 \\ &\quad + \sigma_1 d\,(\theta_1 \wedge \theta_2) - k_1 d\,(\theta_3 \wedge \theta_1) - (t_2 + \varepsilon_3)\, d\,(\theta_2 \wedge \theta_3) = 0. \end{aligned} \tag{7.18}$$

The first three terms in (7.18) are analogous to the expressions (7.5) and give

$$\sigma_{1/3}\theta_3 \wedge \theta_1 \wedge \theta_2 - k_{1/2}\theta_2 \wedge \theta_3 \wedge \theta_1 - (t_2 + \varepsilon_3)_{/1}\, \theta_1 \wedge \theta_2 \wedge_3 .$$

But by using III-(4.3) we have

$$\theta_1 \wedge \theta_2 \wedge \theta_3 = \theta_2 \wedge \theta_3 \wedge \theta_1 = \theta_3 \wedge \theta_1 \wedge \theta_2 \tag{7.19}$$

and hence these three terms become

$$\left\{\sigma_{1/3} - k_{1/2} - (t_2 + \varepsilon_3)_{/1}\right\} \theta_1 \wedge \theta_2 \wedge \theta_3.$$

The last three terms in (7.18) may be readily evaluated by using (7.16)-(7.18). Hence, the resulting coefficient of $\theta_1 \wedge \theta_2 \wedge \theta_3$ becomes

$$\sigma_{1/3} - k_{1/3} - (t_2 + \varepsilon_3)_{/1} - \sigma_1 (k_1 + k_2) + k_1 (\gamma_2 + \sigma_1) + (t_2 + \varepsilon_3)(\gamma_1 - \sigma_2) .$$

This is equal to zero in $\wedge^3 \mathbf{T}^*_P$. Likewise $d(d\theta_2) := d^2\theta_2 = 0$ and $d(d\theta_3) := d^2\theta_3 = 0$ involve the expressions

$$d\sigma_2 \wedge \theta_1 \wedge \theta_2 - d(t_1 + \varepsilon_3) \wedge \theta_3 \wedge \theta_1 + dk_2 \wedge \theta_2 \wedge \theta_3 + \sigma_2 d(\theta_1 \wedge \theta_2) - (t_1 + \varepsilon_3) d(\theta_3 \wedge \theta_1) + k_2 d(\theta_2 \wedge \theta_3) = 0 \tag{7.20}$$

$$-d(t_1 + t_2) \wedge \theta_1 \wedge \theta_2 + d\gamma_1 \wedge \theta_3 \wedge \theta_1 - d\gamma_2 \wedge \theta_2 \wedge \theta_3 - (t_1 + t_2) d(\theta_1 \wedge \theta_2) + \gamma_1 d(\theta_3 \wedge \theta_1) - \gamma_2 d(\theta_2 \wedge \theta_3) = 0 \tag{7.21}$$

which respectively lead to the following coefficients of $\theta_1 \wedge \theta_2 \wedge \theta_3$:

$$\sigma_{2/3} - (t_1 + \varepsilon_3)_{/2} + k_{2/1} - \sigma_2 (k_1 + k_2) + (t_1 + \varepsilon_3)(\gamma_2 + \sigma_1) - k_2 (\gamma_1 - \sigma_2) ;$$

$$-(t_1 + t_2)_{/3} + \gamma_{1/2} - \gamma_{2/1} + (t_1 + t_2)(k_1 + k_2) - \gamma_1 (\gamma_2 + \sigma_1) + \gamma_2 (\gamma_1 - \sigma_2) .$$

Hence, the resulting set $(\mathcal{S}_\mathcal{A})$ $\mathcal{A} =$ I, II, III of Schouten identities become

$$(\mathcal{S}_\mathrm{I}): \quad \sigma_{1/3} - \varepsilon_{3/1} - k_{1/2} - t_{2/1} + k_1\gamma_2 - k_2\sigma_1 + t_2 (\gamma_1 - \sigma_2) + \varepsilon_3 (\gamma_1 - \sigma_2) = 0 \tag{7.22}$$

$$(\mathcal{S}_\mathrm{II}): \quad \sigma_{2/3} - \varepsilon_{3/2} + k_{2/1} - t_{1/2} - k_1\sigma_2 - k_2\gamma_1 + t_1 (\gamma_2 + \sigma_1) + \varepsilon_3 (\gamma_2 + \sigma_1) = 0 \tag{7.23}$$

$$(\mathcal{S}_\mathrm{III}): \quad \gamma_{1/2} - \gamma_{2/1} - (t_1 + t_2)_{/3} + (k_1 + k_2)(t_1 + t_2) - \gamma_1\sigma_1 - \gamma_2\sigma_2 = 0. \tag{7.24}$$

This grouping of terms is convenient for later purposes since when property (T) (recall SECTION 5) is imposed, $t_1 + t_2 = 0$. Note that the permutation algorithm shows that $(\mathcal{S}_\mathrm{I})$ and $(\mathcal{S}_\mathrm{II})$ are conjugate to each other, whereas $(\mathcal{S}_\mathrm{III})$ is self-conjugate.

IV.8 Laplacian of a Scalar Function

In SECTION II-11 we gave an expression involving the Ricci coefficients for the Laplacian Δ of a smooth scalar function F of the ambient coordinates on a domain Ω of $\mathbf{E}_3$. It is now interesting to see how this expression is derived using the general leg calculus and the leg coefficients. This derivation requires an extension of the Cartan calculus and the notion of the *star*, or *adjoint, operator* $*$ which, when applied to the Pfaffian forms, is defined by

$$\begin{aligned} *\theta_1 &:= \theta_2 \wedge \theta_3, \\ *\theta_2 &:= \theta_3 \wedge \theta_1, \\ *\theta_3 &:= \theta_1 \wedge \theta_2; \end{aligned} \tag{8.1}$$

$$\begin{aligned} *(\theta_1 \wedge \theta_2) &:= \theta_3, \\ *(\theta_3 \wedge \theta_1) &:= \theta_2, \\ *(\theta_2 \wedge \theta_3) &:= \theta_1; \end{aligned} \tag{8.2}$$

$$*(\theta_1 \wedge \theta_2 \wedge \theta_3) := 1. \tag{8.3}$$

Hence, we have $** := *^2 = 1$ where 1 denotes the identity operator.

For our purposes we will apply this operator only to the 1-form

$$\omega := dF = F_{/a}\theta_a, \tag{8.4}$$

which yields

$$*\omega = F_{/1}(\theta_2 \wedge \theta_3) + F_{/2}(\theta_3 \wedge \theta_1) + F_{/3}(\theta_1 \wedge \theta_2). \tag{8.5}$$

Hence

$$\begin{aligned} d*\omega &= dF_{/1} \wedge (\theta_2 \wedge \theta_3) + dF_{/2} \wedge (\theta_3 \wedge \theta_1) \\ &\quad + dF_{/3} \wedge (\theta_1 \wedge \theta_2) + F_{/1} d(\theta_2 \wedge \theta_3) \\ &\quad + F_{/2} d(\theta_3 \wedge \theta_1) + F_{/3} d(\theta_1 \wedge \theta_2). \end{aligned} \tag{8.6}$$

Then using (7.5) and (6.16)-(6.18) we have

$$\begin{aligned} d*\omega &= F_{/11} + F_{/22} + F_{/33} - (\gamma_1 - \sigma_2) F_{/1} \\ &\quad - (\gamma_2 + \sigma_1) F_{/2} - (k_1 + k_2) F_{/3}\theta_1 \wedge \theta_2 \wedge \theta_3 \end{aligned}$$

which is a 3-form. We expect ΔF to be a 0-form and this is obtained by using (8.3) and defining

$$\Delta F := *d*\omega = *d*(dF), \tag{8.7}$$

hence

$$\begin{aligned} \Delta F &= F_{/11} + F_{/22} + F_{/33} - (\gamma_1 - \sigma_2) F_{/1} \\ &\quad - (\gamma_2 + \sigma_1) F_{/2} - (k_1 + k_2) F_{/3}. \end{aligned} \tag{8.8}$$

By using (3.1), it is easy to check that (8.8) agrees with II-(11.4) upon expansion.

The operator $*d*$ is called the *co-differential* δ of the exterior differentiation d, and the Laplacian Δ for scalar functions reduces to

$$\Delta = \delta d, \tag{8.9}$$

which is a special case of the more general Hodge-deRham-Laplacian

$$\Delta = \delta d + d\delta \tag{8.10}$$

which is applicable to p-forms when $p > 1$. The properties of (8.10) are discussed in CHOQUET-BRUHAT (1968) (see pages 92-93, 248-250). Co-differentiation lowers the degree of a p-form ($p > 1$), whereas exterior differentiation raises the degree. Hence, when $p = 0$, $d\delta$ is omitted and in this case (8.10) reduces to (8.9). The reader should note that the inverse of the star operator is commonly introduced, but it is not necessary in the present context.

The co-differential of the Pfaffian basis forms is easily evaluated by using (8.1)-(8.3). For example, upon exterior differentiation of the last expression in (8.1) we have

$$d*\theta_3 = d(\theta_1 \wedge \theta_2)$$

which by (6.18) yields

$$d*\theta_3 = -(k_1 + k_2)\,\theta_1 \wedge \theta_2 \wedge \theta_3.$$

Hence

$$*d*\theta_3 = -(k_1 + k_2),$$

so

$$\delta\theta_3 = -(k_1 + k_2),$$

and analogously we have

$$\delta\theta_1 = -(\gamma_1 - \sigma_2)$$
$$\delta\theta_2 = -(\gamma_2 + \sigma_1).$$

Consequently

$$\delta\theta_a = \lambda_a{}^r{}_{,r} = \operatorname{div} \boldsymbol{\lambda}_a \tag{8.11}$$

where div is understood to be the divergence taken with covariant differentiation (compare (8.11) with (6.13)-(6.15)).

IV.9 Lamé Equations and Identities

The Cartan calculus version of the Lamé equations as noted in SECTION III-6 are summarized by the vanishing of the curvature 2-forms III-(6.5) which are equivalent to III-(6.26), viz.

$$\Omega_{12} = 0 \Longleftrightarrow d\omega_{12} = -\omega_{31} \wedge \omega_{23}, \tag{9.1}$$

$$\Omega_{31} = 0 \Longleftrightarrow d\omega_{31} = \omega_{12} \wedge \omega_{23}, \tag{9.2}$$

$$\Omega_{23} = 0 \Longleftrightarrow d\omega_{23} = -\omega_{12} \wedge \omega_{31}. \tag{9.3}$$

By (6.4)-(6.6) each of the connection 1-forms ω_{ab}: ω_{12}, ω_{31} and ω_{23} is a linear combination of the Pfaffian basis forms θ_1, θ_2, θ_3 with the leg coefficients (3.2) appearing as coefficients. Thus (9.1)-(9.3) furnish a set of identities which these leg coefficients must satisfy identically. By virtue of our usage of the leg calculus, these leg coefficients play the role of the primary variables in the theory, and hence these identities are the geometric constraints which reduce $\mathbf{V}_3$ to $\mathbf{E}_3$. We now determine the explicit form of these constraints.

First, by the same methods employed in SECTION 6 exterior differentiation yields

$$\begin{aligned} d\omega_{12} &= \left\{\sigma_{1/2} - \sigma_{2/1} - \sigma_1^2 - \sigma_2^2 + (t_1 + t_2)\,\varepsilon_3\right\} \theta_1 \wedge \theta_2 \\ &\quad + \left\{\varepsilon_{3/1} - \sigma_{1/3} + k_1\sigma_1 + (t_1 + \varepsilon_3)\,\sigma_2 - \varepsilon_3\gamma_1\right\} \theta_3 \wedge \theta_1 \\ &\quad + \left\{\sigma_{2/3} - \sigma_{3/2} - k_2\sigma_2 + (t_1 + \varepsilon_3)\,\sigma_1 + \varepsilon_3\gamma_2\right\} \theta_2 \wedge \theta_3 \end{aligned} \tag{9.4}$$

$$\begin{aligned} d\omega_{31} &= \left\{-t_{2/1} - k_{1/2} + k_1\sigma_1 - t_2\sigma_2 + (t_1+t_2)\,\mu_1\right\}\theta_1\wedge\theta_2 \\ &+\left\{k_{1/3} + \gamma_{1/1} - k_1^2 + (t_1+\varepsilon_3)\,t_2 - \gamma_1^2\right\}\theta_3\wedge\theta_1 \\ &+\left\{t_{2/3} - \gamma_{1/2} - k_1\,(t_2+\varepsilon_3) - k_2t_2 + \gamma_1\gamma_2\right\}\theta_2\wedge\theta_3, \end{aligned} \tag{9.5}$$

$$\begin{aligned} d\omega_{23} &= \left\{t_{1/2} - k_{2/1} - k_2\sigma_2 - t_1\sigma_1 - (t_1+t_2)\,\gamma_2\right\}\theta_1\wedge\theta_2 \\ &+\left\{-t_{1/3} - \gamma_{2/1} + k_1t_1 + (t_1+\varepsilon_3)\,k_2 + \gamma_1\gamma_2\right\}\theta_3\wedge\theta_1 \\ &+\left\{\gamma_{2/2} + k_{2/3} - k_2^2 + (t_2+\varepsilon_3)\,t_1 - \gamma_2^2\right\}\theta_2\wedge\theta_3, \end{aligned} \tag{9.6}$$

while a simple calculation shows that

$$\begin{aligned} \omega_{31}\wedge\omega_{23} &= (-k_1k_2 - t_1t_2)\,\theta_1\wedge\theta_2 \\ &+(t_1\gamma_1 - k_1\gamma_2)\,\theta_3\wedge\theta_1 \\ &+(-t_1\gamma_2 - k_2\gamma_1)\,\theta_2\wedge\theta_3, \end{aligned} \tag{9.7}$$

$$\begin{aligned} \omega_{23}\wedge\omega_{12} &= (t_1\sigma_2 - k_2\sigma_1)\,\theta_1\wedge\theta_2 \\ &+(-\gamma_2\sigma_1 - t_1\varepsilon_3)\,\theta_3\wedge\theta_1 \\ &+(k_2\varepsilon_3 + \gamma_2\sigma_2)\,\theta_2\wedge\theta_3, \end{aligned} \tag{9.8}$$

$$\begin{aligned} \omega_{12}\wedge\omega_{31} &= (t_2\sigma_1 + k_1\sigma_2)\,\theta_1\wedge\theta_2 \\ &+(-k_1\varepsilon_3 - \sigma_1\gamma_1)\,\theta_3\wedge\theta_1 \\ &+(\sigma_2\gamma_1 - t_2\varepsilon_3)\,\theta_2\wedge\theta_3. \end{aligned} \tag{9.9}$$

Hence, forming the combinations (9.1)-(9.3) we obtain the following leg coefficient version of the Lamé equations $(\mathcal{L}_{\mathcal{B}})$, ($\mathcal{B}$ = I, II, ..., IX) which occur respectively as coefficients of $\theta_1\wedge\theta_2$, $\theta_3\wedge\theta_1$, $\theta_2\wedge\theta_3$:

$$(\mathcal{L}_{\mathrm{I}}):\ \sigma_{1/2} - \sigma_{2/1} - \sigma_1^2 - \sigma_2^2 + (t_1+t_2)\,\varepsilon_3 = k_1k_2 + t_1t_2, \tag{9.10}$$

$$(\mathcal{L}_{\mathrm{II}}):\ \varepsilon_{3/1} - \sigma_{1/3} + k_1\sigma_1 + (t_1+\varepsilon_3)\,\sigma_2 - \varepsilon_3\gamma_1 = k_1\gamma_2 - t_1\gamma_1, \tag{9.11}$$

$$(\mathcal{L}_{\mathrm{III}}):\ \sigma_{2/3} - \varepsilon_{3/2} - k_2\sigma_2 + (t_2+\varepsilon_3)\,\sigma_1 + \varepsilon_3\gamma_2 = k_2\gamma_1 + t_2\gamma_2, \tag{9.12}$$

$$(\mathcal{L}_{\mathrm{IV}}):\ -t_{2/1} - k_{1/2} + k_1\sigma_1 - t_1\sigma_2 + (t_1+t_2)\,\gamma_1 = k_2\sigma_1 - t_1\sigma_2, \tag{9.13}$$

$$(\mathcal{L}_{\mathrm{V}}):\ k_{1/3} + \gamma_{1/1} - k_1^2 + (t_1+\varepsilon_3)\,t_2 - \gamma_1^2 = \gamma_2\sigma_1 + t_1\varepsilon_3, \tag{9.14}$$

$$(\mathcal{L}_{\mathrm{VI}}):\ t_{2/3} - \gamma_{1/2} - k_1\,(t_2+\varepsilon_3) - k_2t_2 + \gamma_1\gamma_2 = -k_2\varepsilon_3 - \gamma_2\sigma_2, \tag{9.15}$$

$$(\mathcal{L}_{\mathrm{VII}}): t_{1/2} - k_{2/1} - k_2\sigma_2 - t_1\sigma_1 - (t_1 + t_2)\,\gamma_2 = -t_2\sigma_1 - k_1\sigma_2, \tag{9.16}$$

$$(\mathcal{L}_{\mathrm{VIII}}): \; - t_{1/3} - \gamma_{2/1} + k_1 t_1 + (t_1 + \varepsilon_3)\,k_2 + \gamma_1\gamma_2 = k_1\varepsilon_3 + \sigma_1\gamma_1, \tag{9.17}$$

$$(\mathcal{L}_{\mathrm{IX}}): \gamma_{2/2} + k_{2/3} - k_2^2 + (t_2 + \varepsilon_3)\,t_2 - \gamma_2^2 = -\sigma_2\gamma_1 + t_2\varepsilon_3. \tag{9.18}$$

The equations $(\mathcal{L}_{\mathrm{I}})$, $(\mathcal{L}_{\mathrm{IV}})$ and $(\mathcal{L}_{\mathrm{VII}})$ are self-conjugate, while the remaining equations are conjugate in the pairs

$$\begin{aligned} &(\mathcal{L}_{\mathrm{II}}) \to (\mathcal{L}_{\mathrm{III}}), && (\mathcal{L}_{\mathrm{III}}) \to (\mathcal{L}_{\mathrm{II}}), \\ &(\mathcal{L}_{\mathrm{V}}) \to (\mathcal{L}_{\mathrm{IX}}), && (\mathcal{L}_{\mathrm{VI}}) \to (\mathcal{L}_{\mathrm{VIII}}), \\ &(\mathcal{L}_{\mathrm{VIII}}) \to (\mathcal{L}_{\mathrm{VI}}), && (\mathcal{L}_{\mathrm{IX}}) \to (\mathcal{L}_{\mathrm{V}}). \end{aligned} \tag{9.19}$$

In SECTION II-12, it was noted that in three dimensions since there are six independent components of the Riemann-Christoffel curvature tensor, there must be six independent Lamé equations. Properly speaking these are called *Lamé identities* and we have the following:

Theorem

The nine Lamé identities $(\mathcal{L})$ reduce to six independent identities

$$(\mathcal{L}_{\mathrm{I}}),\ (\mathcal{L}_{\mathrm{V}}),\ (\mathcal{L}_{\mathrm{IX}}) \tag{9.20}$$

and

$$\begin{gathered} (\mathcal{L}_{\mathrm{II}}) \Leftrightarrow (\mathcal{L}_{\mathrm{IV}}),\ (\mathcal{L}_{\mathrm{III}}) \Leftrightarrow (\mathcal{L}_{\mathrm{VII}}) \\ (\mathcal{L}_{\mathrm{VI}}) \Leftrightarrow (\mathcal{L}_{\mathrm{VIII}}). \end{gathered} \tag{9.21}$$

Proof: The above identities are precisely the algebraic symmetries II-(12.7). The first three identities correspond to the components R_{1212}, R_{3131}, R_{2323} and hence are independent, while the last three identities are respectively

$$\begin{aligned} R_{1231} &= R_{3112} \\ R_{1223} &= R_{2312} \\ R_{3123} &= R_{2331}. \end{aligned}$$

Moreover, the latter three can be expressed in terms of the Schouten identities as follows:

$$
\begin{aligned}
(\mathcal{S}_{\mathrm{I}}) &\Leftrightarrow -(\mathcal{L}_{\mathrm{II}}) + (\mathcal{L}_{\mathrm{IV}})\,, \\
(\mathcal{S}_{\mathrm{II}}) &\Leftrightarrow (\mathcal{L}_{\mathrm{III}}) - (\mathcal{L}_{\mathrm{VII}})\,, \\
(\mathcal{S}_{\mathrm{III}}) &\Leftrightarrow -(\mathcal{L}_{\mathrm{VI}}) + (\mathcal{L}_{\mathrm{VIII}})\,.
\end{aligned}
\tag{9.22}
$$

IV.10 Coordinates in the Leg Calculus

The reader will note that up to now we have made very little use of coordinates in either the Ricci or Cartan calculus. Indeed, the only coordinates which we have considered were the so-called ambient coordinates in $\mathbf{E}_3$ or $\mathbf{V}_3$. The question now arises as to what rôle do coordinates play in the leg calculus, or more generally what is the rôle of coordinates in differential geodesy? Forty years ago, both Marussi and Hotine would undoubtably have given the emphatic answer that they were the heart of the theory. Indeed, Marussi originally stated his conditions (SECTION I-2) with coordinates in mind, viz. his *intrinsic coordinates*, and Part II of Hotine's treatise was entirely devoted to the investigation of five types of coordinate systems which he hoped would be useful in differential geodesy. We have summarized this belief in what we call the *Marussi Hypothesis,* which can be restated in two forms:

> *The Strong Form: All geodetic problems can be posed in terms of intrinsic coordinates;*
>
> *The Weak Form: Some geodetic problems can be posed in terms of intrinsic coordinates.*

The Strong Form states that his intrinsic coordinates are the appropriate manner of formulating differential geodesy, and such coordinates can always be found, while the Weak Form asserts that such coordinates need not be available, i.e., need not exist. An alternate version of the Weak Form addresses the problem of determining under what circumstances coordinates do exist. Both statements were perfectly natural to geodesists of their generation, and it is likely that neither Marussi nor Hotine worried much about their validity, or sought to determine whether each of the forms was a realistic hypothesis. Essentially, both forms were vestiges of their classical mode of thinking and the mathematical tools available to them and their contemporaries.

Only partial results relative to the general validity of either form of the Marussi Hypothesis are known. But these suggest that the Strong Form is false, and that the prospects for the Weak Form — in particular in its alternate version — being valid are rather limited. We will defer the question of constructing coordinate systems to CHAPTER X, but it seems clear today that the issues involved are much more difficult than Marussi or Hotine would have suspected.

The first indication that the Strong Form is not valid was the discovery of non-holonomic, or anholonomic coordinate/reference systems by GRAFAREND (1971) and GROSSMAN (1974). The resulting problem, which is known as the *Holonomity Problem*, was one of the most beautiful and challenging questions in differential geodesy. More recently the author has addressed the coordinate construction problem, i.e. the Hotine Problem, in ZUND (1990b).

The leg calculus offers the advantage of de-emphasizing the rôle of coordinate systems in our formulation of differential geodesy. Of course, the theory would considerably simplify if a geodetically attractive coordinate system was known, but this is not required in order to employ the leg calculus. Our hunch is that the primary variables in differential geodesy are the legs, with the secondary variables being the associated set of leg coefficients (3.2), and that the coordinates play a purely subsidiary role. The work of Marussi and Hotine is consistent with this view, and indeed much of the material in Part II of Hotine's treatise supports this view. The reader need not accept this viewpoint since — strictly speaking — our analysis does not depend on it. Hence, a 'coordinate-minded' individual could view the leg calculus as merely an alternate more efficient method of formulating differential geodesy, whereas a 'leg-minded' individual could view our approach as a new formulation of the theory.

Our reasons for pursuing the latter viewpoint are that it does not preclude the former, and that the metric tensor components can be expressed in terms of the leg vectors as in (2.20). Moreover, the analysis given in CHAPTERS II and III show how the geometry of $\mathbf{E}_3$, or $\mathbf{V}_3$, is readily expressible in terms of 3-legs. We suspect that the leg vectors may be more naturally susceptible of being measurable than coordinates, and that mathematically the existence of legs is easier to establish than that of coordinates. In other words, mathematically, vector fields are more *fundamental* than coordinates. Finally, the Reconstruction Theorem proven at the end of this section shows that in a sense there is *no logical distinction* between the leg vectors and their associated leg coefficients. We will elaborate on these views as our presentation unfolds.

We now discuss the important distinction between holonomic and non-holonomic reference systems. This first appeared in classical mechanics and it is helpful to recall its origin, although in differential geodesy the situation is quite different — but not totally unrelated!

In classical mechanics, one is primarily concerned with integrating the equations of motion of a dynamical system. The dynamical system is described by a set of general curvilinear coordinates, called generalized coordinates, q^r, where the superscript ranges over the values $r = 1, \cdots, f$ and f is the number of degrees of freedom of the system. Integration of the equations of motion — most simply given by Lagrange's equations — say

$$F(q^r, \dot{q}^r, \ddot{q}^r; t) = 0, \tag{10.1}$$

where the dot denotes differentiation with respect to the time variable t, then yields the q^r as functions of t. Typically the motion of the system is subject to a system of physical constraints expressed by equations of the form

$$f(q^r, \dot{q}^r; t) = 0 \tag{10.2}$$

or more simply by

$$f(q^r; t) = 0. \tag{10.3}$$

The question naturally arises as to the compatibility of the systems (10.1) and (10.2) or (10.1) and (10.3). In the latter case the constraints (10.3) are said to be *finite*, or *integrable*, and they reduce the number of degrees of freedom of the system, but do not interfere with the integration of (10.1). However, in the former case, (10.2) is said to be *differential* or *non-integrable*. Such constraints do not reduce the number of degrees of freedom of the system and may render integration of (10.1) intractable. In this case the system is *non-holonomic* and it is not possible to exhibit a solution for the q^r, while for integrable constraints the system is said to be *holonomic*. Unfortunately, classical mechanics has essentially focused its attention on holonomic systems and non-holonomic systems have usually been neglected. Indeed, Lagrange himself failed to notice the possibility of non-holonomic systems, and their importance was realized only a century after the appearance of his celebrated *Mécanique analytique*. This delay is understandable since the great successes of classical mechanics were in celestial mechanics where constraints were either trivial or non-existent. Typically, non-holonomic constraints occur in rigid body dynamics as rolling conditions, or in rather complicated situations like the motion of a skate-blade over ice, etc. More generally, constraints involving inequalities can frequently lead to non-holonomic systems.

The reader might well conclude that the above dynamical situation is of no relevance to differential geodesy, but we claim that such a happy misconception is mistaken. Suppose the equations of motion (10.1) are replaced by some equations from theoretical geodesy or from differential geometry. Then the use of the leg calculus in studying these equations is predicated on two assumptions: the leg vectors are *unit* vectors that are identified with tangent vectors to congruences of curves, and as we will see in CHAPTER V, the 3-leg is associated to a surface $\mathbf{S}$ in $\mathbf{E}_3$ in a particular manner. Later in CHAPTER VI, when the 3-leg becomes a *Hotine 3-leg*, the vector $\boldsymbol{\nu}$ will be taken to be proportional to the gradient of the geopotential function N, with $\mathbf{S}$ being an equipotential surface! These requirements may be regarded as *constraints* on our 3-leg, however they are of a different character from those encountered in classical mechanics. Moreover, such constraints are essential if we wish to give geometric interpretations of the leg coefficients (3.2), and to systematically translate vectors/tensors into their leg representations, e.g. as in (2.18), (2.19), etc. Orthogonality is required in order that $\boldsymbol{\lambda}$ and $\boldsymbol{\mu}$ be perpendicular to $\boldsymbol{\nu}$, i.e. be tangents to the surface $\mathbf{S}$ having $\boldsymbol{\nu}$ as its normal vector, however the assumption that $\boldsymbol{\lambda}$ be perpendicular to $\boldsymbol{\mu}$ is a mathematical convenience which could be weakened.

The key result relative to holonomic and non-holonomic leg systems is given by the following theorem which we state without proof and call

<u>The Holonomity Theorem</u>

Let $\{\boldsymbol{\lambda}_a\}$ be a 3-leg on a region Ω in $\mathbf{E}_3$ such that

$$[\boldsymbol{\lambda}_a, \boldsymbol{\lambda}_b] = \mathbf{0} \tag{10.4}$$

for all a, b. Then there exists a local coordinate system x^r on Ω such that each $\boldsymbol{\lambda}_a$ has the form

$$\boldsymbol{\lambda}_a = \delta^r_{\ a} \frac{\partial}{\partial x^r} \tag{10.5}$$

on Ω.

Note that (10.4) is equivalent to

$$\alpha_{abc} = 0 \tag{10.6}$$

for all a, b, c, and as a consequence of (10.5) the Pfaffian basis system $\{\boldsymbol{\theta}_a\}$ is given by

$$\boldsymbol{\theta}_a = \delta_{ar} dx^r. \tag{10.7}$$

Hence, each $\boldsymbol{\theta}_a$ is a perfect/total differential, viz.

$$d\boldsymbol{\theta}_a = 0 \tag{10.8}$$

for $a = 1, 2, 3$. In this case, both leg systems $\{\boldsymbol{\lambda}_a\}$ and $\{\boldsymbol{\theta}_a\}$ are said to be *holonomic*, and the associated local coordinate system

x^r is also *holonomic*. In the contrary case, i.e. when any *one* of (10.4)-(10.8) is not satisfied the leg system is *non-holonomic*, or *anholonomic*, and properly speaking in this case a coordinate system x^r does not exist in the sense that (10.8) is satisfied. Likewise an individual *coordinate* is *holonomic* or *anholonomic*, according to whether or not it can or cannot be expressed in the form (10.7). Hence, by abuse of language we may speak of an *anholonomic coordinate system*, when at least one of the 'coordinates' fails to satisfy (10.8).

The obvious holonomic coordinate system is the familiar Cartesian coordinates system:

$$x^r = (x, y, z),$$

and, as shown in the following example, spherical polar coordinates are anholonomic since *only* the differential of the radial coordinate r coincides with *one* of the Pfaffian basis forms.

The clue to distinguishing between *holonomic* and *non- holonomic* (or *anholonomic*) *systems* in differential geodesy has already appeared in CHAPTER II when we called the α_{abc} coefficients of anholonomity. There these coefficients initially arose as a means of simply computing the Ricci coefficients γ_{abc}, however their real significance was revealed in II-(8.5), i.e.

$$[\boldsymbol{\lambda}_a, \boldsymbol{\lambda}_b] = \alpha_{abc}\boldsymbol{\lambda}_c, \tag{10.9}$$

which when applied to a smooth function F led to the integrability conditions [$\mathrm{F}_{\mathcal{A}}$] $\mathcal{A}$ = I, II, III of the system of leg differential equations II-(7.10). It is interesting to note formally that an analogue of the Lie bracket also occurs in celestial mechanics where it is known as the Poisson bracket, and this bracket, although defined differently, still satisfies a Jacobi-like identity (recall II-(9.3))!

To identify the various holonomic variables GRAFAREND (1971) has introduced the following ingenious device. By virtue of II-(7.4) and (7.7) we have

$$\alpha_{abc} = (\lambda_{cr;s} - \lambda_{cs;r})\,\lambda_a{}^r\lambda_b{}^s \tag{10.10}$$

which directly places the *curl-like* character of $\boldsymbol{\lambda}_c$ in evidence. Then modulo a factor of $\frac{1}{2}$ he defines the skew-symmetric anholonomity matrices:

$$\mathbf{A}_c := \left\| \alpha_{ab\underline{c}} \right\| \tag{10.11}$$

the first two indices are the entries in the matrices for a fixed value of c. By (3.5) we have the following matrices:

$$\mathbf{A}_1 = \left\| \begin{matrix} 0, & -\sigma_1, & -k_1 \\ \sigma_1, & 0, & t_2 + \varepsilon_3 \\ k_1, & -(t_2 + \varepsilon_3), & 0 \end{matrix} \right\| \tag{10.12}$$

$$\mathbf{A}_2 = \left\| \begin{matrix} 0, & -\sigma_2, & -(t_1 + \varepsilon_3) \\ \sigma_2, & 0, & k_2 \\ t_1 + \varepsilon_3, & -k_2, & 0 \end{matrix} \right\| \tag{10.13}$$

$$\mathbf{A}_3 = \left\| \begin{matrix} 0, & t_1 + t_2, & \gamma_1 \\ -(t_1 + t_2), & 0, & \gamma_2 \\ -\gamma_1, & -\gamma_2, & 0 \end{matrix} \right\| . \tag{10.14}$$

The explicit values of the entries in these Grafarend matrices are unimportant — the issue is whether the entries are *all zero* or not — and hence

$$\mathbf{A}_c = \mathbf{0} \tag{10.15}$$

for a value of c is equivalent to the cth leg vector being holonomic. Due to the skew-symmetry of these matrices, each matrix can have at most three non-zero entries, and hence in place of these matrices one can consider a *super anholonomity matrix* $\mathbf{A}$ whose entries are given in (3.5). Then a holonomic leg vector, and consequently a holonomic variable corresponds to a zero column occurring in $\mathbf{A}$.

IV.11 The Reconstruction Theorem and General Viewpoint

We now return to the Lamé equations (recall SECTION 9) and consider their consequences for differential geodesy. It is useful to momentarily return to Riemannian geometry, and reconsider the situation in N-dimensions, viz. we will replace $\mathbf{V}_3$ by an N-dimensional Riemannian space $\mathbf{V}_N$ and consider the reduction of this $\mathbf{V}_N$ to an Euclidean $\mathbf{E}_N$. In the following it will be useful to denote the set of Ricci coefficients by $\{\gamma_{abc}\}$. The converse statement is known as the

Reconstruction Theorem

Given a set $\{\gamma_{abc}\}$, there exists an N-leg $\{\boldsymbol{\lambda}_a\}$ having these $\{\gamma_{abc}\}$ as its Ricci coefficients and for which the Riemann-Christoffel curvature tensor of $\mathbf{V}_N$ has the leg components R_{abcd} given by II-(12.6).

Proof: Consider II-(4.2), i.e.

$$\lambda_{ar,s} = \gamma_{abc}\lambda_{ar}\lambda_{bs} \tag{11.1}$$

which for given $\{\gamma_{abc}\}$ are a set of N^3 quasi-linear partial differential equations for the N-leg $\{\boldsymbol{\lambda}_a\}$. The necessary and sufficient conditions for the existence of solutions of (11.1) is that its integrability conditions be satisfied. These are given by

$$\begin{aligned}\lambda_{ar,s,t} - \lambda_{ar,t,s} &= \{\gamma_{abc/d} - \gamma_{abd/c} \\ &\quad +\gamma_{afd}\gamma_{fbc} - \gamma_{afc}\gamma_{fbd} \\ &\quad +\gamma_{abf}(\gamma_{fcd} - \gamma_{fdc})\}\lambda_{br}\lambda_{cs}\lambda_{dt}.\end{aligned} \tag{11.2}$$

But the right-hand side of this equation is

$$R_{abcd}\lambda_{br}\lambda_{cs}\lambda_{dt} = R^p{}_{rst}\lambda_{ap} \tag{11.3}$$

which is a *mixed version* of II-(12.3). Hence upon *defining* R_{abcd} by II-(12.6), the theorem is proven.

By virtue of this result, the reduction of $\mathbf{V}_N$ to $\mathbf{E}_N$ is given by

$$R_{abcd} = 0, \tag{11.4}$$

which when $N = 3$ are Lamé equations $(\mathcal{L}_\mathcal{B})$ $\mathcal{B} = \mathrm{I}, ..., \mathrm{IX}$. These equations may be considered from two viewpoints. First, if the

3-leg $\{\boldsymbol{\lambda}_a\}$ is known, the $\{\gamma_{abc}\}$ or equivalently the set of leg coefficients (3.2), are also known and hence (11.4) is identically satisfied. Thus, the Lamé equations become the *Lamé identities.* Conversely, if the $\{\gamma_{abc}\}$ are known then (11.4) may be regarded as a system of nine first order quasi-linear partial differential equations for the nine covariant components of the 3-leg vectors of $\{\boldsymbol{\lambda}_a\}$. In this case the Lamé equations are not identities, but *determining equations* for these leg vector components.

These considerations suggest several different rôles for the Lamé equations:

(i) given the $\{\boldsymbol{\lambda}_a\}$, the $\{\gamma_{abc}\}$ are computable and then $(\mathcal{L}_\mathcal{B})$ is identically satisfied;

(ii) given the $\{\gamma_{abc}\}$, the $\{\boldsymbol{\lambda}_a\}$ are determinable and then $(\mathcal{L}_\mathcal{B})$ play the role of determining equations;

(iii) given a partial specification of the $\{\boldsymbol{\lambda}_a\}$ and the $\{\gamma_{abc}\}$ — with this specification being compatible, then the $(\mathcal{L}_\mathcal{B})$ reduce to identities and or determining equations for the remaining $\{\boldsymbol{\lambda}_a\}$ and $\{\gamma_{abc}\}$.

The approach (i) is essentially that employed by Hotine in his treatise (see [pages 70-86]), however, as we have earlier noted he used only a rudimentary form of the leg calculus, and the Lamé equations $(\mathcal{L}_\mathcal{B})$ $\mathcal{B} = \mathrm{I}, \ldots, \mathrm{IX}$ were not completely exhibited. Nevertheless, denoting $k_1 = k_2$ by k we have in our example of Section III-6,

$$\begin{aligned} \lambda^r &= (-k\sec\phi, 0, 0)\,, \\ \mu^r &= (0, -k, 0)\,, \\ \nu^r &= (0, 0, 1)\,; \end{aligned} \tag{11.5}$$

$$\begin{aligned} \lambda_r &= \left((-k\sec\phi)^{-1}\right), 0, 0, \\ \mu_r &= (0, -k^{-1}, 0)\,, \\ \nu_r &= (0, 0, 1)\,; \end{aligned} \tag{11.6}$$

and moreover,

$$\begin{aligned} \omega_r &= -k\sec\phi\lambda_r\,, \\ \phi_r &= -k\mu_r\,, \\ r_r &= \nu_r\,. \end{aligned} \tag{11.7}$$

Equations (11.5) and (11.6) illustrate the situation described in the reconstruction theorem. The details and properties described in these equations, and in particular the significance of (11.7) for coordinate systems of interest to differential geodesy will be discussed in CHAPTER X. Finally, the non-zero values of k and $\sigma := \sigma_2$ given in our example, identically satisfy the Lamé equations (see PROBLEM IV.5).

The approaches (ii) and (iii) are new and useful in eliminating over determined choices of the $\{\boldsymbol{\lambda}_a\}$ and $\{\gamma_{abc}\}$. For example, consider the Lamé equations when the leg coefficients are given by k_1, k_2, σ_2 with $t_1 = t_2 = \sigma_1 = \gamma_1 = \gamma_2 = \varepsilon_3 = 0$ viz.

$$\begin{aligned}
&(\mathcal{L}_{\mathrm{I}}): \sigma_{1/2} - \sigma_1^2 = k_1 k_2,\\
&(\mathcal{L}_{\mathrm{II}}): \ -\sigma_{1/3} + k_1\sigma_1 = 0,\\
&(\mathcal{L}_{\mathrm{III}}): 0 = 0,\\
&(\mathcal{L}_{\mathrm{IV}}): \ -k_{1/2} + k_1\sigma_1 = k_2\sigma_1,\\
&(\mathcal{L}_{\mathrm{V}}): k_{1/3} - k_1^2 = 0,\\
&(\mathcal{L}_{\mathrm{VI}}): 0 = 0,\\
&(\mathcal{L}_{\mathrm{VII}}): \ -k_{2/1} = 0,\\
&(\mathcal{L}_{\mathrm{VIII}}): 0 = 0,\\
&(\mathcal{L}_{\mathrm{IX}}): k_{2/3} - k_2^2 = 0;
\end{aligned} \tag{11.8}$$

with

$$\begin{aligned}
F_{/1} &:= \lambda^r F_{;r} = \lambda^1 F_{;1} + \lambda^2 F_{;2} + \lambda^3 F_{;3},\\
F_{/2} &:= \mu^r F_{;r} = \mu^1 F_{;1} + \mu^2 F_{;2} + \mu^3 F_{;3},\\
F_{/3} &:= \nu^r F_{;r} = \nu^1 F_{;1} + \nu^2 F_{;2} + \nu^3 F_{;3};
\end{aligned} \tag{11.9}$$

for a smooth function F, e.g. k_1, k_2 or σ_1. Then the immediate question is what specializations of the leg vectors are compatible with $k_1 = k_2$ and the value of σ_1 given in our example. Note that in terms of coordinates, does the system (11.8) necessarily demand the identification

$$x^1 = \omega, \quad x^2 = \phi, \quad x^3 = r? \tag{11.10}$$

The answer is not obvious, and requires careful analysis. A natural suspicion, especially since the x^3-curves have zero curvature χ, viz. $\gamma_1 = \gamma_2 = 0$, is that (11.10) is the obvious choice, i.e. x^3 is a linear coordinate, but it is less clear that x^1 and x^2 must be angular coordinates ω and ϕ respectively.

The above situation works only with a subset of the $\{\gamma_{abc}\}$ and seeks the corresponding components of the $\{\boldsymbol{\lambda}_a\}$. A more realistic approach is that of (iii) when one partially specifies *some* of the $\{\gamma_{abc}\}$ and *some* of the $\{\boldsymbol{\lambda}_a\}$ and investigates the demands made on such a choice by the Lamé equations. Naturally one seeks a compatible choice of these quantities and the specification of enough of them that the system $(\mathcal{L}_{\mathcal{B}})$ $\mathcal{B}$ = I, ..., IX determine the remaining quantities. On the other hand, even if the specification is incomplete, it could be useful in that it could indicate when the choices in question were overly restrictive or contradictory.

At the moment these are merely possibilities which are meant to suggest the fruitfulness and importance of the approaches (ii) and (iii) based on considering the Lamé equations *not* as identities, but as determining equations in differential geodesy. As we will later see in CHAPTER VI these equations play a prominent role in the Hotine-Marussi equations, which are the fundamental equations of differential geodesy.

PROBLEMS FOR CHAPTER IV

IV.1 Derive (2.19) and hence show that there is no distinction between the leg components T_{ab} in (2.18) and those in (2.19).

IV.2 Introducing 'covariant' leg derivatives denoted by

$$\xi_{r//1} := \xi_{r,s}\lambda^s, \quad \xi_{r//2} := \xi_{r,s}\mu^s, \quad \xi_{r//3} := \xi_{r,s}\nu^s$$

where $\boldsymbol{\xi}$ is an arbitrary vector, rewrite (6.22)-(6.23) using these derivatives.

N.B. This kind of derivative is not essential, but often useful!

IV.3 Derive the expression (8.8) for the 3-dimensional Laplacian ΔF of an arbitrary smooth function F by covariant differentiation of the leg representation

$$F_r = F_{/1}\lambda_r + F_{/2}\mu_r + F_{/3}\nu_r$$

etc.

IV.4 Write the line element ds^2 of $\mathbf{E}_3$ in spherical polar coordinates $x^r = (\omega, \phi, r)$ where $0 \leq \omega < 2\pi$ is the longitude, $-\frac{\pi}{2} \leq \phi < \frac{\pi}{2}$ the latitude, and $0 < r < \infty$ the radial coordinate, viz.

$$ds^2 = \left(r^2 \cos^2 \phi\right) d\omega^2 + r^2 d\phi^2 + dr^2 .$$

Then exhibit the matrices $\|g_{rs}\|$, $\|g^{rs}\|$; and the contravariant and covariant components of the vectorial 3-leg $\{\boldsymbol{\lambda}, \boldsymbol{\mu}, \boldsymbol{\nu}\}$.

IV.5 Using the line element and results of PROBLEM IV.5, show that

$$\theta_1 = r\cos\phi d\omega, \quad \theta_2 = r d\phi, \quad \theta_3 = dr$$

and show that the connection 1-forms are given by

$$\begin{aligned} \omega_{12} &= -\left(r^{-1}\tan\phi\right)\theta_1, \\ \omega_{31} &= -\left(r^{-1}\right)\theta_2, \\ \omega_{23} &= \left(r^{-1}\right)\theta_3. \end{aligned}$$

Hence, conclude that the leg coefficients are given by

$$\begin{gathered} k_1 = k_2 = -r^{-1}, \quad \sigma_1 = r^{-1}\tan\phi, \\ \gamma_1 = \gamma_2 = t_1 = t_2 = \sigma_2 = \varepsilon_3 = 0. \end{gathered}$$

IV.6 Show that for the connection 1-forms ω_{ab} of PROBLEM IV.5, that the curvature 2-forms Ω_{ab} are identically zero.

IV.7 Prove that the system of PROBLEM IV.5 is anholonomic, i.e. there do not exist functions X and Y such that

$$\begin{aligned} dX &= r\cos\phi d\omega \\ dY &= r d\phi \end{aligned}$$

although θ_3 is a perfect differential.

IV.8 For the ds^2 of PROBLEM IV.4-IV.7 exhibit the *anholonomity matrices* $\mathbf{A}_1$, $\mathbf{A}_2$, $\mathbf{A}_3$ and the *super anholonomity matrix* $\mathbf{A}$.

IV.9 Read GRAFAREND (1971, 1973, and 1975) and BOCCHIO (1975) for an extensive discussion of the anholonomity matrices and their geodetic applications.

IV.10 Read GRAFAREND (1971, 1977, and 1978) for a proposed generalization of Riemannian geometry using the Cartan calculus. An alternate approach to non-symmetric connections — not using the Cartan calculus — is given in BOCCHIO (1970 and 1974).

V

Gaussian Differential Geometry

V.1 Introduction

In this chapter we specialize the general leg calculus of CHAPTER IV to the study of the intrinsic and extrinsic geometry of surfaces and curves when our Riemannian 3-space $\mathbf{V}_3$ is restricted to be an Euclidean 3-space $\mathbf{E}_3$. We will call this geometry Gaussian differential geometry, even though it includes many results obtained earlier by Euler, Monge, and Meusnier. The terminology is justified since the Gaussian viewpoint incorporated these often fragmentary, but beautiful, results in a new and unified light that was only dimly suspected by these workers.

Our immediate goal is to find a geometric interpretation of the leg coefficients/parameters:

$$k_1, k_2, t_1, t_2, \gamma_1, \gamma_2, \sigma_1, \sigma_2, \varepsilon_3$$

where $\mathbf{S}$ denotes a smooth surface that is locally, isometrically and smoothly imbedded in $\mathbf{E}_3$. In this regard, we choose the third congruence $\boldsymbol{\Gamma} = \boldsymbol{\Gamma}_3$ to be normal to $\mathbf{S}$ (in the notation of CHAPTER IV, strictly speaking $\mathbf{S}$ is $\mathbf{S}_3$, but since only this surface is considered prior to SECTION 13, we frequently suppress the numerical index on $\mathbf{S}$). Thus the orthonormal 3-leg $\{\boldsymbol{\lambda}, \boldsymbol{\mu}, \boldsymbol{\nu}\}$ consists of a pair of spatial tangent vectors to $\mathbf{S}$ which have the space components:

$$\boldsymbol{\lambda}: \ \lambda^r \text{ or } \lambda_r,$$
$$\boldsymbol{\mu}: \ \mu^r \text{ or } \mu_r;$$

and a unit normal vector $\boldsymbol{\nu}$ to $\mathbf{S}$ which has the space components

$$\boldsymbol{\nu}: \nu^r \text{ or } \nu_r.$$

Hence, when x^r is an arbitrary ambient curvilinear coordinate system in $\mathbf{E}_3$ in a neighborhood of **S**, we have

$$\nu_r dx^r = 0 \tag{1.1}$$

for all tangential displacements to **S**. By our previous definition of the Pfaffian forms (IV-(2.11))

$$\{\theta_a\}_{a=1}^{3},$$

(1.1) requires that

$$\theta_3 = 0, \tag{1.2}$$

and this condition characterizes the *intrinsic geometry* of **S**. The orthogonality of $\boldsymbol{\Gamma}$ to **S**, i.e., $\boldsymbol{\Gamma} \perp \mathbf{S}$, by Condition 5^0 of SECTION IV-5 requires that

$$t_2 = -t_1. \tag{1.3}$$

Hotine derived this condition relatively early in his analysis, see [Chapter 7, §6 page 40], but it appears that he was unaware of the fact that it arises *ab initio* from the requirement that $\boldsymbol{\Gamma} \perp \mathbf{S}$. Equation (1.3) is a fundamental result — one can regard it as a *constraint* — in our exposition of differential geodesy, and we may call it *property* (T), or simply (T). It is useful and interesting in deducing the meaning of the leg coefficients/parameters to retain both t_1 and t_2 in our analysis — even though we know by virtue of $\boldsymbol{\Gamma} \perp \mathbf{S}$ that (1.3) automatically holds. By doing this we can see how many of our derivations and results explicitly depend on this condition. Thus, in citing our final results, if (1.3) has not been imposed by our analysis, then the specialization to it will be indicated by mentioning property (T), or simply writing (T) following the result. There is no suggestion that by not automatically assuming (1.3) that we are in any way generalizing Hotine's analysis. In effect, our procedure merely indicates where (T) plays a role in the geometric discussion. Later, in CHAPTER VI, when we have successively derived (T) from a number of cogent considerations, we will dispense with it and automatically assume that it is in force.

Our presentation is not a substitute for a previous general familiarity with the tensorial approach to Gaussian differential geometry as may be found in Hotine's treatise [see Part I: Chapters 3-11] or in McCONNELL (1931) (see Part III: Chapters XI-XVI), but rather a reworking of this material from the viewpoint of the leg calculus. Hence, we have restricted our discussion only to those parts of the theory which require some comment before being put in the language of the leg calculus.

V.2 The Framework of Gaussian Differential Geometry

Properly speaking, Gaussian differential geometry begins with the assumption that *ab initio* **S** is given as a parametrized surface. Thus we assume that **S** has a Gaussian representation

$$\mathbf{S} : x^r = x^r\left(u^1, u^2\right) \tag{2.1}$$

where $u^\alpha := (u^1, u^2)$ is an arbitrary parameter (or 'coordinate') system. Henceforth, small Greek indices always refer to surface quantities and range over the values 1, 2 with repeated indices obeying the summation convention. The u^α in (2.1) are completely arbitrary, indeed as arbitrary as the ambient coordinate system x^r in $\mathbf{E}_3$. However, since generally the u^α are curvilinear parameters, there is no presumption that a single parametrization is adequate to parametrize all of **S**. Hence, u^α is a *local* parametrization and when employing (2.1) on **S**, we are actually referring only to that portion, or piece, of **S** for which the parametrization is valid.

The Gaussian geometry of **S** is conveniently described in terms of a set of *basic* quadratic (differential) forms which we denote by I, II, III, ... We now recall the definitions of these forms, where for generality we assume an arbitrary curvilinear coordinate system x^r to serve as an ambient coordinate system in $\mathbf{E}_3$, and a metric tensor with components g_{rs} in this coordinate system.

The first basic form I is defined by the requirement that the distance, or metric, in $\mathbf{E}_3$ induces an isometric distance, or metric, in $\mathbf{S}$, i.e.

$$ds^2 = g_{rs}dx^r dx^s = a_{\alpha\beta}du^\alpha du^\beta := I. \tag{2.2}$$

By virtue of (2.1) we immediately have the following relationship between the coordinate and parameter differentials

$$dx^r = x^r_{\alpha} du^\alpha \tag{2.3}$$

where

$$x^r{}_\alpha := \frac{\partial x^r}{\partial u^\alpha} = x^r{}_{;\alpha}. \tag{2.4}$$

There is no standard name for the quantities (2.4), but since in effect they convert surface indices into space indices, it is tempting to regard them as the mixed components of an *operator*. We propose to call it the *Gauss operator*, but must immediately stress that this operator has no inverse. However, upon exhibiting these mixed components as entries in a 3×2 matrix

$$\|x^r_\alpha\|,$$

one requires that this matrix be of rank two. The motivation for this is as follows. By construction, the space congruences $\boldsymbol{\Gamma}_1$ and $\boldsymbol{\Gamma}_2$ are tangential to $\mathbf{S}$. Let $\boldsymbol{\Gamma}'_1$ and $\boldsymbol{\Gamma}'_2$ denote the respective projections of these congruences on $\mathbf{S}$. These projections are the respective u^α-*parameter curves*, or just the u^α-*curves*,

$$\begin{aligned} \boldsymbol{\Gamma}'_1 &: \ u^1 = \text{variable}, \ u^2 = \text{fixed}; \\ \boldsymbol{\Gamma}'_2 &: \ u^1 = \text{fixed}, \ u^2 = \text{variable}. \end{aligned} \tag{2.5}$$

Then the rank two requirement insures the linear independence of the congruences (as a consequence of the orthogonality of the 3-leg vectors $\boldsymbol{\lambda}$ and $\boldsymbol{\mu}$, in our case we have more: $\boldsymbol{\Gamma}_1$ and $\boldsymbol{\Gamma}_2$ are orthogonal, hence project into orthogonal $\boldsymbol{\Gamma}'_1$ and $\boldsymbol{\Gamma}'_2$ on $\mathbf{S}$), and also the existence of the normal $\boldsymbol{\nu}$ to $\mathbf{S}$. Using (2.3) in (2.2) we immediately have

$$\left(g_{rs}x^r_\alpha x^s_\beta - a_{\alpha\beta}\right) du^\alpha du^\beta = 0 \tag{2.6}$$

for arbitrary parameter differentials. Hence, since both the g_{rs} and $a_{\alpha\beta}$ are symmetric, one has the fundamental result

$$a_{\alpha\beta} = g_{rs} x^r_\alpha x^s_\beta. \tag{2.7}$$

Note that (2.7) is the analytic statement of a local isometric imbedding of **S** in $\mathbf{E}_3$. In classical differential geometry, this is often rather casually and automatically imposed, since the theory supposes that one is *a priori* given a surface **S** in $\mathbf{E}_3$. However, in reality (2.7) is a system of three non-linear first order partial differential equations for the independent variables x^1, x^2, x^3 with the u^1, u^2 being dependent variables. The g_{rs} are regarded as given (since one has chosen a system of x^α in $\mathbf{E}_3$) and the $a_{\alpha\beta}$ are prescribed functions. If (2.1) is known, then (2.7) is a prescription for determining the $a_{\alpha\beta}$, and the situation is unambiguous. On the other hand, differential geodesy does present the possibility of an ambiguous situation. For example, suppose on the basis of observations, or by approximation schemes as in the theory of adjustments, one is led to a particular $a_{\alpha\beta}$. Then does this trial $a_{\alpha\beta}$ correspond to a surface **S** which is locally isometrically imbedded in $\mathbf{E}_3$? The condition that this be true is that (2.7) be satisfied, i.e. in this case (2.7) serves as a compatibility/consistency condition for the $a_{\alpha\beta}$ to be the coefficients of I. Stated in more graphical terms, not every kind of surface can be realized in $\mathbf{E}_3$. A startling example of this non-existence is due to Hilbert, who in 1901 (see WILLMORE (1959) for a discussion) proved that a complete analytic **S** which has no singularities and has constant negative Gaussian curvature K cannot exist in $\mathbf{E}_3$. For our purposes this is a rather pathological example, but nevertheless it indicates the non-trivial character of (2.7). Moreover, as we will see later, the $a_{\alpha\beta}$ do not suffice to determine uniquely a surface **S**.

The second basic form II is defined by

$$\mathrm{II} := b_{\alpha\beta} du^\alpha du^\beta, \tag{2.8}$$

where the $b_{\alpha\beta}$ may be defined by any of the following three expressions:

$$\begin{aligned}
&\text{(i)} \quad b_{\alpha\beta} := -g_{rs} x^r_\alpha \nu^s_\beta \\
&\text{(ii)} \quad b_{\alpha\beta} := g_{rs} x^r_{\alpha\beta} \nu^s \\
&\text{(iii)} \quad b_{\alpha\beta} := -\nu_{rs} x^r_\alpha x^s_\beta
\end{aligned} \tag{2.9}$$

where

$$x^r_{\alpha\beta} := x^r_{\alpha,\beta}$$
$$\nu_{rs} := \nu_{r,s};$$

the comma denotes covariant differentiation with respect to the surface parameters, and space coordinates, respectively. The geometric interpretation of II is obtained as follows. Suppose II $\neq 0$ at a point P of $\mathbf{S}$, and let P and a neighboring point Q on $\mathbf{S}$ have the respective Gaussian parametrizations $x^r(u^1, u^2)$, $x^r(u^1 + h^1, u^2 + h^2)$ where $h^\alpha := du^\alpha$. We will show that II is related to the perpendicular distance δ from Q to the tangent plane $\mathbf{T}_P(\mathbf{S})$ — see Figure 1 — or more precisely

$$\delta = |Q, Q^*| \tag{2.10}$$

where Q^* is the orthogonal projection of Q on $\mathbf{T}_P(\mathbf{S})$.

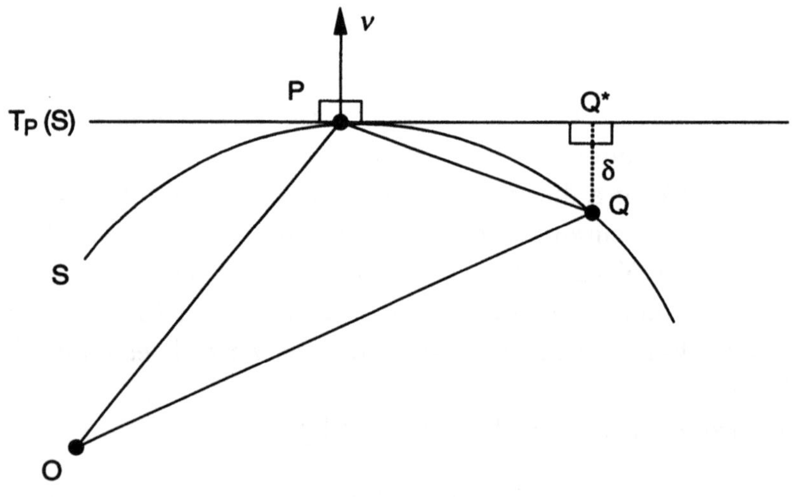

Figure 1

To evaluate (2.10), we consider a Taylor expansion of x^r about P, and denote for simplicity $x^r(Q) = \tilde{x}^r$ with the expressions at P being written as

$$x^r, x^r_\alpha, x^r_{\alpha\beta}, \cdots$$

where the Greek indices denote covariant surface derivatives. Then

$$\tilde{x}^r = x^r + h^\alpha x^r_\alpha + \frac{1}{2!} h^\alpha h^\beta x^r_{\alpha\beta} + \frac{1}{3!} h^\alpha h^\beta h^\gamma x^r_{\alpha\beta\gamma} + \cdots \tag{2.11}$$

and from elementary geometry we know that

$$\delta = \nu_r \left(\tilde{x}^r - x^r\right), \tag{2.12}$$

i.e. $\overrightarrow{QQ^*} = |Q, Q^*|\,\boldsymbol{\nu}$. Now $\nu_r x^r_\alpha = 0$ and hence upon using

$$\begin{aligned} x^r_{\alpha\beta} &= b_{\alpha\beta}\nu^r \\ x^r_{\alpha\beta\gamma} &= b_{\alpha\beta,\gamma}\nu^r + b_{\alpha\beta}\nu^r_\gamma \\ x^r_{\alpha\beta\gamma\delta} &= b_{\alpha\beta,\gamma\delta}\nu^r + b_{\alpha\beta,\gamma}\nu^r_\delta + b_{\alpha\beta,\delta}\nu^r_\gamma + b_{\alpha\beta}\nu^r_{\gamma,\delta} \end{aligned} \tag{2.13}$$

we find that

$$\delta = \frac{1}{2!} h^\alpha h^\beta b_{\alpha\beta} + \frac{1}{3!} h^\alpha h^\beta h^\gamma b_{\alpha\beta,\gamma} + \frac{1}{4!} h^\alpha h^\beta h^\gamma h^\delta b_{\alpha\beta,\gamma\delta} + \cdots. \tag{2.14}$$

Thus, correct to third order terms in h^α, viz. the differentials du^α, we have

$$\delta = \mathrm{II}/2. \tag{2.15}$$

Both of the quadratic forms I and II were introduced by Gauss in 1828 and play a primary role in his formulation of the differential geometry of surfaces in $\mathbf{E}_3$. They are fundamental in the sense that both are required for a complete specification of an $\mathbf{S}$ which is locally isometrically imbedded in $\mathbf{E}_3$. The explicit conditions for this to be the case are that

$$a := \det \|a_{\alpha\beta}\| > 0, \tag{2.16}$$

and $a_{\alpha\beta}$, $b_{\alpha\beta}$ satisfy the equations of Gauss $(\mathcal{G})$ and Codazzi $(\mathcal{C})$. These arise as the integrability conditions of the Gauss formula, i.e. the first equation in (2.13):

$$x^r_{\alpha\beta} = b_{\alpha\beta}\nu^r, \tag{2.17}$$

viz.

$$x^r_{\alpha\beta\gamma} - x^r_{\alpha\gamma\beta} = R^\sigma{}_{\alpha\beta\gamma} x^r_\sigma \tag{2.18}$$

where $R^\sigma{}_{\alpha\beta\gamma}$ is the Riemann curvature tensor of **S**. Note that (2.18) is none other than the difference of the surface covariant derivatives of x^r_α so $R^\sigma{}_{\alpha\beta\gamma}$ occurs naturally as in the tensor calculus. The left-hand side may be immediately evaluated by differentiation of the Gauss formula when the Weingarten formula is employed, i.e.

$$\nu^r{}_\alpha = -a^{\rho\sigma} b_{\rho\alpha} x^r_\sigma. \tag{2.19}$$

Thus we obtain

$$-a^{\rho\sigma}\left(b_{\alpha\beta}b_{\rho\gamma} - b_{\alpha\gamma}b_{\rho\beta}\right)x^r_\sigma + \left(b_{\alpha\beta,\gamma} - b_{\alpha\gamma,\beta}\right)\nu^r = R^\sigma{}_{\alpha\beta\gamma}x^r_\sigma \tag{2.20}$$

and consequently contraction by $g_{rs}x^s_\tau$ and ν^r respectively gives

$$(\mathcal{G}) \qquad R_{\rho\alpha\beta\gamma} = b_{\rho\beta}b_{\alpha\gamma} - b_{\rho\gamma}b_{\alpha\beta}, \tag{2.21}$$

$$(\mathcal{C}) \qquad 0 = b_{\alpha\beta,\gamma} - b_{\alpha\gamma,\beta}. \tag{2.22}$$

Alternative versions of $(\mathcal{G})$ are:

$$(\mathcal{G}') \qquad R_{\alpha\beta\gamma\delta} = K\left(a_{\alpha\gamma}a_{\beta\delta} - a_{\alpha\delta}a_{\beta\gamma}\right), \tag{2.23}$$

$$(\mathcal{G}'') \qquad R_{\alpha\beta\gamma\delta} = K\varepsilon_{\alpha\beta}\varepsilon_{\gamma\delta}; \tag{2.24}$$

where K is the Gaussian (total) curvature; and $(\mathcal{C})$ may be rewritten more concisely as

$$(\mathcal{C}') \qquad \varepsilon^{\beta\gamma}b_{\alpha\beta,\gamma} = 0. \tag{2.25}$$

Since $(\mathcal{G})$ and $(\mathcal{C})$ are surface equations, inspection shows that $(\mathcal{G})$ consists of *one* independent equation, while $(\mathcal{C})$ involves *two* independent equations. The equations given in $(\mathcal{G})$, $(\mathcal{G}')$ and $(\mathcal{G}'')$ then yield the following expressions for K

$$K = R_{1212}/a = b/a \tag{2.26}$$

where

$$b := \det\left\|b_{\alpha\beta}\right\|, \tag{2.27}$$

and

$$\varepsilon_{\alpha\beta} = \sqrt{a}e_{\alpha\beta} \tag{2.28}$$

with $e_{\alpha\beta}$ being the permutation symbol.

The third basic form III is significantly less fundamental than I and II, and is also — at least in spirit — due to Gauss. It concerns the classical problem of mapping a given surface $\mathbf{S}$ onto another surface $\mathbf{S}^*$. Following Hotine [Chapter 11], when $\mathbf{S}^*$ is taken to be a unit 2-sphere $\mathbf{S}_2$ with center 0, we write $\mathbf{S}^* = \overline{\mathbf{S}}$ and denote the image (spherical) quantities by a bar. The construction of this *spherical representation* of $\mathbf{S}$, involves imagining a set of unit normals $\boldsymbol{\nu}$ on $\mathbf{S}$ to be parallely transported to a point 0. Then the endpoints of these normals (all of whose initial points are fixed at 0) will trace out a region on $\mathbf{S}_2$ which is called the *spherical image* of $\mathbf{S}$. Under this mapping I becomes

$$\overline{\mathrm{I}} = \overline{a}_{\alpha\beta}du^\alpha du^\beta \tag{2.29}$$

where the same parametrization u^α is employed on both $\mathbf{S}$ and $\mathbf{S}_2$, i.e. $u^\alpha = \overline{u}^\alpha$ and so $du^\alpha = d\overline{u}^\alpha$. Obviously $\nu_r\nu^r = 1$, so we have $\nu_r\nu^r_\alpha = 0$ and since $\overline{\nu}_r = \nu_r$ this suggests taking

$$\overline{x}^r_\alpha = \nu^r_\alpha. \tag{2.30}$$

Thus

$$\overline{a}_{\alpha\beta} = g_{rs}\overline{x}^r_\alpha\overline{x}^s_\beta = g_{rs}\nu^r_\alpha\nu^s_\beta := c_{\alpha\beta} \tag{2.31}$$

and hence we have

$$\overline{\mathrm{I}} = c_{\alpha\beta}du^\alpha du^\beta := \mathrm{III}$$

and this is the *third basic* form. By using (2.19) we immediately have

$$c_{\alpha\beta} = a^{\rho\sigma}b_{\alpha\rho}b_{\beta\sigma} \tag{2.32}$$

and consequently if

$$c := \det \|c_{\alpha\beta}\|$$

we obtain

$$c = b^2/a > 0. \tag{2.34}$$

The area element on $\mathbf{S}$ is

$$dA := \sqrt{a}du^1du^2 \tag{2.35}$$

and hence on $\mathbf{S}_2$ we have

$$d\overline{A} := \sqrt{\overline{a}}du^1du^2 = \sqrt{c}du^1du^2. \tag{2.36}$$

Thus, the ratio of $d\overline{A}$ and dA is given by

$$\frac{d\overline{A}}{dA} = \frac{\sqrt{c}}{\sqrt{a}} = \frac{b}{a} = K, \tag{2.37}$$

a result anticipated by Rodrigues in 1815 and rigorously establish later by Gauss. Finally, I, II and III are not independent but are connected by the equation

$$\text{III} \ - 2H\text{II} \ + K\text{I} = 0, \tag{2.38}$$

or

$$c_{\alpha\beta} - 2Hb_{\alpha\beta} + Ka_{\alpha\beta} = 0 \tag{2.39}$$

where H is the *Germain* (*mean*) *curvature*

$$2H := a^{\rho\sigma}b_{\rho\sigma}. \tag{2.40}$$

The third basic form, and the idea of representing $\mathbf{S}$ on $\mathbf{S}_2$ is very appealing. But despite the obviousness of representing a spheroid or an ellipsoid on $\mathbf{S}_2$ in geodetic situations, it has found little direct use in differential geodesy. Nevertheless, by virtue of (2.39) III is almost unavoidable in the context of employing $a_{\alpha\beta}$ and $b_{\alpha\beta}$. For our purposes, this equation represents the underlying importance of $c_{\alpha\beta}$.

V.3 Leg Representation of the Gauss Operator and the First Basic Form

Our first task is to derive a leg representation for the *Gauss operator*, (2.4), x^r_α which for tangential surface directions converts space components into surface components by the rules:

$$x^r_\alpha \lambda_r := \lambda_\alpha, \tag{3.1}$$

$$x^r_\alpha \mu_r := \mu_\alpha. \tag{3.2}$$

Since $\boldsymbol{\nu}$ is normal to $\mathbf{S}$, it has zero surface components, viz.

$$x^r_\alpha \nu_r := 0. \tag{3.3}$$

By virtue of this fact, *it is better not to introduce* the quantities ν_α or ν^α which in the case under consideration are identically zero. Later in SECTION 13, when we consider triply-orthogonal systems of surfaces, viz. not merely $\mathbf{S} = \mathbf{S}_3$ but also $\mathbf{S}_1$ and $\mathbf{S}_2$, such quantities do naturally occur and will be discussed. For our present purposes ν_α and ν^α are trivial and hence do not appear in our analysis.

As noted in SECTION 2, there is no inverse Gauss operator, however we do have the following *inverse-like* equations

$$x^r_\alpha \lambda^\alpha = \lambda^r \tag{3.4}$$

$$x^r_\alpha \mu^\alpha = \mu^r. \tag{3.5}$$

Thus, to merely account for its indices, we may expect that the Gauss operator admits a leg representation of the general form

$$x^r_\alpha = (c_1\lambda^r + c_2\mu^r + c_3\nu^r)\,\lambda_\alpha + (c_r\lambda^r + c_5\mu^r + c_6\nu^r)\,\mu^\alpha \tag{3.6}$$

where $c_1, \cdots, c_6$ are scalar coefficients which must be determined. By successively applying (3.1-.3) to (3.6) we obtain

$$\begin{aligned} x^r_\alpha \lambda_r &= \lambda_\alpha \Rightarrow c_1 = 1, c_4 = 0 \\ x^r_\alpha \mu_r &= \mu_\alpha \Rightarrow c_2 = 0, c_5 = 1 \\ x^r_\alpha \nu_r &= 0 \Rightarrow c_3 = 0, c_6 = 0 \end{aligned} \tag{3.7}$$

which gives

$$x^r_\alpha = \lambda^r \lambda_\alpha + \mu^r \mu_\alpha. \tag{3.8}$$

This was stated in a slightly different — perhaps less general — context by Hotine (see his [6.09]). As we will see (3.8) is the *key formula* in the leg calculus formulation of Gaussian differential geometry.

First note that using (3.8), (2.3) becomes

$$dx^r = \lambda^r\,(\lambda_\alpha du^\alpha) + \mu^r\,(\mu_\alpha du^\alpha) \tag{3.9}$$

and hence contraction by λ_r and μ_r respectively yields

$$\begin{aligned} \theta_1 &= \lambda_r dx^r = \lambda_\alpha du^\alpha, \\ \theta_2 &= \mu_r dx^r = \mu_\alpha du^\alpha. \end{aligned} \tag{3.10}$$

This shows that the spatial and surface versions of θ_1 and θ_2 are *identical*, i.e. $\lambda_r dx^r$ and $\lambda_\alpha du^\alpha$ produce the same Pfaffian forms! Likewise, if (3.9) is rewritten as

$$dx^r = \lambda^r \theta_1 + \mu^r \theta_2, \tag{3.11}$$

then by inspection we also have

$$du^\alpha = \lambda^\alpha \theta_1 + \mu^\alpha \theta_2. \tag{3.12}$$

As noted in (1.1-2) we have $\nu_r dx^r = 0$, i.e. $\theta_3 = 0$ which characterizes the restriction of our analysis to the surface $\mathbf{S} = \mathbf{S}_3$.

Thus V-(2.7) immediately gives

$$a_{\alpha\beta} = \lambda_\alpha \lambda_\beta + \mu_\alpha \mu_\beta \tag{3.13}$$

for the surface leg representation of the covariant components of the surface metric of $\mathbf{S}$, and by contracting with $du^\alpha du^\beta$ we have

$$\mathrm{I} = (\theta_1)^2 + (\theta_2)^2 . \tag{3.14}$$

The latter is the Pfaffian leg representation of the first basic form of $\mathbf{S}$. The contravariant and mixed leg representations are given by

$$a^{\alpha\beta} = \lambda^\alpha \lambda^\beta + \mu^\alpha \mu^\beta \tag{3.15}$$

$$\delta^\alpha{}_\beta = \lambda^\alpha \lambda_\beta + \mu^\alpha \mu_\beta \tag{3.16}$$

respectively. It is easy to check that

$$\begin{aligned} \lambda_r \lambda^r = 1 &\Leftrightarrow \lambda_\alpha \lambda^\alpha = 1 \\ \lambda_r \mu^r = 0 &\Leftrightarrow \lambda_\alpha \mu^\alpha = 0 \\ \mu_r \mu^r = 1 &\Leftrightarrow \mu_\alpha \mu^\alpha = 1 \end{aligned} \tag{3.17}$$

and indeed (3.16) is a concise manner of expressing these (surface) orthonormality conditions.

In the following, it is convenient in the leg calculus to agree to mean by the matrix $\|a_{\alpha\beta}\|$ the coefficients of the leg vectors in the expression (3.13). Hence, employing the SCHOUTEN (1954) notation $\stackrel{*}{=}$ to denote an equality holding in a particular reference system, we write

$$\|a_{\alpha\beta}\| \stackrel{*}{=} \left\| \begin{matrix} 1 & 0 \\ 0 & 1 \end{matrix} \right\| \tag{3.18}$$

so that

$$a \stackrel{*}{=} 1. \tag{3.19}$$

This is justified since actually the $a_{\alpha\beta}$ in (3.13) is really an expression of the form

$$a_{\alpha\beta} = a_{AB}\lambda_{A\alpha}\lambda_{B\beta} \tag{3.20}$$

where a_{AB} are scalars and the capital Latin indices are surface leg indices taken over the range 1, 2. Hence (3.18) is actually

$$\|a_{\alpha\beta}\| \stackrel{*}{=} \|a_{AB}\| = \|\delta_{AB}\|\,. \tag{3.21}$$

Our agreement (3.18) is economical since it avoids the introduction of leg indices for the components of tensors! However, it is nevertheless convenient to retain such indices in labelling the surface 2-legs. Hence, we may write

$$\{\ell_A\}_{A=1}^{2} \tag{3.22}$$

to denote the general abstract surface 2-leg which includes both the vectorial 2-leg $\{\boldsymbol{\lambda}_A\}_{A=1}^{2}$ and the Pfaffian 2-leg $\{\theta_A\}_{A=1}^{2}$ as special realizations. Thus, in using the notation (3.22) it is automatically implied that the leg quantities λ_A and θ_A have surface component indices. Likewise, we denote by

$$\{\ell_a\}_{a=1}^{3} \tag{3.23}$$

the spatial 3-leg, which includes both the vectorial 3-leg $\{\boldsymbol{\lambda}_a\}_{a=1}^{3}$ and that Pfaffian 3-leg $\{\theta_a\}_{a=1}^{3}$ as special realizations.

The 2-leg vectors $\{\boldsymbol{\lambda}_A\}_{A=1}^{2}$ are oriented in a right-handed manner, i.e. under a right-handed rotation through ninety degrees — see Figure 2.

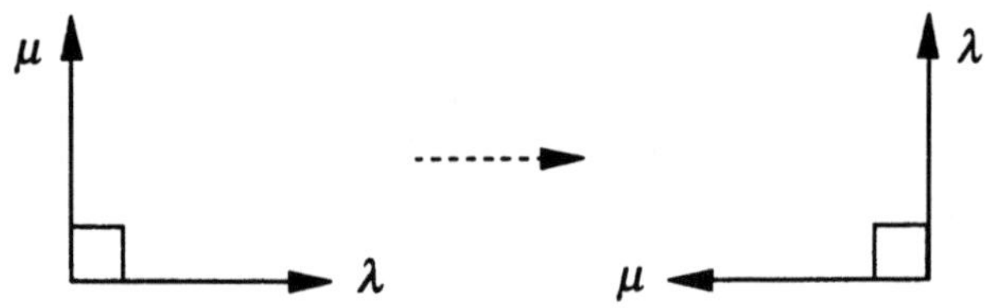

Figure 2

We have

$$\{\boldsymbol{\lambda}, \boldsymbol{\mu}\} \mapsto \{\boldsymbol{\lambda}', \boldsymbol{\mu}'\} = \{\boldsymbol{\mu}, -\boldsymbol{\lambda}\}. \tag{3.24}$$

These are described by the leg representations of the dualizors $\varepsilon^{\alpha\beta}$ and $\varepsilon_{\alpha\beta}$, i.e.

$$\varepsilon^{\alpha\beta} = \lambda^\alpha \mu^\beta - \mu^\alpha \lambda^\beta \tag{3.25}$$

$$\varepsilon_{\alpha\beta} = \lambda_\alpha \mu_\beta - \mu_\alpha \lambda_\beta. \tag{3.26}$$

Note that since we have (2.19), the matrices of these dualizors are equal, i.e.

$$\left\|\varepsilon^{\alpha\beta}\right\| \stackrel{*}{=} \|\varepsilon_{\alpha\beta}\| \stackrel{*}{=} \left\| \begin{matrix} 0 & 1 \\ -1 & 0 \end{matrix} \right\|, \tag{3.27}$$

and

$$\varepsilon^{\alpha\rho} \varepsilon_{\beta\rho} = \delta^\alpha_\beta \tag{3.28}$$

where the right-hand side is given by (3.16). The rotation (3.24) may be concisely expressed by the equations

$$\begin{aligned} \varepsilon^{\alpha\beta} \lambda_\alpha \mu_\beta &= 1 \\ \varepsilon_{\alpha\beta} \lambda^\alpha \mu^\beta &= 1. \end{aligned} \tag{3.29}$$

The proofs of these are easy and relegated to the PROBLEM V.4. More generally suppose $\boldsymbol{\xi}$and $\boldsymbol{\eta}$ are a pair of unit surface vectors and ϑ is the angle between them. Then we have

$$a^{\alpha\beta} \xi_\alpha \eta_\beta = a_{\alpha\beta} \xi^\alpha \eta^\beta = \cos\vartheta \tag{3.30}$$

$$\varepsilon^{\alpha\beta}\xi_\alpha\eta_\beta = \varepsilon_{\alpha\beta}\xi^\alpha\eta^\beta = \sin\vartheta. \tag{3.31}$$

Finally, we note that the right-handedness of $\{\boldsymbol{\lambda}_a\}_{a=1}^{3}$ induces the right-handedness of $\{\boldsymbol{\lambda}_A\}_{A=1}^{2}$.

V.4 The Second Basic Form

We now may immediately obtain the leg representation of $b_{\alpha\beta}$ by surface covariant differentiation of (3.8) which in effect amounts to exhibiting the leg version of the Gauss formula (3.8). Then we have

$$x^r_{\alpha\beta} = (\lambda^r\lambda_\alpha + \mu^r\mu_\alpha) = \left\{\lambda^r_s x^s_\beta\right\}\lambda_\alpha + \lambda^r\lambda_{\alpha\beta} + \left\{\mu^r_s x^s_\beta\right\}\mu_\alpha + \mu^r\mu_{\alpha\beta} \tag{4.1}$$

where $\lambda^r_{\ s} := \lambda^r_{\ ,s}$ and $\lambda_{\alpha\beta} := \lambda_{\alpha,\beta}$ etc. However, by virtue of IV-(6.1, 2) and (3.8) we may readily evaluate the expressions occurring in the curly brackets in (4.1) to get

$$\begin{aligned} x^r_{\alpha\beta} &= \{\lambda_{\alpha\beta} - \sigma_1\mu_\alpha\lambda_\beta - \sigma_2\mu_\alpha\mu_\beta\}\lambda^r \\ &\quad + \{\mu_{\alpha\beta} + \sigma_1\lambda_\alpha\lambda_\beta + \sigma_2\lambda_\alpha\mu_\beta\}\mu^r \\ &\quad + \{k_1\lambda_\alpha\lambda_\beta - t_2\lambda_\alpha\mu_\beta + t_1\mu_\alpha\lambda_\beta + k_2\mu_\alpha\mu_\beta\}\nu^r. \end{aligned} \tag{4.2}$$

But by (3.17) the coefficients of λ^r and μ^r must be zero, while the coefficient of ν^r is $b_{\alpha\beta}$. Hence we have

$$\lambda_{\alpha\beta} = \sigma_1\mu_\alpha\lambda_\beta + \sigma_2\mu_\alpha\mu_\beta \tag{4.3}$$

$$\mu_{\alpha\beta} = -\sigma_1\lambda_\alpha\lambda_\beta - \sigma_2\lambda_\alpha\mu_\beta \tag{4.4}$$

and

$$b_{\alpha\beta} = k_1\lambda_\alpha\lambda_\beta - t_2\lambda_\alpha\mu_\beta + t_1\mu_\alpha\lambda_\beta + k_2\mu_\alpha\mu_\beta.$$

However, $b_{\alpha\beta} = b_{\beta\alpha}$ and hence property (T) is implied by our formulation, so the last equation becomes

$$b_{\alpha\beta} = k_1\lambda_\alpha\lambda_\beta + t_1(\lambda_\alpha\mu_\beta + \mu_\alpha\lambda_\beta) + k_2\mu_\alpha\mu_\beta. \tag{4.5}$$

Equations (4.3), (4.4) and (4.5) occur in Hotine as [4.11] and [7.12] respectively.

We may now deduce the geometric meaning of the leg coefficients/parameters σ_1, σ_2. Let λ^α and μ^α denote the unit tangent vectors to the respective surface (parameter) curves

$$\boldsymbol{\Gamma}_1' := \boldsymbol{\Gamma}'(\boldsymbol{\lambda}), \quad \boldsymbol{\Gamma}_2' := \boldsymbol{\Gamma}_2'(\boldsymbol{\mu})$$

on **S**. Then contraction of (4.3) and (4.4) by λ^β and μ^β respectively yields

$$\lambda^\alpha{}_\beta \lambda^\beta = \sigma_1 \mu^\alpha \tag{4.6}$$

$$\mu^\alpha{}_\beta \mu^\beta = -\sigma_2 \lambda^\alpha \tag{4.7}$$

which are Hotine's [4.07] and [4.09]. The right hand sides of these equations are respectively the *curvature vectors* of $\boldsymbol{\Gamma}_1'$ and $\boldsymbol{\Gamma}_2'$, i.e.

$$k^\alpha = \sigma_1 \mu^\alpha, \quad \ell^\alpha = -\sigma_2 \lambda^\alpha; \tag{4.8}$$

and thus σ_1 and σ_2 are none other than the *geodesic curvatures* of $\boldsymbol{\Gamma}_1'$ and $\boldsymbol{\Gamma}_2'$ respectively. Likewise (4.6) and (4.7) are each one of the surface Frenet equations for $\boldsymbol{\Gamma}_1'$ and $\boldsymbol{\Gamma}_2'$, the two sets of surface Frenet equations being

$$\left.\begin{aligned} \lambda^\alpha{}_\beta \lambda^\beta &= \sigma_1 \mu^\alpha \\ \mu^\alpha{}_\beta \lambda^\beta &= -\sigma_1 \lambda^\alpha \end{aligned}\right\} \text{ for } \boldsymbol{\Gamma}_1' \tag{4.9}$$

$$\left.\begin{aligned} \mu^\alpha{}_\beta \mu^\beta &= -\sigma_2 \lambda^\alpha \\ \lambda^\alpha{}_\beta \mu^\beta &= \sigma_2 \mu^\alpha \end{aligned}\right\} \text{ for } \boldsymbol{\Gamma}_2'. \tag{4.10}$$

Note that these two sets of Frenet equations are related to each other by the permutation algorithm (see IV-(4.10, 11), and also by the rotation rule (3.24) upon an interchange of the numerical indices on σ.

It is interesting to compare (4.6) and (4.7) to the corresponding equations given in SECTION IV-6 for the space congruences $\boldsymbol{\Gamma}_1$ and $\boldsymbol{\Gamma}_2$:

$$\lambda^r{}_s \lambda^s = \sigma_1 \mu^r + k_1 \nu^r \tag{4.11}$$

$$\mu^r{}_s \mu^s = -\sigma_2 \lambda^r + k_2 \nu^r. \tag{4.12}$$

This is consistent with the interpretation given in SECTION IV-5, if we introduce the curvature vectors of $\boldsymbol{\Gamma}_1$ and $\boldsymbol{\Gamma}_2$:

$$\begin{aligned} k^r &:= \sigma_1 \mu^r + k_1 \nu^r \\ \ell^r &:= -\sigma_2 \lambda^r + k_2 \nu^r \end{aligned} \tag{4.13}$$

which have the respective curvatures χ_1, χ_2 given in IV-(5.1). Moreover, by using V-(3.8) we have

$$\begin{aligned} k_\alpha &= x^r_{\alpha} k_r \\ \ell_\alpha &= x^r_{\alpha} \ell_r \end{aligned} \tag{4.14}$$

which are the covariant versions of (4.8).

We have previously said that $\boldsymbol{\Gamma}'_1$ and $\boldsymbol{\Gamma}'_2$ were the projections of $\boldsymbol{\Gamma}_1$ and $\boldsymbol{\Gamma}_2$ respectively onto **S**. Equation (4.14) supports this contention, and identifies the projection as being defined by the Gauss operator (3.8). Moreover, as one can easily check

$$\lambda_{\alpha\beta} = x^r_{\alpha} x^s_{\beta} \lambda_{rs} \tag{4.15}$$

$$\mu_{\alpha\beta} = x^r_{\alpha} x^s_{\beta} \mu_{rs} \tag{4.16}$$

agreeing with (4.3) and (4.4) respectively. Hence, our claim is explicitly proven.

Before leaving these differentiation formulas we note that what is really occurring is the computation of the surface covariant derivative of an arbitrary space vector $\boldsymbol{\xi}$ which is also defined on **S**. Then we have

$$\xi^r{}_\alpha = \xi^r{}_s x^s{}_\alpha \tag{4.17}$$

where $\xi^r{}_\alpha := \xi^r{}_{,\alpha}$ and $\xi^r{}_s = \xi^r{}_{,s}$, and indeed this formula has been employed in the differentiation of two terms in the second equation of (4.1). This equation may be applied to each of the 3-leg vectors, and by virtue of (3.8) we have

$$\xi^r{}_\alpha = (\xi^r{}_s \lambda^s)\, \lambda_\alpha + (\xi^r{}_s \mu^s)\, \mu_\alpha. \tag{4.18}$$

Hence

$$\lambda^r{}_\alpha = (\sigma_1 \mu^r + k_1 \nu^r)\, \lambda_\alpha + (\sigma_2 \mu^r - t_2 \nu^r)\, \mu_\alpha, \tag{4.19}$$

or by (4.13) and (T):

$$\lambda^r{}_\alpha = k^r \lambda_\alpha + (\sigma_2 \mu^r + t_1 \nu^r)\, \mu_\alpha; \tag{4.20}$$

$$\mu^r{}_\alpha = (-\sigma_1 \lambda^r + t_1 \nu^r)\, \lambda_\alpha + (-\sigma_2 \lambda^r + k_2 \nu^r)\, \mu_\alpha, \tag{4.21}$$

or by (4.13) with no use of (T):

$$\mu^r{}_\alpha = (-\sigma_1 \lambda^r + t_1 \nu^r)\, \lambda_\alpha + \ell^r \mu_\alpha; \tag{4.22}$$

and

$$\nu^r{}_\alpha = (-k_1 \lambda^r - t_1 \mu^r)\, \lambda_\alpha + (t_2 \lambda^r - k_2 \mu^r)\, \mu_\alpha, \tag{4.23}$$

or by (4.13) and (T):

$$\nu^r{}_\alpha = -\,(k_1 \lambda^r + t_1 \mu^r)\, \lambda_\alpha - (t_1 \lambda^r + k_2 \mu^r)\, \mu_\alpha. \tag{4.24}$$

We now return to the canonical leg representation (4.5) of the second basic surface tensor $b_{\alpha\beta}$. By a double contraction of (4.5) with the surface 2-leg vectors, we have

$$k_1 = b_{\alpha\beta} \lambda^\alpha \lambda^\beta, \tag{4.25}$$

$$k_2 = b_{\alpha\beta} \mu^\alpha \mu^\beta, \tag{4.26}$$

$$t_1 = b_{\alpha\beta} \lambda^\alpha \mu^\beta = b_{\alpha\beta} \mu^\alpha \lambda^\beta. \tag{4.27}$$

The first two of these equations are — in principle — due to Meusnier who in 1785 identified expressions equivalent to k_1, k_2 as the *normal curvatures* of $\mathbf{S}$ in the respective $\boldsymbol{\lambda}$, $\boldsymbol{\mu}$ directions. The quantity t_1 is of a more recent vintage and probably originated with Bonnet who in 1853 called it the *geodesic torsion* of $\mathbf{S}$ in the $\boldsymbol{\lambda}$ direction, although more precisely it actually refers to the surface congruence $\boldsymbol{\Gamma}'_1$. By our rotation rule (3.24) with an interchange $1 \to 2$ of the numerical index on the geodesic torsion, for the congruence $\boldsymbol{\Gamma}'_2$ we would have

$$t_2 = b_{\alpha\beta} \mu^\alpha \left(-\lambda^\beta\right) \tag{4.28}$$

which equals $-t_1$. Thus, we again have encountered property (T). Note that it is also obtainable from our expression for $b_{\alpha\beta}$ given before (4.5). In terms of the geometry of $\boldsymbol{\Gamma}'_1$ and $\boldsymbol{\Gamma}'_2$, one often states that a fundamental property of the geodesic torsions is that they are related by

$$t_1 + t_2 = 0 \tag{4.29}$$

whenever $\boldsymbol{\Gamma}_1' \perp \boldsymbol{\Gamma}_2'$. This is how Hotine (see [page 40]) deduced this result, i.e. our property (T).

An alternate approach to t_1 and t_2 may be given by introducing the *Euler tensor*

$$h_{\alpha\beta} := \varepsilon^{\rho\sigma} a_{\alpha\rho} b_{\beta\sigma}. \tag{4.30}$$

By employing the leg representations (3.25), (3.13) and (4.5) we immediately obtain the following leg representation for $h_{\alpha\beta}$

$$h_{\alpha\beta} = t_1 (\lambda_\alpha \lambda_\beta - \mu_\alpha \mu_\beta) + k_2 \lambda_\alpha \mu_\beta - k_1 \mu_\alpha \lambda_\beta \tag{4.31}$$

which shows that this tensor has both non-trivial symmetric and skew-symmetric parts (this is also evident from (4.30)). Then the geodesic torsions of $\boldsymbol{\Gamma}_1'$ and $\boldsymbol{\Gamma}_2'$ in the respective directions λ and μ are given by

$$\begin{aligned} t_1 &= h_{\alpha\beta} \lambda^\alpha \lambda^\beta \\ -t_1 &= h_{\alpha\beta} \mu^\alpha \mu^\beta = t_2, \end{aligned} \tag{4.32}$$

with the normal curvatures $\mathbf{S}$ being given by

$$k_2 = h_{\alpha\beta} \lambda^\alpha \mu^\beta, \tag{4.33}$$

$$-k_1 = h_{\alpha\beta} \mu^\alpha \lambda^\beta. \tag{4.34}$$

The Pfaffian version of the second basic form is obtained by contraction of (4.5) with $du^\alpha du^\beta$, i.e.

$$\begin{aligned} \mathrm{II} &= k_1 (\theta_1)^2 + t_1 (\theta_1 \theta_2 + \theta_2 \theta_1) + k_2 (\theta_2)^2 , \\ \text{or} \quad & \\ \mathrm{II} &= k_1 (\theta_1)^2 + 2 t_1 \theta_1 \theta_2 + k_2 (\theta_2)^2 . \end{aligned} \tag{4.35}$$

Recalling the expressions IV-(6.5, 6) for the connection 1-forms ω_{31} and ω_{23}, (4.35) is equivalent to

$$\mathrm{II} = \theta_1 \omega_{31} - \theta_2 \omega_{23} \tag{4.36}$$

since by virtue of (T) and the fact that $\theta_3 = 0$ on $\mathbf{S}$ we have

$$\theta_1 (k_1 \theta_1 + t_1 \theta_2) - \theta_2 (-t_1 \theta_1 - k_2 \theta_2) = k_1 (\theta_1)^2 + 2 t_1 \theta_1 \theta_2 + k_2 (\theta_2)^2 .$$

Thus, we may write

$$\|b_{\alpha\beta}\| \stackrel{*}{=} \left\| \begin{matrix} k_1 & t_1 \\ t_1 & k_2 \end{matrix} \right\| \tag{4.37}$$

and hence

$$b = \det \|b_{\alpha\beta}\| \stackrel{*}{=} k_1 k_2 - t_1^2. \tag{4.38}$$

In (2.26) we noted that $K = b/a$ and thus by (3.19) we have the following leg calculus expression for the Gaussian curvature

$$K \stackrel{*}{=} k_1 k_2 - t_1^2. \tag{4.39}$$

The traditional manner of introducing the *Gauss* (*total*) *curvature* K and the *Germain* (*mean*) *curvature* H is to consider the determinantal expansion of the quantity $b_{\alpha\beta} - \tau a_{\alpha\beta}$ where τ is a parameter. By using the leg calculus expressions (4.37) and (3.18) we have

$$\|b_{\alpha\beta} - \tau a_{\alpha\beta}\| \stackrel{*}{=} \left\| \begin{matrix} k_1 - \tau, & t_1 \\ t_1, & k_2 - \tau \end{matrix} \right\| \tag{4.40}$$

and consequently

$$\det \|b_{\alpha\beta} - \tau a_{\alpha\beta}\| = (k_1 - \tau)(k_2 - \tau) - t_1^2. \tag{4.41}$$

Hence, the requirement $\det \|b_{\alpha\beta} - \tau a_{\alpha\beta}\| = 0$ leads to the quadratic τ-equation

$$\tau^2 - (k_1 + k_2)\tau + \left(k_1 k_2 - t_1^2\right) = 0 \tag{4.42}$$

which customarily one rewrites as

$$\tau^2 - 2H\tau + K = 0 \tag{4.43}$$

where

$$2H \stackrel{*}{=} k_1 + k_2 \tag{4.44}$$

and K is given by (4.39). The corresponding tensorial versions of these quantities are given by

$$2H := a^{\rho\sigma} b_{\rho\sigma}, \tag{4.45}$$

$$2K := \varepsilon^{\alpha\beta}\varepsilon^{\rho\sigma} b_{\alpha\rho} b_{\beta\sigma}. \tag{4.46}$$

It is easy to check that by using (3.15), (3.25), and (4.5) we obtain (4.44) and (4.39).

Finally, we note that in terms of the Cartan calculus, one has the following definitions of H and K:

$$2H\theta_1 \wedge \theta_2 := \omega_{31} \wedge \theta_2 + \omega_{23} \wedge \theta_1, \tag{4.47}$$

$$d\omega_{12} := K\theta_1 \wedge \theta_2. \tag{4.48}$$

The former is easily checked, simply employ the surface versions of IV-(6.5) using (T) and $\theta_3 = 0$. However, the latter requires some discussion which will be deferred to SECTION 8.

V.5 The Third and Fourth Basic Forms

The leg representation of the third basic tensor $c_{\alpha\beta}$ may now be determined by using either (2.31) or (2.32). Using the former, i.e.

$$c_{\alpha\beta} = g_{rs}\nu^r_\alpha \nu^s_\beta,$$

and the expression (4.23) which retains t_2 we obtain

$$\begin{aligned} c_{\alpha\beta} = (k_1^2 + t_1^2)\, \lambda_\alpha\lambda_\beta + (-k_1 t_2 + k_2 t_1)\,(\lambda_\alpha\mu_\beta + \mu_\alpha\lambda_\beta) \\ + (k_2^2 + t_2^2)\, \mu_\alpha\mu_\beta. \end{aligned} \tag{5.1}$$

Note that this expression for $c_{\alpha\beta}$ is symmetric, and unlike the case encountered in SECTION 4 for $b_{\alpha\beta}$, there is *no* requirement that property (T) be satisfied. However, imposing (T) and recalling (4.44) we have

$$\begin{aligned} c_{\alpha\beta} = (k_1^2 + t_1^2)\, \lambda_\alpha\lambda_\beta + 2Ht_1\,(\lambda_\alpha\mu_\beta + \mu_\alpha\lambda_\beta) \\ + (k_2^2 + t_1^2)\, \mu_\alpha\mu_\beta \end{aligned} \tag{5.2}$$

which we take to be the canonical leg representation of $c_{\alpha\beta}$. Hence, we have

$$\|c_{\alpha\beta}\| \overset{*}{=} \left\| \begin{matrix} k_1^2 + t_1^2, & (k_1 + k_2)\, t_1 \\ (k_1 + k_2)\, t_1, & k_2^2 + t_1^2 \end{matrix} \right\| \tag{5.3}$$

and so

$$\det \|c_{\alpha\beta}\| = K^2 \stackrel{*}{=} \left(k_1 k_2 - t_1^2\right)^2 . \tag{5.4}$$

By analogy with (4.25-.27) we have

$$k_1^2 + t_1^2 = c_{\alpha\beta}\lambda^\alpha\lambda^\beta, \tag{5.5}$$

$$k_2^2 + t_1^2 = c_{\alpha\beta}\mu^\alpha\mu^\beta, \tag{5.6}$$

$$2Ht_1 = c_{\alpha\beta}\lambda^\alpha\mu^\beta, \tag{5.7}$$

which may be regarded as $\overline{k}_1^2$, $\overline{k}_2^2$, $\overline{t}_1$ where the bar (as in SECTION 2) denotes the spherical quantity. However, such a practice is not common and is probably best avoided.

Contracting (5.2) with $du^\alpha du^\beta$ we obtain the third basic form

$$\text{III} = (k_1^2 + t_1^2)\,(\theta_1)^2 + 2Ht_1\,(\theta_1\theta_2 + \theta_2\theta_1) + (k_2^2 + t_1^2)\,(\theta_2)^2$$

or

$$\text{III} = (k_1^2 + t_1^2)\,(\theta_1)^2 + 4Ht_1\theta_1\theta_2 + (k_2^2 + t_1^2)\,(\theta_2)^2 . \tag{5.8}$$

As in SECTION 4, the above expression can be rewritten in terms of the connection 1-forms, as

$$\text{III} = (\omega_{31})^2 + (\omega_{23})^2 . \tag{5.9}$$

The so-called fourth basic tensor is defined by

$$d_{\alpha\beta} := \frac{1}{2}a^{\rho\sigma}\left(\varepsilon_{\alpha\rho}b_{\sigma\beta} + \varepsilon_{\beta\rho}b_{\sigma\alpha}\right) \tag{5.10}$$

and occurs in the theory of the bending, i.e. deformation, of surfaces. Strictly speaking, it is a very special construction which is even less fundamental than $c_{\alpha\beta}$. Clearly the expression (5.10) defines a symmetric surface tensor, and we recognize it as being the symmetric part of the Euler tensor (4.30). This is immediately evident from (4.30), i.e.

$$d_{\alpha\beta} = \frac{1}{2}\left(h_{\alpha\beta} + h_{\beta\alpha}\right), \tag{5.11}$$

and by using (4.31) — or evaluating (5.10) in terms of the obvious leg representations — we have

$$d_{\alpha\beta} = t_1\lambda_\alpha\lambda_\beta + \frac{(k_2 - k_1)}{2}\lambda_\alpha\mu_\beta + \frac{(k_2 - k_1)}{2}\mu_\alpha\lambda_\beta - t_1\mu_\alpha\mu_\beta \tag{5.12}$$

for the canonical leg representation of $d_{\alpha\beta}$.

Hence

$$\|d_{\alpha\beta}\| \stackrel{*}{=} \left\| \begin{matrix} t_1, & \dfrac{(k_2-k_1)}{2} \\ \dfrac{(k_2-k_1)}{2}, & -t_1 \end{matrix} \right\| \tag{5.13}$$

and

$$\det \|d_{\alpha\beta}\| \stackrel{*}{=} 2K - k_1^2 - k_2^2 - 2t_1^2. \tag{5.14}$$

By analogy with (4.25-.27) we have

$$t_1 = d_{\alpha\beta}\lambda^\alpha\lambda^\beta, \tag{5.15}$$

$$-t_1 = d_{\alpha\beta}\mu^\alpha\mu^\beta, \tag{5.16}$$

$$\frac{k_2-k_1}{2} = d_{\alpha\beta}\lambda^\alpha\mu^\beta. \tag{5.17}$$

Contraction of (5.12) with $du^\alpha du^\beta$ yields

$$\begin{aligned} \text{IV} &= t_1(\theta_1)^2 + \frac{(k_2-k_1)}{2}(\theta_1\theta_2+\theta_2\theta_1) - t_1(\theta_2)^2 \\ \text{or} \\ \text{IV} &= t_1(\theta_1)^2 + (k_2-k_1)\,\theta_1\theta_2 - t_1(\theta_2)^2 \end{aligned} \tag{5.18}$$

which corresponds to

$$\text{IV} = -\omega_{31}\theta_2 - \omega_{23}\theta_1. \tag{5.19}$$

V.6 Congruence and Surface Eigenstructure Theorems

Further insight into the space congruences $\boldsymbol{\Gamma}_1$, $\boldsymbol{\Gamma}_2$, $\boldsymbol{\Gamma}:=\boldsymbol{\Gamma}_3$ and the surface congruences $\boldsymbol{\Gamma}_1'$ and $\boldsymbol{\Gamma}_2'$ and the set of basic leg coefficients/parameters

$$k_1, k_2, t_1, \gamma_1, \gamma_2, \sigma_1, \sigma_2, \varepsilon_3$$

can be obtained by considering the eigenstructure, i.e. the eigenvectors and eigenvalues of the leg representations of the covariant derivatives of the vectors of $\{\boldsymbol{\lambda}_a\}_{a=1}^3$ and $\{\boldsymbol{\lambda}_A\}_{A=1}^2$.

We first consider the spatial case and note that by virtue of property (T) the basic equations of the leg calculus, (recall IV-(6.1, 2, 3)) become

$$\begin{aligned}\lambda_{rs} &= \sigma_1\mu_r\lambda_s + \sigma_2\mu_r\mu_s + \varepsilon_3\mu_r\nu_r \\ &\quad +k_1\nu_1\lambda_s + t_1\nu_r\mu_s - \gamma_1\nu_r\nu_s,\end{aligned} \tag{6.1}$$

$$\begin{aligned}\mu_{rs} &= -\sigma_1\lambda_r\lambda_s - \sigma_2\lambda_r\mu_s - \varepsilon_3\lambda_r\nu_s \\ &\quad +t_1\nu_r\lambda_s + k_2\nu_r\mu_s - \gamma_2\nu_r\nu_s,\end{aligned} \tag{6.2}$$

$$\begin{aligned}\nu_{rs} &= -k_1\lambda_r\lambda_s - t_1\lambda_r\mu_s + \gamma_1\lambda_r\nu_s \\ &\quad -t_1\mu_r\lambda_s - k_2\mu_r\mu_s + \gamma_2\mu_r\nu_s.\end{aligned} \tag{6.3}$$

Then via the basic orthogonality conditions II-(2.6) we have

$$\begin{aligned}\lambda_{rs}\lambda^s &= \sigma_1\mu_r + k_1\nu_r = k_r \\ \lambda_{rs}\mu^s &= \sigma_2\mu_r + t_1\nu_r \\ \lambda_{rs}\nu^s &= \varepsilon_3\mu_r - \gamma_1\nu_r,\end{aligned} \tag{6.4}$$

$$\begin{aligned}\mu_{rs}\lambda^s &= -\sigma_1\lambda_r + d_1\nu_r \\ \mu_{rs}\mu^s &= -\sigma_2\lambda_r + k_2\nu_r = \ell_r \\ \mu_{rs}\nu^s &= -\varepsilon_3\lambda_r - \gamma_2\nu_r,\end{aligned} \tag{6.5}$$

$$\begin{aligned}\nu_{rs}\lambda^s &= -k_1\lambda_r - t_1\mu_r \\ \nu_{rs}\mu^s &= -t_1\lambda_r - k_2\mu_r \\ \nu_{rs}\nu^s &= \gamma_1\lambda_r + \gamma_2\mu_r,\end{aligned} \tag{6.6}$$

where the curvature vectors **k** and $\boldsymbol{\ell}$ were introduced in (4.13). These equations then immediately yield the

<u>Space Congruence Eigenstructure Theorem</u>

For the space congruences $\boldsymbol{\Gamma}_1$, $\boldsymbol{\Gamma}_2$ and $\boldsymbol{\Gamma}=\boldsymbol{\Gamma}_3$ in $\mathbf{E}_3$ we have:

1^0 λ_{rs} has eigenvector λ^s with zero eigenvalue if and only if $\sigma_1 = 0$, $k_1 = 0$, i.e. $\chi_1 = 0$; and in this case $\boldsymbol{\Gamma}_1$ is a congruence of straight lines in $\mathbf{E}_3$;

2^0 μ_{rs} has eigenvector μ^s with zero eigenvalue if and only if $\sigma_2 = 0$, $k_2 = 0$, i.e. $\chi_2 = 0$, and in this case $\boldsymbol{\Gamma}_2$ is a congruence of straight lines in $\mathbf{E}_3$;

3^0 ν_{rs} has eigenvector ν^s with zero eigenvalue if and only if $\gamma_1 = 0$, $\gamma_2 = 0$, i.e. $\chi = \chi_3 = 0$, and in this case $\boldsymbol{\Gamma}$ is a congruence of straight lines in $\mathbf{E}_3$.

Recall the curvatures χ_a of the $\boldsymbol{\Gamma}_a$ were introduced in SECTION II-5 and later in IV-(5.1) (the latter employing the language of leg coefficients/parameters.

We also have two spatial eigenstructure theorems which characterize the vanishing of t_1 and ε_3 respectively.

Theorem A. The condition $t_1 = 0$ is equivalent to each of the following statements:

i) λ_{rs} has eigenvector μ^s with eigenvalue σ_2,
ii) μ_{rs} has eigenvector λ^s with eigenvalue $-\sigma_1$,
iii) ν_{rs} has eigenvector λ^s with eigenvalue $-k_1$,
iv) ν_{rs} has eigenvector μ^s with eigenvalue $-k_2$.

Theorem B. The condition $\varepsilon_3 = 0$ is equivalent to each of the following statements:

i) λ_{rs} has eigenvector ν^s with eigenvalue $-\gamma_1$,
ii) μ_{rs} has eigenvector ν^s with eigenvalue $-\gamma_2$.

Each of these theorems follows immediately from an inspection of (6.3-.6).

Statements iii) and iv) of Theorem A are algebraic generalizations of expressions of the *Rodrigues formulas* (see McCONNELL (1931), Examples XVI, #8, page 216), since by (4.24) we have

$$\begin{aligned} \nu^r{}_\alpha \lambda^\alpha &= -k_1 \lambda^r - t_1 \mu^r \\ \nu^r{}_\alpha \mu^\alpha &= -t_1 \lambda^r - k_2 \mu^r \end{aligned} \tag{6.7}$$

which are respectively identical with $\nu^r{}_s\lambda^s$ and $\nu^r{}_s\mu^s$. The surface analogue of (6.7) is obtained by defining the *Weingarten operator*

$$w^\alpha{}_\beta := -a^{\alpha\rho}b_{\rho\beta} \tag{6.8}$$

and noting that then (4.24) may be rewritten as

$$\nu^r{}_\alpha = w_\alpha{}^\sigma x^r_\sigma. \tag{6.9}$$

The Weingarten operator is obviously linear and has the following leg representation

$$w^\alpha{}_\beta = -k_1\lambda^\alpha\lambda_\beta - t_1\lambda^\alpha\mu_\beta - t_1\mu^\alpha\lambda_\beta - k_2\mu^\alpha\mu_\beta \tag{6.10}$$

and hence

$$\begin{aligned} w^\alpha{}_\beta\lambda^\beta &= -k_1\lambda^\alpha - t_1\mu^\alpha := \rho^\alpha \\ w^\alpha{}_\beta\mu^\beta &= -t_1\lambda^\alpha - k_2\mu^\alpha := \sigma^\alpha \end{aligned} \tag{6.11}$$

where the surface vectors ρ^α and σ^α are the *Rodrigues vectors* of **S**. Then $w^\alpha{}_\beta$ clearly has the eigenvectors λ^α and μ^α with the corresponding eigenvalues $-k_1$, and $-k_2$ if and only if $t_1 = 0$. This is the usual case of the Rodrigues formulas. Finally, we note that

$$-k_1 = w_{\alpha\beta}\lambda^\alpha\lambda^\beta, \tag{6.12}$$

$$-k_2 = w_{\alpha\beta}\mu^\alpha\mu^\beta, \tag{6.13}$$

$$-t_1 = w_{\alpha\beta}\lambda^\alpha\mu^\beta, \tag{6.14}$$

which is analogous to our previous equations (4.25-.27) and (4.32-.34). It is interesting to observe that

$$\varepsilon^{\alpha\beta}\rho_\alpha\sigma_\beta = k_1k_2 - t_1^2 = K \tag{6.15}$$

via using (3.29).

To complete our space analysis we note that by virtue of the leg representation of g_{rs}, we also have

$$g_{rs}\lambda^s = \lambda_r, \quad g_{rs}\mu^s = \mu_r, \quad g_{rs}\nu^s = \nu_r; \tag{6.16}$$

which can be interpreted as saying that g_{rs} admits the 3-leg vectors as eigenvectors, with each eigenvector having a unit eigenvalue.

The corresponding surface congruence and surface eigenstructure theory is much richer than the spatial theory. By (4.6) and (4.7) we have the

Surface Congruence Eigenstructure Theorem

For the surface congruences $\boldsymbol{\Gamma}_1'$ and $\boldsymbol{\Gamma}_2'$ on S we have:

1^0 $\lambda_{\alpha\beta}$ has eigenvector λ^β with zero eigenvalue, if and only if $\sigma_1 = 0$, i.e., $\boldsymbol{\Gamma}_1'$ is a surface geodesic congruence;

2^0 $\mu_{\alpha\beta}$ has eigenvalue μ^β with zero eigenvalue, if and only if $\sigma_2 = 0$, i.e., $\boldsymbol{\Gamma}_2'$ is a surface geodesic congruence.

By virtue of (3.13) we have the obvious surface analogue of (6.17), i.e.

$$a_{\alpha\beta}\lambda^\beta = \lambda_\alpha, \qquad a_{\alpha\beta}\mu^\beta = \mu_\alpha. \tag{6.17}$$

Likewise we may examine the eigenstructure of the surface tensors $b_{\alpha\beta}$, $h_{\alpha\beta}$, $c_{\alpha\beta}$, $d_{\alpha\beta}$ and $w_{\alpha\beta}$. Clearly by (4.5) we have

$$\begin{aligned} b_{\alpha\beta}\lambda^\beta &= k_1\lambda_\alpha + t_1\mu_\alpha, \\ b_{\alpha\beta}\mu^\beta &= t_1\lambda_\alpha + k_2\mu_\alpha; \end{aligned} \tag{6.18}$$

while (4.31) gives

$$\begin{aligned} h_{\alpha\beta}\lambda^\beta &= t_1\lambda_\alpha - k_1\mu_\alpha, \\ h_{\alpha\beta}\mu^\beta &= k_2\lambda_\alpha - t_1\mu_\alpha; \end{aligned} \tag{6.19}$$

and (5.12) yields

$$\begin{aligned} c_{\alpha\beta}\lambda^\beta &= \left(k_1^2 + t_1^2\right)\lambda_\alpha + 2Ht_1\mu_\alpha, \\ c_{\alpha\beta}\mu^\beta &= 2Ht_1\lambda_\alpha + (k_2^2 + t_1^2)\mu_\alpha; \end{aligned} \tag{6.20}$$

and finally (5.12) gives

$$d_{\alpha\beta}\lambda^\beta = \frac{(k_2 - k_1)}{2}\lambda_\alpha - t_1\mu_\alpha; \tag{6.21}$$

with the eigenvector/eigenvalue equation for the Weingarten operator (tensor) being given in (6.11-.12). These equations lead to a surface eigenstructure theorem, however before stating it one must make a brief digression to introduce the important concept of principal directions and lines of curvature on **S**.

The notion of a *principal direction* is purely of an algebraic character, and defined by the requirement that $\boldsymbol{\lambda}$ and $\boldsymbol{\mu}$ be eigenvectors of $b_{\alpha\beta}$. By (6.18), this condition obviously reduces to

$$t_1 = 0 \tag{6.22}$$

which by SECTION 4, means that the geodesic torsions of both $\boldsymbol{\Gamma}'_1$ and $\boldsymbol{\Gamma}'_2$ vanish. In this case the congruences $\boldsymbol{\Gamma}'_1$ and $\boldsymbol{\Gamma}'_2$ are said to consist of *lines of curvature* of **S**. An alternate approach is to define a principal direction of **S** to be an arbitrary surface vector field $\boldsymbol{\xi}$ such that

$$h_{\alpha\beta}\xi^\alpha\xi^\beta = 0. \tag{6.23}$$

Hence, since $\boldsymbol{\xi}$ can be expressed as a linear combination of the vectors $\boldsymbol{\lambda}$ and $\boldsymbol{\mu}$, we must have (4.32) which leads to (6.22).

When (6.22) holds, by (4.39) the Gauss curvature K reduces to a product of curvatures which are known as principal curvatures and to indicate this distinction we write

$$\begin{matrix} k_1 = \kappa_1 \\ k_2 = \kappa_2 \end{matrix} \quad \text{(when } t_1 = 0\text{)}. \tag{6.24}$$

There is no corresponding structural simplification of the Germain curvature H — recall (4.44) — but for purposes of reference we write

$$K \stackrel{*}{=} \kappa_1\kappa_2, \tag{6.25}$$

$$2H \stackrel{*}{=} \kappa_1 + \kappa_2 \tag{6.26}$$

when $t_1 = 0$.

The choice of $\boldsymbol{\lambda}$ and $\boldsymbol{\mu}$ being principal directions is a common one in Gaussian differential geometry. Indeed, most texts — BLASCHKE & LEICHTWEISS (1970) being the prime exception — automatically assume that the roots of (4.41) define a pair of principal directions on **S**. Such an assumption is warranted — in a general setting — since it does simplify the analysis. However, by doing so one has taken the otherwise arbitrary parameter system u^α (often called conveniently a *parameter net*) to consist of lines of curvature of **S**. In differential geodesy, such an assumption is not *a priori* justifiable and as Hotine did in his treatise, it is better to retain the arbitrariness of $\boldsymbol{\Gamma}'_1$ and $\boldsymbol{\Gamma}'_2$ on **S**. Hence, unless specified otherwise, we will assume that (6.22) is not automatically satisfied, and so $\boldsymbol{\lambda}$ and $\boldsymbol{\mu}$ are not automatically eigenvectors of $b_{\alpha\beta}$.

One further caveat is in order. In Gaussian differential geometry, the vanishing of H is an admissible condition and surfaces satisfying this requirement are known as *minimal surfaces*. They have many attractive and intriguing propers — see NITSCHE (1975) for a detailed exposition — however, in differential geodesy they are generally of no interest. Clearly by (4.44), $H = 0$ yields

$$k_1 = -k_2$$

and, thus, by (4.37), we would have

$$K \stackrel{*}{=} -\left(k_1^2 + t_1^2\right).$$

However, in differential geodesy the condition $K < 0$ is contrary to our implicit and intuitive assumption that our **S** — which will eventually be identified with an equipotential surface — is 'spherelike', i.e. has *both* k_1 and k_2 of the same sign, and this is clearly impossible when $H = 0$. Thus, we will exclude the case $H = 0$ from our consideration and always assume that $H \neq 0$. Note that the above discussion is not essentially changed when (6.22) holds.

We may now announce the

<u>Surface Eigenstructure Theorem</u>

A smooth surface **S** has the following eigenstructure:

1^0 $b_{\alpha\beta}$ has eigenvector λ^β with eigenvalue k_1, if and only if $t_1 = 0$, and in this case $k_1 = \kappa_1$ and λ^β is a principal direction of **S** so $\boldsymbol{\Gamma}'_1$ consists of lines of curvature;

2^0 $b_{\alpha\beta}$ has eigenvector μ^β with eigenvalue k_2 if and only if $t_1 = 0$, and in this case $k_2 = \kappa_2$ and μ^β is a principal direction of **S** so Γ_2' consists of lines of curvature of **S**;

3^0 $h_{\alpha\beta}$ has eigenvector λ^β with eigenvalue t_1 if and only if $k_1 = 0$;

4^0 $h_{\alpha\beta}$ has eigenvector μ^β with eigenvalue $-t_1$ if and only if $k_2 = 0$;

5^0 $c_{\alpha\beta}$ has eigenvector λ^β with eigenvalue $k_1^2 + t_1^2$ if and only if $t_1 = 0$, and in this case the eigenvalue reduces to $k_1^2 = \kappa_1^2$;

6^0 $c_{\alpha\beta}$ has eigenvector μ^β with eigenvalue $k_2^2 + t_1^2$ if and only if $t_1 = 0$, and in this case the eigenvalue reduces to $k_2^2 = \kappa_2^2$;

7^0 $d_{\alpha\beta}$ has eigenvector λ^β with eigenvalue t_1 if and only if $k_1 = k_2$;

8^0 $d_{\alpha\beta}$ has eigenvector μ^β with eigenvalue $-t_1$ if and only if $k_1 = k_2$;

9^0 $w_{\alpha\beta}$ has eigenvector λ^β with eigenvalue $-k_1$ if and only if $t_1 = 0$;

10^0 $w_{\alpha\beta}$ has eigenvector μ^β with eigenvalue $-k_2$ if and only if $t_1 = 0$.

Note that in this theorem, cases 1^0 and 2^0; 5^0 and 6^0; 7^0 and 8^0; 9^0 and 10^0 respectively occur simultaneously. The first of these insures that if **S** admits one principal direction, it necessarily has a second principal direction. Furthermore, these situations lead to two additional characterizations of the conditions $t_1 = 0$ and $k_1 = k_2$ respectively:

Theorem C. The condition $t_1 = 0$ is characterized by each of the following statements:

i) $b_{\alpha\beta}$ has a pair of eigenvectors, i.e. principal directions, λ^β and μ^β with the respective eigenvalues κ_1 and κ_2;

ii) $c_{\alpha\beta}$ has a pair of eigenvectors λ^β and μ^β with the respective eigenvalues κ_1^2 and κ_2^2;

iii) $w_{\alpha\beta}$ has a pair of eigenvectors λ^β and μ^β with the respective eigenvalues $-\ k_1$ and $-k_2$ which reduce respectively to $-\kappa_1$ and $-\kappa_2$.

Theorem D. The condition $k_1 = k_2$ is characterized by $d_{\alpha\beta}$ having a pair of eigenvectors λ^β and μ^β which have the respective eigenvalues t_1 and $-t_1$.

V.7 Two Digressions

In SECTION 4 we applied a pair of Gauss operators to λ_{rs} and μ_{rs}, and obtained $\lambda_{\alpha\beta}$ and $\mu_{\alpha\beta}$ respectively — see (4.15) and (4.16). We now examine what happens if the same procedure were applied to ν_{rs}. Suppose we disregarded our agreement and introduced a quantity ν_α defined by

$$\nu_\alpha := x^r_\alpha \nu_r \tag{7.1}$$

which is identically zero. Then by surface covariant differentiation we would obtain

$$\nu_{\alpha\beta} \equiv 0 = x^r_{\alpha\beta}\nu_r + x^r_\alpha x^s_\beta \nu_{rs}. \tag{7.2}$$

But each term on the right hand side is non-zero. By the Gauss formula (3.8) the first term is equal to $b_{\alpha\beta}$, while

$$x^r_\alpha x^s_\beta \nu_{rs} = -k_1\lambda_\alpha\lambda_\beta + t_2\lambda_\alpha\mu_\beta - t_1\mu_\alpha\lambda_\beta - k_2\mu_\alpha\mu_\beta,$$

which by (T) is

$$x^r_\alpha x^s_\beta \nu_{rs} = -k_1\lambda_\alpha\lambda_\beta - t_1\left(\lambda_\alpha\mu_\beta + \mu_\alpha\lambda_\beta\right) - k_2\mu_\alpha\mu_\beta. \tag{7.3}$$

Hence by (7.3) we have

$$x^r_\alpha x^s_\beta \nu_{rs} = -b_{\alpha\beta} \tag{7.4}$$

and as anticipated $\nu_{\alpha\beta} \equiv 0$ since (7.2) is actually $\nu_{\alpha\beta} = b_{\alpha\beta} - b_{\alpha\beta}$. This result explains why in SECTION 4 we made the decision *not to introduce* the quantities ν_α and ν^α.

Unfortunately, Hotine did not follow this path and in his treatise on several occasions did employ $\nu_{\alpha\beta}$, see his [12.024] [page 73] and [12.078] [page 78] which are $\nu_{\alpha\beta} = -b_{\alpha\beta}$ and $b_{\alpha\beta} = -\nu_{\alpha\beta}$ respectively; and ν^{α} appears in his [12.081] and [12.082] [page 78]. If the ν_{α} and $\nu_{\alpha\beta}$ are interpreted as the quantities defined by (7.1) and (7.2) or (7.3), with $\nu^{\alpha} := a^{\alpha\beta}\nu_{\beta}$ then his equations are wrong! However, as Hotine's discussion suggests — see in particular [page 78] — what he presumably meant by ν_{α} and $\nu_{\alpha\beta}$ was the two-dimensional parts of ν_r and ν_{rs}. If this is the case, and his subsequent usage on [page 81] [see subsection 76], on [page 82]: [12.126], [12.129]-[12.131], on [page 86]: [12.155], [12.158]-[12.161], etc., seem to imply such an interpretation, then his choice of notation was merely poor. In this case, he should have written ν_{α} as ν_A, and $\nu_{\alpha\beta}$ as ν_{AB} with the understanding that

$$\nu_A = \nu_r \text{ when } r = A$$

$$\nu_{AB} = \nu_{rs} \text{ when } r, s = A, B$$

with the capital Latin indices ranging over the subset of values 1, 2 of 1, 2, 3. This maneuver would have prevented a confusion of ν_{AB} with our $\nu_{\alpha\beta}$, but it would not have explained the meaning of b_{AB} since there is no b_{rs}!

All of this is somewhat confusing and probably not worth the effort to sort out, since — as far as we can ascertain — Hotine's ill advised practice of automatically identifying the two-dimensional indices A, B with surface indices α, β did not lead him to any results which are essential for differential geodesy. Hence, as we said in SECTION 4, the whole issue is best avoided, and having explained what is involved in the issue, in the remainder of our work we will do so.

Our second digression concerns what happens in the basic equations of the leg calculus — recall SECTION IV-6 — when the equations IV-(6.4-9) are specialized to a surface $\mathbf{S} : \theta_3 = 0$. Then clearly we obtain the connection 1-forms

$$\omega_{12} = -\sigma_1\theta_1 - \sigma_2\theta_2, \tag{7.5}$$

$$\omega_{31} = k_1\theta_1 - t_2\theta_2, \tag{7.6}$$

$$\omega_{23} = -t_1\theta_1 - k_2\theta_2, \tag{7.7}$$

and from the first set of structural equations we have

$$d\theta_1 = \sigma_1\theta_1 \wedge \theta_2, \tag{7.8}$$

$$d\theta_2 = \sigma_2\theta_1 \wedge \theta_2, \tag{7.9}$$

$$0 = -(t_1 + t_2)\,\theta_1 \wedge \theta_2, \tag{7.10}$$

(the last equation being $d\theta_3$, but $d\theta_3 \equiv 0$ when $\theta_3 = 0$). Notice that (7.10) once again leads us to property (T).

There is only one non-trivial Lie bracket in surface theory and it may be deduced by specialization of IV-(7.16) by simply omitting the $\boldsymbol{\nu}$ terms, or directly by exterior differentiation of (3.12) and using $d(du^\alpha) = d^2u^\alpha = 0$. When $\boldsymbol{\lambda}$, $\boldsymbol{\mu}$ are taken to be surface vectors, i.e.,

$$\boldsymbol{\lambda} = \lambda^\alpha \frac{\partial}{\partial u^\alpha}, \quad \boldsymbol{\mu} = \mu^\alpha \frac{\partial}{\partial u^\alpha} \tag{7.11}$$

then this Lie bracket is

$$[\boldsymbol{\lambda}, \boldsymbol{\mu}] = -\sigma_1\boldsymbol{\lambda} - \sigma_2\boldsymbol{\mu}. \tag{7.12}$$

Exterior differentiation of $d\theta_1$ and $d\theta_2$ where these are understood to be the surface expressions given in (7.8) and (7.9) lead to no surface Schouten identities, viz. intrinsic Schouten identities. Finally, only two of the leg coefficients/parameters are purely intrinsic quantities — i.e. independent of $\boldsymbol{\nu}$ — and these are

$$\begin{aligned} \gamma_{121} = \sigma_1 &= \lambda_{rs}\mu^r\lambda^s = -\mu_{rs}\lambda^r\lambda^s \\ &= \lambda_{\alpha\beta}\mu^\alpha\lambda^\beta = -\mu_{\alpha\beta}\lambda^\alpha\lambda^\beta, \end{aligned} \tag{7.13}$$

$$\begin{aligned} \gamma_{122} = \sigma_2 &= \lambda_{rs}\mu^r\mu^s = -\mu_{rs}\lambda^r\mu^s \\ &= \lambda_{\alpha\beta}\mu^\alpha\mu^\beta = -\mu_{\alpha\beta}\lambda^\alpha\mu^\beta. \end{aligned} \tag{7.14}$$

These follow from (4.15-.16), or contraction of (4.3-.4) by the surface components of $\boldsymbol{\lambda}$ and $\boldsymbol{\mu}$.

V.8 The Basic Equations of the Surface Leg Calculus

We many now list the basic equations which occur in the surface leg calculus, and in doing so we will employ property (T).

The three connection 1-forms are

$$\omega_{12} = -\sigma_1\theta_1 - \sigma_2\theta_2, \tag{8.1}$$

$$\omega_{31} = k_1\theta_1 + t_1\theta_2, \tag{8.2}$$

$$\omega_{23} = -t_1\theta_1 - k_2\theta_2; \tag{8.3}$$

and the first Cartan structural equations yield two non-trivial equations

$$d\theta_1 = \sigma_1\theta_1 \wedge \theta_2, \tag{8.4}$$

$$d\theta_2 = \sigma_2\theta_1 \wedge \theta_2. \tag{8.5}$$

By virtue of the local isometric imbedding of $\mathbf{S}$ in $\mathbf{E}_3$, we have the following specializations of the general 3-leg equations in $\mathbf{E}_3$, viz. the Schouten identities IV-(7.22-.24):

$$(\mathcal{S}_{\mathrm{I}}) \quad \begin{aligned} &\sigma_{1/3} - \varepsilon_{3/1} - k_{1/2} + t_{1/1} \\ &= -k_1\gamma_2 + k_2\sigma_1 + t_1(\gamma_1 - \sigma_1) - \varepsilon_3(\gamma_1 - \sigma_2), \end{aligned} \tag{8.6}$$

$$(\mathcal{S}_{\mathrm{II}}) \quad \begin{aligned} &\sigma_{2/3} - \varepsilon_{3/2} + k_{2/1} - t_{1/2} \\ &= k_1\sigma_2 + k_2\gamma_1 - t_1(\gamma_2 + \sigma_2) - \varepsilon_3(\gamma_2 + \sigma_1), \end{aligned} \tag{8.7}$$

$$(\mathcal{S}_{\mathrm{III}}) \quad \gamma_{1/2} - \gamma_{2/1} = \gamma_1\sigma_1 + \gamma_2\sigma_2; \tag{8.8}$$

and the Lamé equations IV-(9.10-.18):

$$(\mathcal{L}_{\mathrm{I}}) \quad \sigma_{1/2} - \sigma_{2/1} = \sigma_1^2 + \sigma_2^2 + k_1k_2 - t_1^2, \tag{8.9}$$

$$(\mathcal{L}_{\mathrm{II}}) \quad \varepsilon_{3/1} - \sigma_{1/3} = k_1(\gamma_2 - \sigma_1) - t_1(\gamma_1 + \sigma_2) + \varepsilon_3(\gamma_1 - \sigma_2), \tag{8.10}$$

$$(\mathcal{L}_{\mathrm{III}}) \quad \varepsilon_{3/2} - \sigma_{2/3} = -k_2(\gamma_1 + \sigma_2) + t_1(\gamma_2 - \sigma_1) + \varepsilon_3(\gamma_2 + \sigma_1), \tag{8.11}$$

$$(\mathcal{L}_{\mathrm{IV}}) \quad t_{1/1} - k_{1/2} = (k_2 - k_1)\sigma_1 - 2t_1\sigma_2, \tag{8.12}$$

$$(\mathcal{L}_{\mathrm{V}}) \quad k_{1/3}+\gamma_{1/1}=k_1^2+t_1\,(t_1+2\varepsilon_3)+\gamma_1^2+\gamma_2\sigma_1, \tag{8.13}$$

$$(\mathcal{L}_{\mathrm{VI}}) \quad t_{1/3}+\gamma_{1/2}=k_1\,(t_1-\varepsilon_3)+k_2\,(t_1+\varepsilon_3)+\gamma_2\,(\gamma_1+\sigma_2)\,, \tag{8.14}$$

$$(\mathcal{L}_{\mathrm{VII}}) \quad t_{1/2}-k_{2/1}=(k_2-k_1)\,\sigma_1+2t_1\sigma_1, \tag{8.15}$$

$$(\mathcal{L}_{\mathrm{VIII}}) \quad t_{1/3}+\gamma_{2/1}=k_1\,(t_1-\varepsilon_3)+k_2\,(t_1+\varepsilon_3)+\gamma_1\,(\gamma_2-\sigma_1)\,, \tag{8.16}$$

$$(\mathcal{L}_{\mathrm{IX}}) \quad k_{2/3}+\gamma_{2/2}=k_2^2+t_1\,(t_1-2\varepsilon_3)+\gamma_2^2-\gamma_1\sigma_2. \tag{8.17}$$

By virtue of the Lamé equations, as expressed in terms of the Cartan calculus in SECTION III-6, we have $\Omega_{ab}=0$ and hence

$$d\omega_{12}=K\theta_1\wedge\theta_2, \tag{8.18}$$

$$d\omega_{31}=\omega_{12}\wedge\omega_{23}, \tag{8.19}$$

$$d\omega_{23}=-\omega_{12}\wedge\omega_{31}. \tag{8.20}$$

As noted at the end of SECTION 4, (8.18) requires some discussion. Upon exterior differentiation of (8.1) and by using (8.4) and (8.5) we have

$$d\omega_{12}=\left(\sigma_{1/2}-\sigma_{2/1}-\sigma_1^2-\sigma_2^2\right)\theta_1\wedge\theta_2. \tag{8.21}$$

Then by the right-hand side of (8.18) it follows that

$$(\mathcal{G}) \quad K=\sigma_{1/2}-\sigma_{2/1}-\sigma_1^2-\sigma_2^2 \tag{8.22}$$

which is the leg version of a classical *formula of Liouville* (see EISENHART (1947) Ex. 12, page 193). On the other hand, by virtue of $(\mathcal{L}_{\mathrm{I}})$ the left-hand side of this equation is equal to $k_1k_2-t_1^2$, i.e. our leg expression (4.39). Hence (8.22) is equivalent to the *Gauss equation* $(\mathcal{G})$, or $(\mathcal{G}')$, or $(\mathcal{G})$ given in SECTION 2. This equation is interesting in that since σ_1 and σ_2 are intrinsic leg coefficients/parameters — recall (7.13) and (7.14) — K must also be an intrinsic quantity. This was the great discovery of Gauss in 1828, but it is not *a priori* evident from all the forms of the Gauss equation, e.g. $(\mathcal{G}')$. Note also that both the flatness of $\mathbf{E}_3$ and the local isometric imbedding of $\mathbf{S}$ in $\mathbf{E}_3$ are required to deduce (8.22).

We now turn to equations (8.19) and (8.20). For the former, exterior differentiation of (8.2) yields

$$d\omega_{31} = \left(t_{1/1} - k_{1/2} + k_1\sigma_1 + t_1\sigma_2\right)\theta_1 \wedge \theta_2,$$

and

$$\omega_{12} \wedge \omega_{23} = (-t_1\sigma_2 + k_2\sigma_1)\,\theta_1 \wedge \theta_2;$$

hence we have

$$(\mathcal{C}_{\mathrm{I}}) \quad k_{1/2} - t_{1/1} = (k_1 - k_2)\,\sigma_1 + 2t_1\sigma_2 \tag{8.23}$$

which is the *first Codazzi equation*. Likewise,

$$d\omega_{23} = \left(t_{1/2} - k_{2/1} - t_1\sigma_1 - k_2\sigma_2\right)\theta_1 \wedge \theta_2,$$

and

$$-\omega_{12} \wedge \omega_{31} = (-k_1\sigma_2 + t_1\sigma_1)\,\theta_1 \wedge \theta_2;$$

hence we have

$$(\mathcal{C}_{\mathrm{II}}) \quad k_{2/1} - t_{1/2} = (k_1 - k_2)\,\sigma_2 - 2t_1\sigma_1 \tag{8.24}$$

which is the *second Codazzi equation.*

Equations (8.23) and (8.24) are due to Hotine who derived them — in a slightly different notation — by using the leg representations of our $(\mathcal{C})$ or $(\mathcal{C}')$ from SECTION 2 (see [Chap. 8, subsections 8, 9 and 13 pages 44-45]). Modulo a change of signs

$$(\mathcal{C}_{\mathrm{I}}) \Leftrightarrow (\mathcal{L}_{\mathrm{IV}}), \tag{8.25}$$

$$(\mathcal{C}_{\mathrm{II}}) \Leftrightarrow (\mathcal{L}_{\mathrm{VII}}); \tag{8.26}$$

and by the Theorem in SECTION IV-9

$$(\mathcal{L}_{\mathrm{II}}) \Leftrightarrow (\mathcal{L}_{\mathrm{IV}}), \tag{8.27}$$

$$(\mathcal{L}_{\mathrm{III}}) \Leftrightarrow (\mathcal{L}_{\mathrm{VII}}); \tag{8.28}$$

hence

$$(\mathcal{C}'_{\mathrm{I}}) \quad \varepsilon_{3/1} - \sigma_{1/3} = k_1 (\gamma_2 - \sigma_1) - t_1 (\gamma_1 + \sigma_2) + \varepsilon_3 (\gamma_1 - \sigma_2), \tag{8.29}$$

$$(\mathcal{C}'_{\mathrm{II}}) \quad \varepsilon_{3/2} - \sigma_{2/3} = k_1 (\gamma_1 + \sigma_2) + t_1 (\gamma_2 - \sigma_1) + \varepsilon_3 (\gamma_2 + \sigma_1), \tag{8.30}$$

are alternate equivalent versions of $(\mathcal{C}_{\mathrm{I}})$ and $(\mathcal{C}_{\mathrm{II}})$ respectively. Indeed, the differences between these Lamé equations are two of the Schouten identities, viz. formally we have

$$(\mathcal{L}_{\mathrm{II}}) - (\mathcal{L}_{\mathrm{IV}}) = (\mathcal{S}_{\mathrm{I}}), \tag{8.31}$$

$$(\mathcal{L}_{\mathrm{III}}) - (\mathcal{L}_{\mathrm{VII}}) = (\mathcal{S}_{\mathrm{II}}). \tag{8.32}$$

V.9 The Frenet Equations

In SECTION 4, i.e. (4.9) and (4.10), we obtained the surface Frenet equations for the congruences $\boldsymbol{\Gamma}'_1$ and $\boldsymbol{\Gamma}'_2$ on **S**. We now consider the Frenet equations for the space congruence $\boldsymbol{\Gamma}=\boldsymbol{\Gamma}_3$ which is normal to **S**. We already know from SECTION II-5, or more precisely from SECTION IV-5 that the curvature χ of $\boldsymbol{\Gamma}$ is given by

$$\chi = \sqrt{\gamma_1^2 + \gamma_2^2}, \tag{9.1}$$

and our immediate task is now to determine a leg coefficient/parameter expression for the *torsion* τ of $\boldsymbol{\Gamma}$.

The *Frenet vectorial* 3-leg considers of the following three vectors associated to $\boldsymbol{\Gamma}{:=}\boldsymbol{\Gamma}(\boldsymbol{\ell})$:

$\boldsymbol{\ell}$: *tangent,*
m : *principal normal,*
n : *binormal,*

defined along $\boldsymbol{\Gamma}$, or more precisely along some (open) arc of $\boldsymbol{\Gamma}$. Note that the vector $\boldsymbol{\ell}$ is not to be confused with our curvature vector (4.8), and $\boldsymbol{\ell}$ is introduced only to agree with Hotine's notation for the Frenet equations, see [4.06]. Indeed, this new $\boldsymbol{\ell}$ will quickly disappear from our analysis. Then following Hotine, the *Frenet equations* for $\boldsymbol{\Gamma}(\boldsymbol{\ell})$ are

$$\ell_{rs}\ell^s = \chi m_r, \tag{9.2}$$

$$m_{r s}\ell^s = -\chi \ell_r + \tau n_r, \tag{9.3}$$

$$n_{r s}\ell^s = -\tau m_r. \tag{9.4}$$

Since ℓ^r are the components of a unit tangent vector, $\ell^r = dx^r/ds$ where s is the arc length along $\boldsymbol{\Gamma}(\boldsymbol{\ell})$, the left-hand sides of (9.2)-(9.4) become the *absolute/intrinsic derivatives* of ℓ_r, m_r and n_r along $\boldsymbol{\Gamma}(\boldsymbol{\ell})$ (see McCONNELL (1931), Chap. XIII pages 156-160). But following Hotine we will not explicitly employ such terminology, or introduce an appropriate notation, for these derivatives in our analysis. This has the advantage of allowing us to avoid specifying the arc length in our work, however, it is there since ℓ^r are components of a tangent vector.

We now specialize the Frenet equations by choosing $\boldsymbol{\ell} = \boldsymbol{\nu}$ where $\boldsymbol{\nu}$ is the unit normal to S and hence the unit tangent to $\boldsymbol{\Gamma}=\boldsymbol{\Gamma}(\boldsymbol{\nu})$, i.e.

$$\nu_{r s}\nu^s = \chi m_r, \tag{9.5}$$

$$m_{r s}\nu^s = -\chi \nu_r + \tau n_r, \tag{9.6}$$

$$n_{r s}\nu^s = -\tau m_r. \tag{9.7}$$

The vectors m and n are unit vectors which are orthogonal to $\boldsymbol{\nu}$, but neither m nor n need be equal to $\boldsymbol{\lambda}$ or $\boldsymbol{\mu}$.

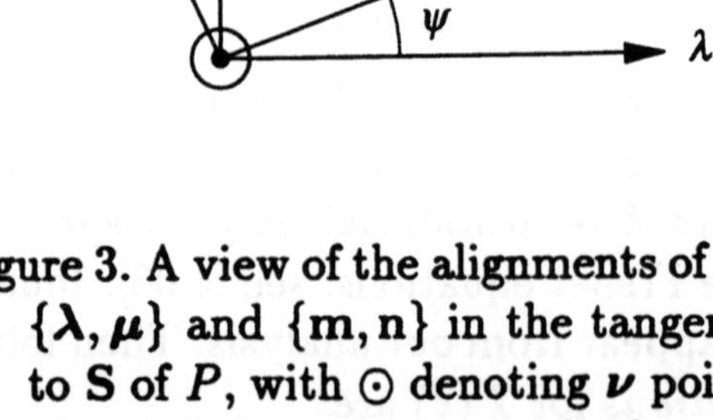

Figure 3. A view of the alignments of the vector pairs $\{\boldsymbol{\lambda}, \boldsymbol{\mu}\}$ and $\{\mathbf{m}, \mathbf{n}\}$ in the tangent plane $\mathbf{T}_P(\mathbf{S})$ to S of P, with ⊙ denoting $\boldsymbol{\nu}$ pointing out of the plane of the paper towards the reader.

Since both **m** and **n** are tangent vectors in the tangent plane to **S** the pairs $\{\mathbf{m}, \mathbf{n}\}$ and $\{\boldsymbol{\lambda}, \boldsymbol{\mu}\}$ are related to each other by a rotation about $\boldsymbol{\nu}$. Denoting the angle between m and λ by ψ we have

$$m_r = \cos\psi\lambda_r + \sin\psi\mu_r, \tag{9.8}$$

$$n_r = -\sin\psi\lambda_r + \cos\psi\mu_r. \tag{9.9}$$

Now by the last equation in the set (6.6) we have

$$\chi m_r = \gamma_1\lambda_r + \gamma_2\mu_r, \tag{9.10}$$

so

$$\chi m_r\lambda^r = \gamma_1 \Rightarrow \chi\cos\psi = \gamma_1, \tag{9.11}$$

$$\chi m_r\mu^r = \gamma_2 \Rightarrow \chi\sin\psi = \gamma_2. \tag{9.12}$$

Hence, we have

$$\chi = \gamma_1\lambda_r m^r + \gamma_2\mu_r m^r \tag{9.13}$$

or more conveniently,

$$\chi = \gamma_1\cos\psi + \gamma_2\sin\psi. \tag{9.14}$$

Note that both (9.11) and (9.12) satisfy (9.1) and (9.14), although (9.1) remains the nicest expression.

The torsion τ is now obtained by contracting (9.7) with ν^r,

$$-\tau = n_{rs}m^r\nu^s, \tag{9.15}$$

however, to do so one must express the right-hand side in terms of the Frenet 3-leg $\{\boldsymbol{\nu}, \mathbf{m}, \mathbf{n}\}$. Clearly,

$$n_{rs} = -\cos\psi\lambda_r\psi_s - \sin\psi\lambda_{rs} - \sin\psi\mu_r\psi_s + \cos\psi\mu_{rs} \tag{9.16}$$

and thus using the leg expressions (6.1) and (6.2) for λ_{rs} and μ_{rs}, the contraction indicated in (9.15) yields

$$\tau = \varepsilon_3 + \psi_{/3} \tag{9.17}$$

where "/3" denotes the leg derivative

$$\psi_{/3} := \psi_r\nu^r. \tag{9.18}$$

This result is analogous to a formula given by Bonnet in 1853, (i.e. [7.07] or McCONNELL (1931), equation (26), page 215). However, this result has a particular significance for us since it exhibits a connection between the leg coefficient/parameter ε_3 with the normal congruence $\boldsymbol{\Gamma}$ of **S**.

V.10 The Meusnier and Bonnet Formulas

In this and the subsequent section we derive various formulas relating the Frenet 3-leg quantities, i.e. the curvature χ and torsion τ to some of the leg coefficients/parameters associated with the canonical 3-leg $\{\boldsymbol{\lambda}, \boldsymbol{\mu}, \boldsymbol{\nu}\}$ of CHAPTER IV.

We now choose the Frenet 3-leg $\{\boldsymbol{\ell}, \mathbf{m}, \mathbf{n}\}$ such that $\boldsymbol{\ell}$ coincides with the spatial tangent vector $\boldsymbol{\lambda}$ to $\mathbf{S}$ as in the following figure

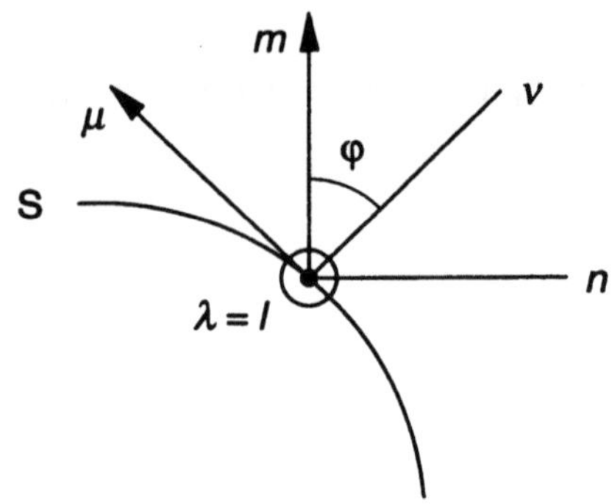

Figure 4. A side, or cross-sectional, view of S, showing the alignments of the vector pairs $\{\boldsymbol{\mu}, \boldsymbol{\nu}\}$ and $\{\mathbf{m}, \mathbf{n}\}$ with $\boldsymbol{\lambda}$ pointing into the plane of the paper away from the reader.

with φ the angle between $\mathbf{m}$ and $\boldsymbol{\nu}$. Thus, we have

$$m_r = \sin\varphi \mu_r + \cos\varphi \nu_r, \tag{10.1}$$

$$n_r = -\cos\varphi \mu_r + \sin\varphi \nu_r. \tag{10.2}$$

Then by the first Frenet equation (9.2), it follows that

$$\lambda_{rs}\lambda^s = \sigma_1 \mu_r + k_1 \nu_r = \tilde{\chi} m_r \tag{10.3}$$

where the curvature is denoted by $\tilde{\chi}$ (to avoid any possible confusion with χ). Hence we write

$$\tilde{\chi} m_r \nu^r = k_1 \Rightarrow k_1 = \tilde{\chi}\cos\varphi, \tag{10.4}$$

$$\tilde{\chi} m_r \mu^r = \sigma_1 \Rightarrow \sigma_1 = \tilde{\chi} \sin \varphi. \tag{10.5}$$

These are called the *Meusnier formulas* (see [7.03-.04], or McCONNELL (1931), page 210), and obviously

$$\tilde{\chi}^2 = k_1^2 + \sigma_1^2, \tag{10.6}$$

a relationship previously given in II-(5.1) since $\tilde{\chi} = \chi_1$ in this case.

Consider the third Frenet equation

$$n_{rs}\lambda^s = -\tilde{\tau} m_r, \tag{10.7}$$

upon differentiation of (10.2) we have

$$\begin{aligned} n_{rs}\lambda^s &= \varphi_{/1} \cos \varphi \nu_r + \varphi_{/1} \sin \varphi \mu_r \\ &\quad + \sin \varphi \{-k_1 \lambda_4 - t_1 \mu_r\} \\ &\quad - \cos \varphi \{-\sigma_1 \lambda_r + t_1 \nu_r\} \end{aligned} \tag{10.8}$$

and

$$-\tilde{\tau} m_r = -\tilde{\tau} \sin \varphi \mu_r - \tilde{\tau} \cos \varphi \nu_r. \tag{10.9}$$

Thus, equating (10.8) and (10.9) from (10.7) we obtain

$$\lambda_r : -k_1 \sin \varphi + \sigma_1 \cos \varphi = 0, \tag{10.10}$$

$$\mu_r : \sin \varphi \left(\varphi_{/1} - t_1\right) = -\tilde{\tau} \sin \varphi, \tag{10.11}$$

$$\nu_r : \cos \varphi \left(\varphi_{/1} - t_1\right) = -\tilde{\tau} \cos \varphi. \tag{10.12}$$

Equation (10.10) is identically satisfied by the Meusnier formulas, and for general φ, i.e. those $0 < \varphi < \pi/2$, (10.11) and (10.12) both give

$$\tilde{\tau} = t_1 - \varphi_{/1} \tag{10.13}$$

which is the *Bonnet formula* (see [7.08], or McCONNELL (1931), page 215).

V.11 Geodesic Curvature and Torsion in an Arbitrary Direction, and Euler's Formula

On the surface **S** the surface 2-leg $\{\boldsymbol{\lambda}, \boldsymbol{\mu}\}$ is aligned in arbitrary orthogonal directions, however, having chosen these directions in a particular situation, it may be required to determine the geodesic curvature $\tilde{\sigma}$ and torsion $\tilde{t}$ in another direction. We now derive such expressions when $\{\boldsymbol{\xi}, \boldsymbol{\eta}\}$ is any other surface 2-leg obtained from $\{\boldsymbol{\lambda}, \boldsymbol{\mu}\}$ by a rotation about the normal ν to **S** through an angle ϑ, i.e. ϑ is the angle between $\boldsymbol{\lambda}$ and $\boldsymbol{\xi}$.

Clearly, as shown in the accompanying figure, we have

$$\xi_\alpha = \cos\vartheta\,\lambda_\alpha + \sin\vartheta\,\mu_\alpha, \tag{11.1}$$

$$\eta_\alpha = -\sin\vartheta\,\lambda_\alpha + \cos\vartheta\,\mu_\alpha, \tag{11.2}$$

so

$$\begin{aligned} \xi_\alpha\lambda^\alpha &= \cos\vartheta, & \xi_\alpha\mu^\alpha &= \sin\vartheta \\ \eta_\alpha\lambda^\alpha &= -\sin\vartheta, & \eta_\alpha\mu^\alpha &= \cos\vartheta \end{aligned} \tag{11.3}$$

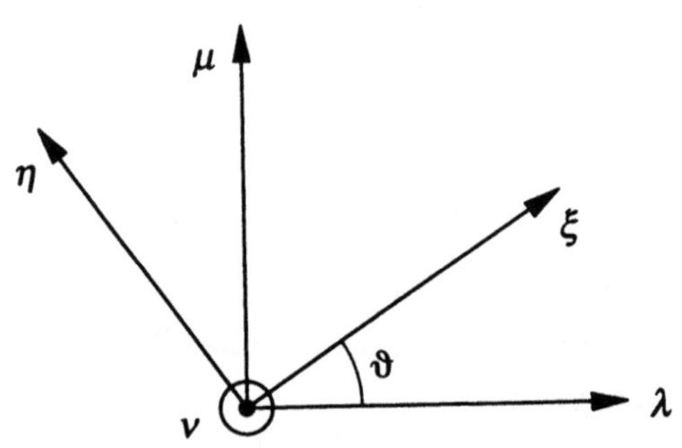

Figure 5

By covariant differentiation of (11.1) we have

$$\begin{aligned}\xi_{\alpha\beta} &= \eta_\alpha \vartheta_\beta + \cos\vartheta \left\{\sigma_1 \mu_\alpha \lambda_\beta + \sigma_2 \mu_\alpha \mu_\beta\right\} \\ &\quad + \sin\vartheta \left\{-\sigma_1 \lambda_\alpha \lambda_\beta - \sigma_2 \lambda_\alpha \mu_\beta\right\}\end{aligned} \tag{11.4}$$

where we have used (4.3) and (4.4) for $\lambda_{\alpha\beta}$ and $\mu_{\alpha\beta}$ respectively. If we now take $\boldsymbol{\xi}$ to be unit tangent to a surface curve C, then $\boldsymbol{\eta}$ is a normal to C and by the first surface Frenet equation — the first equation in the set (4.9) — we have

$$\xi_{\alpha\beta}\xi^\beta = \tilde{\sigma}\eta_\alpha, \tag{11.5}$$

where $\tilde{\sigma}$ is the geodesic curvature of C, viz. the required curvature in the direction $\boldsymbol{\xi}$. Since

$$\vartheta_\beta \xi^\beta = \vartheta_{/1}\cos\vartheta + \vartheta_{/2}\sin\vartheta \tag{11.6}$$

explicit evaluation of (11.5) leads to the following equations which are respectively the coefficients of λ_α and μ_α:

$$\begin{aligned}-\tilde{\sigma}\sin\vartheta &= -\sin\vartheta\left(\vartheta_{/1}\cos\vartheta + \vartheta_{/2}\sin\vartheta\right) \\ &\quad +\sigma_1 \sin^2\vartheta - \sigma_2 \sin\vartheta\cos\vartheta,\end{aligned} \tag{11.7}$$

$$\begin{aligned}\tilde{\sigma}\cos\vartheta &= \cos\vartheta\left(\vartheta_{/1}\cos\vartheta + \vartheta_{/2}\sin\vartheta\right) \\ &\quad -\sigma_1 \cos\vartheta\sin\vartheta + \sigma_2\cos^2\vartheta.\end{aligned} \tag{11.8}$$

Both of these expressions for a general ϑ, i.e. $0 < \vartheta < \pi/2$, yield the same result, i.e.

$$\tilde{\sigma} = \vartheta_{/1}\cos\vartheta + \vartheta_{/2}\sin\vartheta - \sigma_1\sin\vartheta + \sigma_2\cos\vartheta \tag{11.9}$$

which is the required equation. A slightly different expression is quoted in Hotine [12.068], and the classical versions are given by EISENHART (1947) (see Exercise 13, page 193) and McCONNELL (1931) (see Example XIV-8, page 191).

By definition, the geodesic torsion $\tilde{t}$ in the direction ξ is given by

$$\tilde{t} := \varepsilon^{\rho\sigma} a_{\alpha\rho} b_{\beta\sigma} \xi^\alpha \xi^\beta \tag{11.10}$$

which is equivalent to

$$\tilde{t} = h_{\alpha\beta}\xi^\alpha\xi^\beta \tag{11.11}$$

where $h_{\alpha\beta}$ is the Euler tensor introduced in (4.30). (For details of the classical approach see EISENHART (1947), equation (44.11)). Since the surface 2-leg representation (4.31) is available, an immediate evaluation of (11.11) gives

$$\tilde{t} = (k_2 - k_1) \sin\vartheta \cos\vartheta + t_1 \left(\cos^2\vartheta - \sin^2\vartheta\right) \tag{11.12}$$

which is the desired expression for the geodesic torsion in the $\boldsymbol{\xi}$ direction. This is a generalization of the classical result

$$\tilde{t} = (\kappa_2 - \kappa_1) \sin\vartheta \cos\vartheta \tag{11.13}$$

which occurs when the curve C is a line of curvature, i.e. $t_1 = 0$ (see McCONNELL (1931), Example 1, page 215). Hotine, [7.08], computed (11.12) from the expression

$$\tilde{t} = b_{\alpha\beta}\xi^\alpha \eta^\beta \tag{11.14}$$

which again yields (11.12).

To obtain the Euler formula for the normal curvature $\tilde{k}$ of $\mathbf{S}$ in a direction $\boldsymbol{\xi}$on $\mathbf{S}$, we need only evaluate the leg representation of the expression

$$\tilde{k} := b_{\alpha\beta}\xi^\alpha \xi^\beta \tag{11.15}$$

by using (4.5) and the obvious contravariant form of (11.1). The result

$$\tilde{k} := k_1 \cos^2\vartheta + k_2 \sin^2\vartheta + 2t_1 \sin\vartheta \cos\vartheta \tag{11.16}$$

may be regarded as a generalization of Euler's formula (see McCONNELL (1931), Example 2, page 212) which holds when $\boldsymbol{\lambda}$ and $\boldsymbol{\mu}$ are not principal directions. Of course, when $\boldsymbol{\lambda}$ and $\boldsymbol{\mu}$ are principal directions $t_1 = 0$, and (11.16) reduces to

$$\tilde{k} = \kappa_1 \cos^2\vartheta + \kappa_2 \sin^2\vartheta \tag{11.17}$$

which is the usual Euler formula of 1760. Equation (11.16) was also given by Hotine [12.049], for a different choice of the angle ϑ.

V.12 Euler's Osculating Paraboloid and Dupin's Indicatrix of Curvature

In this section we will consider how the leg coefficients k_1, k_2 and t_1 can be given an intuitive geometric interpretation. Logically this will be a preliminary to CHAPTER VIII when in an analogous manner we will introduce the Marussi quadric as an aid to visualizing the meaning of the Marussi tensor. The situation is somewhat similar to that occurring in rigid body dynamics where Cauchy in 1827 described the components of the inertia tensor (also known as the ellipsoid of inertia) and subsequently Poinsot in 1857 gave a kinematic representation of the motion (see WHITTAKER (1904), Chapters 5 and 6).

The basic idea is to take the entries in a square matrix as the coefficients of a quadratic form whose variables are some unspecified quantities (in algebra these are called indeterminates). Then this quadratic form can be regarded as defining a quadric surface in space, or a conic in a plane. Both of these quadrics and conics are *abstract* in the sense that the variables *per se* are not the familiar ambient coordinates, but nevertheless they may be thought of as geometric loci in *abstract* Euclidean $\mathcal{E}_3$ and $\mathcal{E}_2$. A somewhat more precise identification does occur in Gaussian differential geometry, but then formally the coefficients in question are considered to be coordinate independent although strictly speaking they are coordinate dependent.

In $\mathcal{E}_3$ we employ a system of Cartesian-like indeterminates: X, Y and Z which are *coordinates* along orthogonal axes. In SECTION 4, recall (4.37), we gave a leg representation of $b_{\alpha\beta}$, and we now take them to be coefficients in a binary quadratic form

$$k_1X^2 + 2t_1XY + k_2Y^2.$$

Then upon introducing the third coordinate Z by

$$Z - \frac{1}{2}\left(k_1X^2 + 2tXY + k_2Y^2\right) = 0 \qquad (12.1)$$

we obtain an osculating paraboloid in $\mathcal{E}_3$. The numerical factor of $\frac{1}{2}$ is introduced to obtain agreement with the second basic form II. In the classical theory, e.g. in POGORELOV (1955), Chapter 5, the construction is given in detail, and the axis of the paraboloid is aligned along the unit normal to **S**, and the quadric defined by (12.1) is said to be an *osculating paraboloid* Φ which touches **S** at P. In fact, Φ may be regarded as being a quadratic approximation to **S** in the sense that at P:

$$\mathrm{I}\,|_S = \mathrm{I}\,|_\Phi \quad \text{and} \quad \mathrm{II}\,|_S = \mathrm{II}\,|_\Phi\,. \tag{12.2}$$

The usefulness of Φ is that it allows us to classify points on **S** in the following manner: a point P on **S** is

(i) *elliptic* whenever Φ is an *elliptic paraboloid* at P;

(ii) *hyperbolic* whenever Φ is a *hyperbolic paraboloid* at P;

(iii) *parabolic* whenever Φ is a *parabolic cylinder* at P;

(iv) *flat/planar* whenever Φ reduces to the *tangent plane* $\mathbf{T}_P(S)$ to **S** at P.

These correspond to the respective specializations of the leg coefficients:

$$\begin{array}{ll} \text{(i)} & k_1k_2 - t_1^2 > 0, \\ \text{(ii)} & k_1k_2 - t_1^2 < 0, \\ \text{(iii)} & k_1k_2 - t_1^2 = 0, \\ \text{(iv)} & k_1 = k_2 = t_1 = 0, \end{array} \tag{12.4}$$

which are statements about the Gauss curvature K; so in (iii) and (iv) $K \equiv 0$. If $\boldsymbol{\lambda}$ and $\boldsymbol{\mu}$ are principal directions on **S**, we have $k_1 = \kappa_1$, $k_2 = \kappa_2$, and writing sgn for *signum*, then

$$\begin{array}{ll} \text{(i)} & \operatorname{sgn}\kappa_1 = \operatorname{sgn}\kappa_2, \\ \text{(ii)} & \operatorname{sgn}\kappa_1 \neq \operatorname{sgn}\kappa_2, \\ \text{(iii)} & \text{either } \kappa_1 \text{ or } \kappa_2 \text{ (but not both) is zero,} \\ \text{(iv)} & \text{both } \kappa_1 \text{ and } \kappa_2 \text{ are identically zero.} \end{array} \tag{12.5}$$

Thus, the shape of Φ is determined by the values of the leg coefficients k_1, k_2, t_1 at a point P of **S**.

A related construction is the *indicatrix of curvature* Ψ suggested by Dupin in 1813. This is a planar locus which essentially amounts to rewriting Euler's formula (11.17) in a convenient manner when $\boldsymbol{\lambda}$ and $\boldsymbol{\mu}$ are principal directions. Introducing the principal radius of curvature

$$\tilde{\rho} := \frac{1}{\tilde{\kappa}} \tag{12.6}$$

into (11.17), we have

$$\frac{1}{\tilde{\rho}} = \kappa_1 \cos^2 \vartheta + \kappa_2 \sin^2 \vartheta. \tag{12.7}$$

Then putting

$$\begin{aligned} X &:= \sqrt{|\tilde{\rho}|} \cos \vartheta, \\ Y &:= \sqrt{|\tilde{\rho}|} \sin \vartheta, \end{aligned} \tag{12.8}$$

the indicatrix of curvature Ψ becomes

$$\kappa_1 X^2 + \kappa_2 Y^2 = 1. \tag{12.9}$$

The equations in (12.9) resemble ordinary polar coordinates where ϑ is taken to be the angle between the directions $\boldsymbol{\xi}$ and $\boldsymbol{\lambda}$, and the absolute value of $\tilde{\rho}$ is required to insure that the radial variable is well-defined. One could replace κ_1 and κ_2, by their radii of curvatures, however (12.9) is preferable since our goal is to obtain an interpretation of κ_1 and κ_2.

The previous classification (12.3) of points on **S** is geometrically given as follows: a point on **S** is

(i) elliptic whenever Ψ is an ellipse,
(ii) hyperbolic whenever Ψ is a hyperbola,
(iii) parabolic whenever Ψ reduces to a pair of parallel straight lines,
(iv) flat/planar whenever Ψ reduces to a single point.
(12.10)

These respectively correspond to the following values of κ_1 and κ_2:

$$\begin{array}{ll} \text{(i)} & \operatorname{sgn}\kappa_1 = \operatorname{sgn}\kappa_2, \\ \text{(ii)} & \operatorname{sgn}\kappa_1 \neq \operatorname{sgn}\kappa_2, \\ \text{(iii)} & \text{either } \kappa_1 \text{ or } \kappa_2 \text{ (but not both) is zero,} \\ \text{(iv)} & \text{both } \kappa_1 \text{ and } \kappa_2 \text{ are identically zero.} \end{array} \tag{12.11}$$

In the classical theory this may be visualized by a plane Π parallel to $\mathbf{T}_P(\mathbf{S})$, where Π is close to $\mathbf{T}_P(\mathbf{S})$ and on the surface side of $\mathbf{T}_P(\mathbf{S})$. Then Ψ is none other than the intersection of Π with $\mathbf{T}_P(\mathbf{S})$ in a neighborhood of P on $\mathbf{S}$. In case (iv) Π coincides with $\mathbf{T}_P(\mathbf{S})$.

V.13 Triply Orthogonal Systems and Dupin's Theorem

We now conclude our discussion of Gaussian differential geometry by considering a triply-orthogonal system $\mathcal{O}_3$ of surfaces $\{\mathbf{S}_a\}_{a=1}^{3}$ defined by their triply-orthogonal systems of normal congruences $\{\boldsymbol{\Gamma}_a\}_{a=1}^{3}$. Hence, properly speaking instead of considering individual surfaces, say $\mathbf{S}_a$, $\mathbf{S}'_a$, $\mathbf{S}''_a$, ... for a specified value of the index a, by S_a we mean a *family* of a-surfaces $\boldsymbol{\Sigma}_a$, i.e. for a fixed a,

$$\boldsymbol{\Sigma}_a = \{\mathbf{S}_a, \mathbf{S}'_a, \mathbf{S}''_a, \cdots\}.$$

Our considerations will exhibit the permutability/symmetry implicit in our canonical enumeration of the Ricci coefficients (SECTION II-5) and establish the correspondence between the leg coefficients for different families of congruences and surfaces. In particular, this will reveal a further meaning of the leg coefficient ε_3 in our usual case $\boldsymbol{\Gamma} \perp \mathbf{S}$ i.e. $\boldsymbol{\Gamma}_3 \perp \mathbf{S}_3$ or more generally $\boldsymbol{\Gamma}_3 \perp \boldsymbol{\Sigma}_3$.

In order to obtain an overview of the situation we will not automatically single out the case $a = 3$ in our analysis, but rather see what happens to the nine leg coefficients when the condition $\boldsymbol{\Gamma}_a \perp \mathbf{S}_a$ is imposed in *each* case $a = 3, 1, 2$. This will allow us to consider the geometry on *each* surface $\mathbf{S}_a$, i.e. exhibit their basic forms I_a, II_a, III_a, IV_a and curvatures K_a, H_a. Then imposing the simultaneous requirements

$$\boldsymbol{\Gamma}_1 \perp \boldsymbol{\Sigma}, \quad \boldsymbol{\Gamma}_2 \perp \boldsymbol{\Sigma}_2, \quad \boldsymbol{\Gamma}_3 \perp \boldsymbol{\Sigma}_3$$

leads to the notion of an $\mathcal{O}_3$.

Our first step is to re-examine the canonical identification scheme of SECTION II-5 and translate it into the leg coefficient language. By a permutation of the indices $3 \to 1$, $1 \to 2$, and $2 \to 3$, by virtue of the identification IV-(3.1) we have the following display:

$$\begin{array}{ccc}
\boldsymbol{\Gamma}_3 & \boldsymbol{\Gamma}_1 & \boldsymbol{\Gamma}_2 \\
\gamma_{131} := k_1 & \gamma_{212} = -\sigma_2 & \gamma_{323} = \gamma_2 \\
\gamma_{232} := k_2 & \gamma_{313} = \gamma_1 & \gamma_{121} = \sigma_1 \\
 & & \\
\gamma_{231} := t_1 & \gamma_{312} = t_2 & \gamma_{123} = \varepsilon_3 \\
\gamma_{312} := t_2 & \gamma_{123} = \varepsilon_3 & \gamma_{231} = t_1 \\
 & & \\
\gamma_{123} := \varepsilon_3 & \gamma_{231} = t_1 & \gamma_{312} = t_2 \\
 & & \\
\gamma_{313} := \gamma_1 & \gamma_{121} = \sigma_1 & \gamma_{232} = k_2 \\
\gamma_{323} := \gamma_2 & \gamma_{131} = k_1 & \gamma_{212} = -\sigma_2 \\
 & & \\
\gamma_{121} := \sigma_1 & \gamma_{232} = k_2 & \gamma_{313} = \gamma_1 \\
\gamma_{122} := \sigma_2 & \gamma_{233} = -\gamma_2 & \gamma_{311} = k_1
\end{array} \tag{13.1}$$

in which we have used the skew-symmetry of the Ricci coefficients, i.e. $\gamma_{abc} = -\gamma_{bac}$, in some of the expressions occurring in the columns for $\boldsymbol{\Gamma}_1$ and $\boldsymbol{\Gamma}_2$. In the first column we have taken the canonical identifications for the Ricci and leg coefficients employed in IV-(3.1). Apart from the sign changes the entries in these columns are identical but permuted and the display indicates the meaning of these $\boldsymbol{\Gamma}_3$ coefficients for $\boldsymbol{\Gamma}_1$ and $\boldsymbol{\Gamma}_2$.

For example, the individual $\mathbf{S}_a$, $a = 3, 1, 2$ are defined by the respective Pfaffian forms with

$$\begin{aligned}
&\mathbf{S}_3 : \theta_3 = 0 \text{ having tangent vectors } \boldsymbol{\lambda}, \boldsymbol{\mu}; \\
&\mathbf{S}_1 : \theta_1 = 0 \text{ having tangent vectors } \boldsymbol{\mu}, \boldsymbol{\nu}; \\
&\mathbf{S}_2 : \theta_2 = 0 \text{ having tangent vectors } \boldsymbol{\nu}, \boldsymbol{\lambda};
\end{aligned} \tag{13.2}$$

so by SECTION 4 we have the following results:

If k_1, k_2 are the *normal curvatures* [1] of $\mathbf{S}_3$ in the respective $\boldsymbol{\lambda}$, $\boldsymbol{\mu}$ directions; then

$-\sigma_2$, γ_1 are the *normal curvatures* of $\mathbf{S}_1$ in the respective $\boldsymbol{\mu}$, $\boldsymbol{\nu}$ directions, and

γ_2, σ_1 are the *normal curvatures* of $\mathbf{S}_2$ in the respective $\boldsymbol{\nu}$, $\boldsymbol{\lambda}$ directions.

This suggests that the respective Germain curvatures are

$$\mathbf{S}_3 : H_3 := (k_1 + k_2)/2, \tag{13.3}$$

$$\mathbf{S}_1 : H_1 := (-\sigma_2 + \gamma_1)/2, \tag{13.4}$$

$$\mathbf{S}_2 : H_2 := (\gamma_2 + \sigma_1)/2. \tag{13.5}$$

The latter two identifications can be proven by examining the respective covariant divergences of $\lambda^r{}_r$ and $\mu^r{}_r$ previously evaluated in IV-(6.13) and IV-(6.14), and observing that

$$\underset{a}{\lambda}{}^r{}_r = -2H_a. \tag{13.6}$$

In CHAPTER VI we will further investigate (13.6) when $a = 1$.

Thus, formally using our permutation scheme we have for $\mathbf{S}_3$:

$$\mathrm{I}_3 := (\theta_1)^2 + (\theta_2)^2, \tag{13.7}$$

$$\mathrm{II}_3 = \theta_1\omega_{31} - \theta_2\omega_{23}, \tag{13.8}$$

$$\mathrm{III}_3 = (\omega_{31})^2 + (\omega_{23})^2, \tag{13.9}$$

$$\mathrm{IV}_3 = -\omega_{31}\theta_2 - \omega_{23}\theta_1, \tag{13.10}$$

$$d\omega_{12} = K_3\theta_1 \wedge \theta_2, \tag{13.11}$$

$$2H_3\theta_1 \wedge \theta_3 = \omega_{31} \wedge \theta_2 + \omega_{23} \wedge \theta_1; \tag{13.12}$$

for $\mathbf{S}_1$:

$$\mathrm{I}_1 = (\theta_2)^2 + (\theta_3)^2, \tag{13.13}$$

$$\mathrm{II}_1 = \theta_2\omega_{12} - \theta_3\omega_{31}, \tag{13.14}$$

---

[1] *Note that in SECTION IV-5 (see item 3^0) the negative quantities were called the tendencies of Γ_3 in the directions of Γ_1 and Γ_2.*

$$\mathrm{III}_1 = (\omega_{12})^2 + (\omega_{31})^2, \tag{13.15}$$

$$\mathrm{IV}_1 = -\omega_{12}\theta_3 - \omega_{31}\theta_2, \tag{13.16}$$

$$d\omega_{23} = K_1\theta_2 \wedge \theta_3, \tag{13.17}$$

$$2H_1\theta_2 \wedge \theta_3 = \omega_{12} \wedge \theta_3 + \omega_{31} \wedge \theta_2; \tag{13.18}$$

and for $\mathbf{S}_2$:

$$\mathrm{I}_2 = (\theta_3)^2 + (\theta_1)^2, \tag{13.19}$$

$$\mathrm{II}_2 = \theta_3\omega_{23} - \theta_1\omega_{12}, \tag{13.20}$$

$$\mathrm{III}_2 = (\omega_{23})^2 + (\omega_{12})^2, \tag{13.21}$$

$$\mathrm{IV}_2 = -\omega_{23}\theta_1 - \omega_{12}\theta_3, \tag{13.22}$$

$$d\omega_{31} = K_2\theta_3 \wedge \theta_1, \tag{13.23}$$

$$2H_2\theta_3 \wedge \theta_1 = \omega_{23} \wedge \theta_1 + \omega_{12} \wedge \theta_3. \tag{13.24}$$

The coefficients of the first basic forms I_a ($a = 1, 2, 3$), i.e. the surface metric tensors, have the respective leg representations:

$$\overset{(3)}{a}_{\alpha\beta} = \lambda_\alpha\lambda_\beta + \mu_\alpha\mu_\beta, \tag{13.25}$$

$$\overset{(1)}{a}_{\alpha\beta} = \mu_\alpha\mu_\beta + \nu_\alpha\nu_\beta, \tag{13.26}$$

$$\overset{(2)}{a}_{\alpha\beta} = \nu_\alpha\nu_\beta + \lambda_\alpha\lambda_\beta; \tag{13.27}$$

and the corresponding Gauss operators are given by

$$x^r_\alpha = \lambda^r\lambda_\alpha + \mu^r\mu_\alpha, \tag{13.28}$$

$$y^r_\alpha = \mu^r\mu_\alpha + \nu^r\nu_\alpha, \tag{13.29}$$

$$z^r_\alpha = \nu^r\nu_\alpha + \lambda^r\lambda_\alpha. \tag{13.30}$$

The orthogonality, or normality, conditions indicated in SECTION II-5 (see item 4^0) then yield

$$\boldsymbol{\Gamma}_3 \perp \mathbf{S}_3 \Leftrightarrow t_1 + t_2 = 0 \Rightarrow t_2 = -t_1, \tag{13.31}$$

$$\boldsymbol{\Gamma}_1 \perp \mathbf{S}_1 \Leftrightarrow t_2 + \varepsilon_3 = 0 \Rightarrow t_2 = -\varepsilon_3, \tag{13.32}$$

$$\boldsymbol{\Gamma}_2 \perp \mathbf{S}_2 \Leftrightarrow \varepsilon_3 + t_1 = 0 \Rightarrow t_1 = -\varepsilon_3. \tag{13.33}$$

In the final part of these assertions we have chosen to eliminate one of the leg coefficients in favor of another leg coefficient. In (13.31) the choice is the obvious one, however in the latter two we have chosen to retain ε_3 since ultimately it will turn out to be the coefficient of primary interest to us.

The general leg representations of the basic tensors appearing in II_a, III_a, and IV_a ($a = 1, 2$) *cannot* be deduced by an inspection of (4.5), (5.2) and (5.12) and using our permutation rule, since these expressions made use of Property (T). However, these can be easily obtained by using the Cartan versions of these forms and employing the general expressions of the connection 1-forms IV-(6.4-.6) with the appropriate specialization (13.3-.5). For example, for $\mathbf{S}_1$ we have

$$\omega_{12} = -\sigma_2\theta_2 - \varepsilon_3\theta_3, \quad \omega_{31} = \varepsilon_3\theta_2 - \gamma_1\theta_3 \tag{13.34}$$

where we have used (13.32) and the 1-form ω_{23} does not occur in (13.14-.16). Then we immediately have

$$\mathrm{II}_1 = -\sigma\,(\theta_2)^2 - \varepsilon_3\,(\theta_2\theta_3 + \theta_3\theta_2) + \gamma_1\,(\theta_3)^2\,, \tag{13.35}$$

$$\begin{aligned} \mathrm{III}_1 \; &= \{\sigma_2^2 + \varepsilon_3^2\}\,(\theta_2)^2 + (\sigma_2 - \gamma_1)\,\varepsilon_3\,(\theta_2\theta_3 + \theta_3\theta_2) \\ &\quad + \{\gamma_1^2 + \varepsilon_3^2\}\,(\theta_3)^2\,, \end{aligned} \tag{13.36}$$

$$\mathrm{IV}_1 = \varepsilon_3\,(\theta_2)^2 - \left\{\frac{\sigma_2 + \gamma_1}{2}\right\}(\theta_2\theta_3 + \theta_3\theta_2) - \varepsilon_3\,(\theta_3)^2 \tag{13.37}$$

and hence

$$\overset{(1)}{b}_{\alpha\beta} = -\sigma_2\mu_\alpha\mu_\beta - \varepsilon_3\,(\mu_\alpha\nu_\beta + \nu_\alpha\mu_\beta) + \gamma_1\nu_\alpha\nu_\beta, \tag{13.38}$$

$$\begin{aligned} \overset{(1)}{c}_{\alpha\beta} \; &= (\sigma_2^2 + \varepsilon_3^2)\,\mu_\alpha\mu_\beta + (\sigma_2 - \gamma_1)\,\varepsilon_3\,(\mu_\alpha\nu_\beta + \nu_\alpha\mu_\beta) \\ &\quad + (\gamma_1^2 + \varepsilon_3^2)\,\nu_\alpha\nu_\beta, \end{aligned} \tag{13.39}$$

$$\overset{(1)}{d}_{\alpha\beta} = \varepsilon_3\mu_\alpha\mu_\beta - \left\{\frac{(\sigma_2 + \gamma_1)}{2}\right\}(\mu_\alpha\nu_\beta + \nu_\alpha\mu_\beta) - \varepsilon_3\nu_\alpha\nu_\beta. \tag{13.40}$$

The Gauss curvature may be determined by

$$K_1 \overset{*}{=} \det \left\| \overset{(1)}{b}_{\alpha\beta} \right\| \overset{*}{=} -\sigma_2\gamma_1 - \varepsilon_3^2, \tag{13.41}$$

or via (13.17), and the Germain curvature H_1 i.e. (13.4) via $\lambda^r{}_r$, or (13.18). As a check of I_1, II_1, III_1, K_1 and H_1, we note that they must satisfy the $a = 1$ version of (2.38), viz.

$$\mathrm{III}_1 - 2H_1\mathrm{II}_1 + K_1\mathrm{I}_1 = 0. \tag{13.42}$$

For purposes of reference we note that

$$K_2 = \mu_2\sigma_1 - \varepsilon_3^2. \tag{13.43}$$

We now come to the definitive interpretation of the leg coefficient ε_3, and in particular, of the condition $\varepsilon_3 = 0$. By SECTION II-5 (item 6^0) and SECTION IV-5 (item 4^0) we know that $\boldsymbol{\Gamma}_a \perp \mathbf{S}_a$. More precisely $\boldsymbol{\Gamma}_a \perp \boldsymbol{\Sigma}_a$, *simultaneously* for $a = 1,2,3$ requires that

$$\boldsymbol{\Gamma}_1 \perp \boldsymbol{\Sigma}_1 \Leftrightarrow \varepsilon_3 + t_2 = 0, \tag{13.44}$$

$$\boldsymbol{\Gamma}_2 \perp \boldsymbol{\Sigma}_2 \Leftrightarrow \varepsilon_3 + t_1 = 0, \tag{13.45}$$

$$\boldsymbol{\Gamma}_3 \perp \boldsymbol{\Sigma}_3 \Leftrightarrow t_1 + t_2 = 0, \tag{13.46}$$

so (13.46) is Property (T). Hence, these three conditions are enough to guarantee that $\varepsilon_3 = 0$. However, from the Ricci calculus $\varepsilon_3 = 0$ is a necessary and sufficient condition for $\{\boldsymbol{\Sigma}_a\}_{a=1}^3$ to be an $\mathcal{O}_3$ only when the congruences $\{\boldsymbol{\Gamma}_a\}_{a=1}^3$ are *canonical*, i.e. $\boldsymbol{\Gamma}_1$ and $\boldsymbol{\Gamma}_2$ are canonical with respect to $\boldsymbol{\Gamma}_3$, and by SECTION IV-5 (see item 5^0) this will be the case only when

$$t_2 = t_1. \tag{13.47}$$

Thus, we must have

$$\varepsilon_3 = 0 \text{ and } t_1 = 0 \text{ and } t_2 = 0 \tag{13.48}$$

which cause the third, fourth and fifth rows in the display (13.1) to be zero. Geometrically, this is the requirement that the geodesic torsions of the congruences defined by the intersections of the families of the surfaces of $\mathcal{O}_3$ to be zero, viz.

$$\boldsymbol{\Sigma}_1 \cap \boldsymbol{\Sigma}_2 := \boldsymbol{\Gamma}_3, \quad \boldsymbol{\Sigma}_3 \cap \boldsymbol{\Sigma}_2 := \boldsymbol{\Gamma}_1, \quad \boldsymbol{\Sigma}_3 \cap \boldsymbol{\Sigma}_1 := \boldsymbol{\Gamma}_2.$$

Thus, (13.48) really insures and is obviously needed to insure, that the curves of $\boldsymbol{\Gamma}_3$, $\boldsymbol{\Gamma}_1$ and $\boldsymbol{\Gamma}_2$ are simultaneously lines of curvature of each pair of the intersections of the families $\boldsymbol{\Sigma}_1$, $\boldsymbol{\Sigma}_2$ and $\boldsymbol{\Sigma}_3$. This is essentially the content of a remarkable result:

Dupin's Theorem

If three surfaces mutually intersect each other orthogonally, then the curves of intersection are lines of curvature on each surface.

Later in SECTION 3 of CHAPTER X we will determine an expression corresponding to (13.48) which arises when one asks that a surface $\mathbf{S} = \mathbf{S}_3$ defined by a given implicit equation $N = 0$, or $N =$ constant, be a member of a $\mathcal{O}_3$.

PROBLEMS FOR CHAPTER V

V.1 Let $x^r(u^1,u^2)$ be a Gaussian representation of a surface $\mathbf{S}$ in $\mathbf{E}_3$ having $\nu^r(u^1,u^2)$ as contravariant components of the unit normal. If z^r is a curvilinear coordinate system in $\mathbf{E}_3$ defined by

$$z^r := x^r\left(u^1,u^2\right) + h\nu^r\left(u^1,u^2\right)$$

where h is a parameter independent of the u^α, and in $\mathbf{E}_3$ we have

$$ds^2 = g_{rs}dz^r dz^s$$

then show that

$$ds^2 = \mathrm{I} - 2h\mathrm{II} + h^2\mathrm{III} + dh^2.$$

Interpret this result (consider the meaning of $h = 0$ and $h \neq 0$) and observe that upon taking

$$z^r = \left(u^1,u^2,h\right)$$

we have

$$\|g_{rs}\| = \left\| \begin{matrix} g_{\alpha\beta} & & \begin{matrix}0\\0\end{matrix} \\ 0 & 0 & 1 \end{matrix} \right\|$$

where

$$(*) \qquad g_{\alpha\beta} := a_{\alpha\beta} - 2hb_{\alpha\beta} + h^2 c_{\alpha\beta}.$$

V.2 Evaluate the expression (*) in PROBLEM V.1 for the leg representations of $a_{\alpha\beta}$, $b_{\alpha\beta}$ and $c_{\alpha\beta}$, when

$$\theta_1 := \lambda_\alpha du^\alpha, \quad \theta_2 := \mu_\alpha du^\alpha, \quad \theta_3 := dh$$

and show that

$$\begin{aligned} ds^2 &= \{1 - 2hk_1 + h^2(k_1^2 + t_1^2)\}(\theta_1)^2 \\ &\quad 4ht_1\{-1 + hH\}\theta_1\theta_2 \\ &\quad + \{1 - 2hk_2 + h^2(k_2^2 + t_1^2)\}(\theta_2)^2 + (\theta_3)^2 . \end{aligned}$$

Evaluate ds^2 when $u^\alpha = (\omega, \phi)$ and

$$x^r(\omega, \phi) = (r_0 \cos\omega \cos\phi, r_0 \sin\omega \cos\phi, r_0 \sin\phi)$$

where r_0 = constant, and show that $k_1 = k_2 = -r_0^{-1}$, $\sigma_1 = r_0^{-1}\tan\phi$ with all other leg coefficients being zero.

Hence

$$ds^2 = \left(1 + \frac{h}{r_0}\right)^2 (\theta_1)^2 + \left(1 + \frac{h}{r_0}\right)^2 (\theta_2)^2 + dh^2,$$

and defining $h := r - r_0$, then $h = 0$ gives $\mathbf{S} = \mathbf{S}_2$ (a 2-sphere of radius r_0), while $h \neq 0$ corresponds to points on $\mathbf{S} \neq \mathbf{S}_2$. Show that in the former case

$$ds^2 = r_0^2 \cos^2\phi d\omega^2 + r_0^2 d\phi^2$$

while in the latter case

$$ds^2 = \left(\frac{r}{r_0}\right)^2 (\theta_1)^2 + \left(\frac{r}{r_0}\right)^2 (\theta_2)^2 + dh^2.$$

V.3 Derive the tensorial expression (4.46) for the Gauss curvature K. Hint: recall the determinantal expression (2.26)!

V.4 Prove (3.25-.29) in general *without* making any specialization on the components of the 2-leg vectors $\boldsymbol{\lambda}$ and $\boldsymbol{\mu}$.

Generalize your argument to include derivations of (3.30-.31).

V.5 In Hotine's treatise, see [pages 43-44], he observed that upon raising the covariant indices of $b_{\alpha\beta}$ by using the contravariant components of the surface metric tensor, i.e. by forming the quantity

$$a^{\alpha\rho}a^{\beta\sigma}b_{\rho\sigma},$$

one obtains a surface tensor. Show that one obtains a surface tensor. Show that this tensor, which we denote by $b^{\alpha\beta}_{\cdot\cdot}$, is *not* the *inverse tensor* of $b_{\alpha\beta}$, i.e. one does not have

$$b^{\alpha\rho}_{\cdot\cdot}\, b_{\beta\rho} = \delta^{\alpha}_{\beta}.$$

However, upon defining

$$\tilde{b}^{\alpha\beta} := \frac{a}{b}\varepsilon^{\alpha\rho}\varepsilon^{\beta\sigma}b_{\rho\sigma}$$

where a and b are the determinants of $\|a_{\alpha\beta}\|$ and $\|b_{\alpha\beta}\|$ respectively, show that we then have

$$\tilde{b}^{\alpha\rho}b_{\beta\rho} = \delta^{\alpha}_{\beta}.$$

[Note, the quantity $\tilde{b}^{\alpha\beta}$ is called the *reciprocal tensor* of $b_{\alpha\beta}$, and is employed by Hotine — without introducing a specific notation for it — in the initial sections of his [Chapter 8]].

V.6 Develop a general theory of reciprocal tensors as defined for the tensor $b_{\alpha\beta}$ in PROBLEM V.5, i.e. for an arbitrary surface tensor $t_{\alpha\beta}$ having a non-zero determinant t, define the *reciprocal tensor* of $t_{\alpha\beta}$ by

$$\tilde{t}^{\alpha\beta} := \frac{1}{t}T^{\alpha\beta}$$

where $T^{\alpha\beta}$ is the cofactor of $t_{\alpha\beta}$. Show that

$$\tilde{t}^{\alpha\rho} t_{\beta\rho} = \delta^{\alpha}_{\beta}$$

and apply your results to $c_{\alpha\beta}$, etc.

ZUND (1988a)

V.7 Show that in general a vector $\mathbf{A}$ in $\mathbf{E}_3$ having contravariant components A^r admits the following representation relative to a given smooth surface $\mathbf{S}$ having unit normal $\boldsymbol{\nu}$:

$$(*) \qquad A^r = A^\alpha x^r_\alpha + \mathcal{A}\nu^r.$$

In $(*)$ the A^α are the *surface components* of $\mathbf{A}$ and $\mathcal{A}$ is a scalar.

V.8 PROBLEM V.7 suggests the following terminology: a vector $\mathbf{A}$ is *general* when both $A^\alpha \neq 0$ and $\mathcal{A} \neq 0$, *special* when only $\mathcal{A} \neq 0$, and *tangential* to $\mathbf{S}$ when only $A^\alpha \neq 0$. Use this terminology to analyze the nature of the components appearing in equations (2.3), (3.1-.5).

Prove that relative to the representation $(*)$ of PROBLEM V.7 one has

$$A_r A^r = A_\alpha A^\alpha + \mathcal{A}^2,$$

and hence $A_r A^r = A_\alpha A^\alpha$ if and only if $\mathcal{A} = 0$.

V.9 In SECTION 3 we have considered the Gauss operator x^r_α relative to a smooth surface $\mathbf{S}$ in $\mathbf{E}_3$. In addition to this operator one can also introduce the projection operator

$$(*) \qquad \pi^r{}_s := \delta^r_s - \nu^r \nu_s$$

relative to $\mathbf{S}$. Show that this operator satisfies the equations

$$\pi^r{}_m \pi^m{}_s = \pi^r{}_s, \quad \pi^r{}_r = 2$$

and that an alternate expression for (∗) is given by

$$\pi^r{}_s = g_{sm} a^{\rho\sigma} x^r_{;\rho} x^m_{;\sigma}.$$

ZUND (1993)

V.10 Show that the operator defined by (∗) in PROBLEM V.9, when applied to a general space vector **A** yields

$$\pi^r{}_m A^m = A^r - (\nu_m A^m)\,\nu^r.$$

Identify this expression, (it is the tangential part of **A**, i.e. $x^r_{;\alpha} A^\alpha$), and deduce that for *any* special space vector one has

$$\pi^r{}_m A^m = 0.$$

[Note that these results together with the equations in PROBLEM V.9 justify our interpretation of $\pi^r{}_s$ as a projection operator.]

ZUND (1993)

VI

Basic Equations of Differential Geodesy

VI.1 Introduction

In this chapter we will apply the leg calculus as developed in CHAPTERS IV and V to the geometry of the geopotential field of the Earth. This will yield the basic equations of differential geodesy. Our discussion is essentially a major reworking of the material contained in Hotine's treatise, i.e. [Chapters 12 and 20], however, we make *no use* of the (ω, ϕ, N) coordinate system which he employed in his analysis. Consequently, our approach is more general and many of his derivations may be recast in a more lucid and less restrictive form, and our results are applicable to any choice of a local coordinate system involved in the local isometric imbedding of the family of equipotential surfaces $\boldsymbol{\Sigma} := \{\mathbf{S}, \mathbf{S}', \mathbf{S}'', \cdots\}$ in $\mathbf{E}_3$ in which $\mathbf{S} := \mathbf{S}_3$, and $\boldsymbol{\Gamma} := \boldsymbol{\Gamma}_3$ is a normal congruence of curves, $\boldsymbol{\Gamma} \perp \boldsymbol{\Sigma}$, having the third leg vector $\boldsymbol{\nu}$ as its unit tangent.

VI.2 Newtonian Gravitation and the Geopotential Field

The starting point of theoretical geodesy is classical Newtonian potential theory. This involves determining the potential function V in various physical situations, i.e. by the *Laplace equation*:

$$\Delta V = 0 \tag{2.1}$$

in empty space, or the *Poisson equation*:

$$\Delta V = 4\pi G\rho \tag{2.2}$$

in a mass distribution of density ρ in space where G denotes the Newtonian gravitation constant. In both equations (2.1) and (2.2), Δ denotes the 3-*dimensional Laplacian operator* with respect to an ambient local coordinate system x^r on a region $\mathcal{R}$, i.e. an open connected subset, of Euclidean 3-space $\mathbf{E}_3$.

These equations express the Newtonian law of gravitation in the exterior and interior of a non-rotating source in $\mathcal{R}$. The problems of determining solutions of these equations are known as the *exterior* and *interior geodetic boundary value problems,* respectively, and much of the activity in theoretical geodesy is devoted to their solution subject to appropriate conditions. The literature on these problems is extensive; however, strictly speaking, they are not the concern of differential geodesy — at least in the Marussi-Hotine approach to the subject. In their work, the potential function V, or more precisely, the geopotential function N (in Hotine's notation which we will employ), is assumed to be *known.* Then the task of differential geodesy is concerned with describing the geometry of curves and surfaces, i.e. plumb lines and equipotential surfaces in the Earth's gravitational field. In practice this is done only for the exterior problem because of the obvious difficulty of specifying a realistic — and generally accepted — candidate for the density function ρ for the interior problem. In the following discussion we will restrict our considerations to the exterior problem.

For the purposes of differential geodesy, since the Earth is rotating, we must consider not V but the geopotential function N which is the sum of V and a centrifugal potential Φ in a local coordinate system rotating about the axis of the Earth with a uniform angular velocity $\tilde{\omega}$. Typically, then, the ambient coordinate system x^r in $\mathcal{R}$ is taken to be a Cartesian system y^r with the third coordinate aligned along the Earth's polar axis. Then (2.1) is replaced by the modified Laplace equation:

$$\Delta N = -2\tilde{\omega}^2 \tag{2.3}$$

in the exterior problem. This equation assumes that the axis of the Earth is invariable, and that $\tilde{\omega}$ is a constant. Neither of these assumptions is precisely true, see LAMBECK (1988), however, both are convenient in setting out a preliminary formulation of theoretical geodesy. The former assumption will always be made in our work, i.e. we neglect all procession effects, but as we will see, the leg calculus readily accommodates the possibility of a variable time dependence of $\tilde{\omega}$ in (2.3) (see PROBLEM VI.1). Hence, although our primary emphasis will be of the classical constant $\tilde{\omega}$ case, we leave open the possibility of a time dependent $\tilde{\omega}$. Usually the exterior problem for N is solved not in the Cartesian coordinate system but in a convenient coordinate system z^r which is related to it by a regular transformation. Typically, this would be some kind of spherical/ellipsoidal based coordinate system and N would involve an expansion in terms of spherical harmonics. The precise nature of this z^r system need not concern us, and the essential feature is that the function N be expressed in some curvilinear coordinate system. For convenience — and to avoid a proliferation of notation — we will henceforth denote z^r by x^r. This agrees with Hotine's usage, and we assume that in the following analysis N is a *known* function in a *known* curvilinear coordinate system x^r. Then, the family of equipotential surfaces Σ, or *N-surfaces* in Hotine's terminology, are given by

$$N\left(x^1, x^2, x^3\right) = c \tag{2.4}$$

where c is a constant whose values enumerate the various surfaces $\mathbf{S}$, $\mathbf{S}'$, $\mathbf{S}''$, ... of Σ. For a given choice of c, (2.4) defines a unique N-surface which we assume is locally isometrically and smoothly imbedded in some region $\mathcal{R}$ of $\mathbf{E}_3$. We imagine that, by varying c, $\mathcal{R}$ is filled by a family Σ which remains external to the topographic surface of the Earth. Such a picture is appealing on both geometric and physical grounds and is one of the basic tenets of potential theory. For the moment we will accept it, or rather avoid it, by restricting our consideration to a single N-surface $\mathbf{S}$. In CHAPTER VII, we will examine the family Σ in more detail when we consider the *Extension Problem* which is concerned with the geometric questions which arise in passing from $\mathbf{S}$ to a neighboring $\mathbf{S}'$.

VI.3 Marussi Ansatz and the Hotine 3-Leg

We now take the first steps in adapting the leg calculus to the geopotential field as defined by (2.5). Essentially this amounts to relating the function N to the vectorial 3-leg $\{\boldsymbol{\lambda}, \boldsymbol{\mu}, \boldsymbol{\nu}\}$ and this is achieved in *two* steps both of which are done simultaneously in the analysis by Marussi and Hotine. We proceed more cautiously and thereby arrive at a more general viewpoint.

The two steps respectively are the *Marussi Ansatz*:

$$(\mathcal{A}) \qquad x^3 := N, \tag{3.1}$$

and the *basic gradient equation*:

$$(\mathcal{B}) \qquad N_r = n\nu_r, \tag{3.2}$$

where N_r denotes the gradient of N with respect to the curvilinear coordinate system x^r; n is the magnitude of N_r, i.e.

$$n^2 := g^{rs} N_r N_s, \tag{3.3}$$

which physically is the *local gravity*; and $\boldsymbol{\nu}$ is the third vector of the 3-leg.

Conditions $(\mathcal{A})$ and $(\mathcal{B})$ are logically independent and have a totally different character: $(\mathcal{A})$ is merely a *definition* of the third local coordinate in the curvilinear system x^r, whereas $(\mathcal{B})$ is a general mathematical result which states that the gradient of an arbitrary function is collinear with the normal to the surface defined by setting the function equal to a constant. Mathematically, the factor n in (3.2) is required to insure that ν_r are the components of a unit vector.

In our work, we will not impose $(\mathcal{A})$ since it evidently imposes *some* restriction on the generality of our otherwise arbitrary curvilinear coordinate system on $\mathcal{R}$ in $\mathbf{E}_3$. Such a restriction, although it may be natural, is not in keeping with our goal of determining and surveying the conceptual framework of differential geodesy; however, it was adopted by Marussi and Hotine in virtually all of their work. Indeed, the Ansatz $(\mathcal{A})$ essentially characterizes the structure of Marussi's intrinsic geodesy and

Hotine's exposition of this theory in his treatise. While $(\mathcal{A})$ is an attractive assumption, it is not mandatory and in our view it unnecessarily restricts the generality of their approach. Our choice of not adopting $(\mathcal{A})$ *ab initio* means that our considerations will be more general than those given in Chapters 12-20 of Hotine (1969), and most of the papers in the monographs MARUSSI (1985) and HOTINE (1991). In particular, our analysis does not preclude $(\mathcal{A})$ but rather indicates the general situation which is only hinted at in their pioneering investigations. Later in CHAPTER X we will return to this issue when we consider the possibilities for choosing geodetic coordinate systems, and this will suggest an approach to differential geodesy which is different from that proposed by Marussi and Hotine.

The condition $(\mathcal{B})$ is not of a restrictive character, but is nevertheless crucial since it involves a specialization of the vectorial 3-leg by adapting the third leg vector $\boldsymbol{\nu}$ to the gradient of the geopotential function N. This move then converts our previous considerations of CHAPTERS II-V from pure differential geometry to differential geodesy. It entails *no change* in the theory given in those chapters, but merely makes a physical choice of SECTION IV-13, to be an equipotential surface of the geopotential field. All that has changed is that the function N is required to be a solution of the modified Laplace equation (2.3), and of course we must investigate the consequences arising from $(\mathcal{B})$, etc.

First we review the geometric setting of differential geodesy. This consists of a smooth surface $\mathbf{S}$ defined by the function N, i.e. (2.4), for a given choice of the constant c and we assume that this surface is locally isometric and smoothly imbedded in $\mathbf{E}_3$. The degree of this smoothness is not specified, but, on the basis of Gaussian differential geometry, for everything to be well-defined the quantities involved should typically be continuously differentiable of class C^2 and class C^3 piecewise. For consistency, the degrees of smoothness of the surface *per se* must be equal to that of the imbedding which defines $\mathbf{S}$ as an object in $\mathbf{E}_3$, since otherwise some of the geometric properties of $\mathbf{S}$ would be lost when the surface is realized in $\mathbf{E}_3$. These conditions are discussed in ZUND (1992), and can be weakened or strengthened in special circumstances. The first two leg vectors are tangents to **S**, i.e.

$$\boldsymbol{\Gamma}_1 := \boldsymbol{\Gamma}(\boldsymbol{\lambda}), \quad \boldsymbol{\Gamma}_2 := \boldsymbol{\Gamma}(\boldsymbol{\mu}) \tag{3.4}$$

are *tangential congruences* to **S**, however these congruences do not lie in **S**. As shown in CHAPTER V, $\boldsymbol{\Gamma}_1$ and $\boldsymbol{\Gamma}_2$ define the *surface congruences* $\boldsymbol{\Gamma}_1'$ and $\boldsymbol{\Gamma}_2'$ on **S**. The resulting 3-leg with the third leg vector $\boldsymbol{\nu}$ being adapted to **S** by (3.2), viz. with

$$\boldsymbol{\Gamma} = \boldsymbol{\Gamma}_3 = \boldsymbol{\Gamma}(\boldsymbol{\nu}) \tag{3.5}$$

being a *normal congruence* to **S**, is called the *Hotine* 3-*leg*. This 3-leg can be written in either vectorial or Pfaffian forms as

$$\{\boldsymbol{\lambda}_a\}_{a=1}^{3} := \{\boldsymbol{\lambda}, \boldsymbol{\mu}, \boldsymbol{\nu}\}, \tag{3.6}$$

or

$$\{\theta_1\}_{a=1}^{3} := \{\theta_1, \theta_2, \theta_3\}; \tag{3.7}$$

where

$$\lambda := \lambda^r \frac{\partial}{\partial x^r}, \quad \mu := \mu^r \frac{\partial}{\partial x^r}, \quad \nu := \nu^r \frac{\partial}{\partial x^r}; \tag{3.8}$$

or

$$\theta_1 := \lambda_r dx^r, \quad \theta_2 := \mu_r dx^r, \quad \theta_3 := \nu_r dx^r; \tag{3.9}$$

and by (3.2)

$$\nu_r := \frac{1}{n} N_r, \quad \nu^r := \frac{1}{n} g^{rs} N_s \tag{3.10}$$

where $n \neq 0$ on **S** and in $\mathcal{R}$. The corresponding Hotine 2-leg on **S** is defined by

$$\{\lambda_A\}_{A=1}^{2} := \{\boldsymbol{\lambda}, \boldsymbol{\mu}\}, \tag{3.11}$$

$$\{\theta_A\}_{A=1}^{2} := \{\theta_1, \theta_2\}, \tag{3.12}$$

where

$$\boldsymbol{\lambda} := \lambda^\alpha \frac{\partial}{\partial u^\alpha}, \quad \boldsymbol{\mu} := \mu^\alpha \frac{\partial}{\partial u^\alpha}, \tag{3.13}$$

and

$$\begin{aligned} \theta_1 &:= \lambda_\alpha du^\alpha = \lambda_r dx^r, \\ \theta_2 &:= \mu_\alpha du^\alpha = \mu_r dx^r; \end{aligned} \tag{3.14}$$

where $u^\alpha := (u^1, u^2)$ are the local Gaussian parameters (or coordinates) on a parametrized coordinate neighborhood $\mathcal{U} := \mathbf{S} \cap \mathcal{R}$ of **S**.

The N-surface **S** is characterized by the condition

$$\theta_3 = 0 \tag{3.15}$$

and the Pfaffian version of $(\mathcal{B})$ is given by

$$dN = n\theta_3 \tag{3.16}$$

in $\mathcal{R}$ and on $\mathcal{U}$ of **S** we have

$$dN = 0 \tag{3.17}$$

by (2.4). Note that in (3.16) the local gravity n, which previously occurred as a *normalization factor* to insure that the ν_r defined by $(\mathcal{B})$ is a unit vector, is now a *multiplicative factor* — actually an *integrating factor* in the language of differential equations — which is necessary for the left hand side to be a *total/perfect differential.*

VI.4 Hotine Leg Calculus and N-Integrability Conditions

We now consider the first consequences of the basic gradient equation $(\mathcal{B})$, i.e. (3.2). In terms of the leg calculus $(\mathcal{B})$ admits two equivalent expressions: a vectorial representation

$$N_r = N_{/1}\lambda_r + N_{/2}\mu_r + N_{/3}\nu_r \tag{4.1}$$

and a Pfaffian representation

$$dN = N_{/1}\theta_1 + N_{/2}\theta_2 + N_{/3}\theta_3 \tag{4.2}$$

where the slash "/" denotes a leg derivative. By (3.2) and (3.16) we immediately see that these representations are equivalent to the following set of leg differential equations:

$$(\mathcal{B})\qquad \begin{aligned} N_{/1} &= 0, && (4.3)\\ N_{/2} &= 0 && (4.4)\\ N_{/3} &= n. && (4.5) \end{aligned}$$

This set of differential equations is consistent only when certain compatibility conditions called the *N-integrability conditions* are identically satisfied, i.e. they are identities which the quantities appearing in them must satisfy.

In order to obtain these conditions we first solve (3.16) for

$$\theta_3 = n^{-1} dN. \tag{4.6}$$

Then by exterior differentiation we have

$$d\theta_3 = -n^2 dn \wedge dN + n^{-1} d(dN),$$

and since $d(dN) = d^2N = 0$, this reduces to

$$d\theta_3 = -n^{-2} dn \wedge dN. \tag{4.7}$$

The differential of the local gravity n admits a Pfaffian representation which is analogous to that exhibited in (4.2), and upon substitution of this expression on the right-hand side of (4.7) and simplifying we obtain

$$\begin{aligned} -n^2 d\theta_3 \;&= \left(n_{/1}N_{/2} - n_{/2}N_{/1}\right)\theta_1 \wedge \theta_2 \\ &\quad + \left(n_{/3}N_1 - n_1 N_{/3}\right)\theta_3 \wedge \theta_1 \\ &\quad + \left(n_{/2}N_{/3} - n_{/3}N_{/2}\right)\theta_2 \wedge \theta_3. \end{aligned} \tag{4.8}$$

The left-hand side of this equation has previously been given in IV-(6.9) *without* imposition of property (T), and it is

$$d\theta_3 = -\left(t_1 + t_2\right)\theta_1 \wedge \theta_2 + \gamma_1 \theta_3 \wedge \theta_1 - \gamma_2 \theta_2 \wedge \theta_3.$$

We will use this expression since as we will see it leads to another proof of property (T) which is to be expected since $(\mathcal{B})$ is obviously a statement of the fact $\boldsymbol{\Gamma} \perp \mathbf{S}$. Hence, employing

IV-(6.9) in (4.8) and equating the respective coefficients of the linearly independent exterior products, we have

$$\begin{aligned} n_{/1}N_{/2} - n_{/2}N_{/1} &= n^2\,(t_1 + t_2) \\ n_{/1}N_{/3} - n_{/3}N_{/1} &= n^2\gamma_1 \\ n_{/2}N_{/3} - n_{/3}N_{/2} &= n^2\gamma_2. \end{aligned}$$

Since $n \neq 0$, the set $\{\mathcal{B}\}$ yields the set of N-integrability conditions which we collectively denote by $\{N\}$ and individually by $(N_{\mathcal{A}})$ $\mathcal{A}$ = I, II, III (recall the notation employed in II-(7.15) and SECTION III-7):

$$(\mathrm{N_I}) \quad t_1 + t_2 = 0, \tag{4.9}$$

$$\{N\}: \qquad (\mathrm{N_{II}}) \quad n_{/1} = n\gamma_1, \tag{4.10}$$

$$(\mathrm{N_{III}}) \quad n_{/2} = n\gamma_2. \tag{4.11}$$

Thus $(\mathrm{N_I})$ is precisely property (T) which we henceforth will assume in the remainder of our work, while $(\mathrm{N_{II}})$ and $(\mathrm{N_{III}})$ are *new* equations which couple the leg coefficients γ_1, γ_2 with the leg derivatives of the local gravity. Equations $(\mathrm{N_{II}})$ and $(\mathrm{N_{III}})$, but not $(\mathrm{N_I})$, were derived by Hotine in a different manner in his treatise. His derivation will be given in SECTION 5.

As noted previously in CHAPTERS IV and V, see IV-(5.1) and V-(9.1), the principal curvature χ of the normal congruence $\boldsymbol{\Gamma}$ is given by

$$\chi^2 = \gamma_1^2 + \gamma_2^2.$$

Hence, in terms of the local gravity by $(\mathrm{N_{II}})$ and $(\mathrm{N_{III}})$ we have the physical relation

$$\chi^2 = \left((\log n)_{/1}\right)^2 + \left((\log n)_{/2}\right)^2. \tag{4.12}$$

Finally, we conclude with a few important — although obvious — comments about the importance of $(\mathcal{B})$ in differential geodesy. Up to now we have essentially regarded $(\mathcal{B})$ as purely a mathematical result. This is indeed the case, but the choice of the geopotential N in $(\mathcal{B})$ — which holds for *any* smooth function of the local coordinates x^r on $\mathcal{R}$ — converts it into a physical equation. The left-hand side of $(\mathcal{B})$ is the *local gravity vector*, while the right-hand side identifies this vector with the tangents of the normal congruence $\boldsymbol{\Gamma}$. Thus, in our case $(\mathcal{B})$ involves a *geometrization* of the geopotential gravity field. We have called $(\mathcal{B})$ the basic gradient equation, however, in a sense it is the *fundamental equation* of *differential geodesy* for without it we have no geometric picture of the gravity field. Likewise, since $(\mathcal{B})$ is a *vector equation* which is valid in all local curvilinear coordinate systems on $\mathcal{R}$, it is not surprising that it is equivalent to three scalar equations, viz. (4.9)-(.411) or simply $\{N\}$, and that these lead to important geometric properties of $\boldsymbol{\Gamma}$, i.e. property (T) and (4.12).

VI.5 Marussi Tensor

The tensor of gravity gradients, or the *Marussi tensor* as Hotine called it in his writings, was introduced in MARUSSI (1947). Although the second partial derivatives of the geopotential had been previously considered in geodesy, e.g. by BRUNS (1878), it was left to Marussi to recognize that the set of these derivatives could be regarded as the covariant components of a second order symmetric tensor in an arbitrary curvilinear coordinate system x^r in $\mathbf{E}_3$. With characteristic modesty he always called it the *Eötvös tensor*; however, due to Hotine's work it is now more commonly known as the Marussi tensor. Marussi's derivation of it was based on the homographic calculus, and much of his analysis was specialized to the Ansatz $(\mathcal{A})$ and his local astronomical coordinate system which was essentially identical with Hotine's (ω, ϕ, N) system. Subsequently, in HOTINE (1957a) (see paper #3 of HOTINE (1991)) and in Chapter 12 of the treatise, Hotine gave a purely tensor-theoretic discussion. This approach — as far as it went — was general but almost immediately tied to the (ω, ϕ, N) system. In this section we will show how the leg calculus permits a completely general coordinate-independent derivation of the Marussi tensor which reveals its principal differential-geodetic properties. Later in CHAPTER VIII we will present a general algebraic theory of this tensor.

The Marussi tensor is the primary structural feature of differential geodesy, and a basic task of the subject is to develop its theory in a manner which is unobscured by the choice of a particular local coordinate system. Before proceeding to the derivation of this tensor let us briefly indicate why we may expect that it will be of interest to us. First, it is obtained by covariant differentiation of ($\mathcal{B}$), and hence the left-hand side involves the covariant gradient of the gravity vector. Now following Hotine, we have written N_r for the partial derivative $N_{;r}$, which is identical with the covariant derivative $N_{,r}$ because N is a scalar function, i.e.,

$$N_r := N_{;r} = N_{,r}\,. \tag{5.1}$$

Thus, upon computing the covariant derivative, we have

$$N_{r,s} := N_{,rs} := N_{;rs} - \Gamma^p_{rs} N_{,p} \tag{5.2}$$

which is symmetric in the indices r and s, and is a tensorial equation. Covariant differentiation of the right hand side of ($\mathcal{B}$) then yields a term involving $n_s := n_{,s}$ and also a term $\nu_{rs} := \nu_{r,s}$. The latter is especially important, since by SECTION V-6, we know that it is one of the basic equations of the leg calculus and as such contains a set of leg coefficients which occur in our congruence and surface eigenstructure theorems.

We now proceed with the formal derivation which is essentially identical to that given by Hotine in his treatise [page 73]. Covariant differentiation of ($\mathcal{B}$) clearly yields

$$N_{r,s} := N_{rs} = \nu_r n_s + n\nu_{rs} \tag{5.3}$$

and, as noted above, we have

$$N_{rs} = N_{sr}\,. \tag{5.4}$$

Thus, by (5.3) we must have

$$\nu_r n_s + n\nu_{rs} = \nu_s n_r + n\nu_{sr}\,. \tag{5.5}$$

Contraction of this equation with ν^s ($\nu_{sr}\nu^s = 0$ since $(\nu_s\nu^s)_r = 0$) gives

$$\nu_r n_s \nu^s + n\nu_{rs}\nu^s = n_r\,. \tag{5.6}$$

Upon solving for $n\nu_{rs}\nu^s$ we obtain

$$n\nu_{rs}\nu^s = \left(n_r - n_{/3}\nu_r\right),$$

where $n_s\nu^s := n_{/3}$, and by recalling the leg representation of n_r, we get

$$n\nu_{rs}\nu^s = \left(n_{/1}\lambda_r + n_{/2}\mu_r\right). \tag{5.7}$$

Choosing $n > 0$, which is consistent with the choice of the sign of N and $\boldsymbol{\nu}$ being a zenith-pointing normal of the N-surface $\mathbf{S}$, we have

$$\nu_{rs}\nu^s = (\log n)_{/1}\,\lambda_r + (\log n)_{/2}\,\mu_r \tag{5.8}$$

which is Hotine's [12.020]. Since $\boldsymbol{\nu}$ is the unit tangent of the normal congruence $\boldsymbol{\Gamma}$, Hotine then *defined*

$$\begin{aligned} \gamma_1 &:= (\log n)_{/1} \\ \gamma_2 &:= (\log n)_{/2} \end{aligned} \tag{5.9}$$

where γ_1, γ_2 are the respective curvatures of $\boldsymbol{\Gamma}$ in the tangential $\boldsymbol{\lambda}$, $\boldsymbol{\mu}$ directions. But as we have seen in the previous section, the two equations given in (5.9) are precisely (N_{II}) and (N_{III}) respectively. Hotine's procedure was correct, but unnecessary since these identifications are consequences of the N-integrability conditions and the specializations need not be taken as definitions. However, such an approach was necessary for him since he did not investigate the integrability conditions of N, or apparently have the leg representation V-(6.3) at his disposal. Contraction of V-(6.3) by ν^s and using (N_{II}) and (N_{III}) also immediately yields (5.8).

Because a leg representation of ν_{rs} is available to us from CHAPTER V, we may directly write out a leg representation of N_{rs}. Equation V-(6.3) employs property (T), but it is instructive to see what happens if we use the more general expression IV-(6.3), i.e.

$$\begin{aligned} \nu_{rs} &= -k_1\lambda_r\lambda_s + t_2\lambda_r\mu_s + \gamma_1\lambda_r\nu_s \\ &\quad -t_1\mu_r\lambda_s - k_2\mu_r\mu_s + \gamma_2\mu_r\nu_s. \end{aligned}$$

Then we have the following representation

$$\begin{aligned} N_{rs} &= -nk_1\lambda_r\lambda_s + nt_2\lambda_r\mu_s + n\gamma_1\lambda_r\nu_s \\ &\quad -nt_1\mu_r\lambda_s - nk_2\mu_r\mu_s + n\gamma_2\mu_r\nu_s \\ &\quad +n_{/1}\nu_r\lambda_s + n_{/2}\nu_r\mu_s + n_{/3}\nu_r\nu_s. \end{aligned} \tag{5.10}$$

Curiously, this result with $t_2 = -t_1$ was given in equation (18.1) of HOTINE (1957a) in a slightly different notation with an identification of $n_{/3}$ which we will obtain in SECTION 7. It was not repeated in his treatise, and this seems to have been a remarkable oversight on his part. Returning to (5.10) we now observe that, by the symmetry (5.4),

$$\begin{aligned} \lambda_r\mu_s \text{ term } &: nt_2 = -nt_1 \Rightarrow t_2 = -t_1 \\ \lambda_r\nu_s \text{ term } &: n\gamma_1 = n_{/1} \Rightarrow n_{/1} = n\gamma_1 \\ \mu_r\nu_s \text{ term } &: n\gamma_2 = n_{/2} \Rightarrow n_{/2} = n\gamma_2 \end{aligned}$$

which we recognize as property (T), ($\mathrm{N_{II}}$), and ($\mathrm{N_{III}}$)! At first sight this appears to be a Marussi tensor derivation of these conditions, but this is illusory. The Marussi tensor was obtained by covariant differentiation of ($\mathcal{B}$), and, *a priori* in, order for ($\mathcal{B}$) to be meaningful its integrability conditions must be satisfied; these are identical with the above 'derived' conditions. Nevertheless, it is nice that the above symmetry argument *reproduces* the conditions $\{N\} = \{\mathrm{N_I}, \mathrm{N_{II}}, \mathrm{N_{III}}\}$ in a consistent manner, and this probably explains to some extent why Hotine did not explicitly consider the N-integrability conditions; he had already trivialized them by his geometric proof of property (T) (recall our discussion in SECTION V-4) and his '*definitions*' (5.9).

Hence, the canonical leg representation for N_{rs} becomes

$$\begin{aligned} N_{rs} &= -nk_1\lambda_r\lambda_s - nt_1\lambda_r\mu_s + n\gamma_1\lambda_r\nu_s \\ &\quad -nt_1\mu_r\lambda_s - nk_2\mu_r\mu_s + n\gamma_1\mu_r\nu_s \\ &\quad +n\gamma_1\nu_r\lambda_s + n\gamma_2\nu_r\mu_s + n_{/3}\nu_r\nu_s \end{aligned} \tag{5.11}$$

which is clearly symmetric as required. Note that by virtue of (5.3) and the symmetry (5.4), or more simply by (5.11), we have

$$N_{rs}\nu^r = n_s, \tag{5.12}$$

and

$$N_{rs}\nu^s = n_r. \tag{5.13}$$

In his treatise, see [12.162] on [page 86], Hotine introduced five leg coefficients which he called *curvature parameters*

$$k_1, k_2, t_1, \gamma_1, \gamma_2$$

via

$$N_{rs}\lambda^r\lambda^s := -nk_1, \tag{5.14}$$

$$N_{rs}\mu^r\mu^s := -nk_2, \tag{5.15}$$

$$N_{rs}\lambda^r\mu^s = N_{rs}\mu^r\lambda^s := -nt_1, \tag{5.16}$$

$$N_{rs}\lambda^r\nu^s = N_{rs}\nu^r\lambda^s := n\gamma_1, \tag{5.17}$$

$$N_{rs}\mu^r\nu^s = N_{rs}\nu^r\mu^s := n\gamma_2, \tag{5.18}$$

$$N_{rs}\nu^r\nu^s := n_{/3}. \tag{5.19}$$

However, (5.17) and (5.18), were not new and had previously been defined in his [12.020], (see our discussion of (5.9)) or [12.028] (which contains a misprint!). These equations are formally equivalent to (5.11) and having these in hand, he could easily have written out the leg representation of N_{rs} had he chose to do so. Hotine's analysis was tied to his (ω, ϕ, N) system by $(\mathcal{A})$ and the gradients of the (ω, ϕ) Gaussian parameters (or coordinates) and $\log n$ were given by his [12.046-12.048] [page 75], viz.

$$\omega_r = \sec\phi\{-k_1\lambda_r - t_1\mu_r + \gamma_1\nu_r\},$$

$$\phi_r = -t_1\lambda_r - k_2\mu_r + \gamma_2\nu_r,$$

$$(\log n)_r = \gamma_1\lambda_r + \gamma_2\mu_r + (\log n)_{/3}\nu_r,$$

respectively; or by his expressions in [12.037] for the contravariant components of $\boldsymbol{\lambda}$, $\boldsymbol{\mu}$, $\boldsymbol{\nu}$ involving (ω, ϕ) and n [page 74].

On the other hand, our derivation of (5.11) is exclusively based upon covariant differentiation of $(\mathcal{B})$, the N-integrability conditions $\{N\} = \{\mathrm{N_I}, \mathrm{N_{II}}, \mathrm{N_{III}}\}$, and the leg representation V-(6.3) of ν_{rs}. This did not involve $(\mathcal{A})$ and is completely coordinate-independent. Indeed, since N_{rs} are the covariant components of a symmetric second order tensor **N**, by using the abstract notion of a tensor as a symmetric bilinear functional (as discussed in SECTION I-4), and allowing this functional to operate on pairs of vectors of the Hotine 3-leg we have:

$$\mathbf{N}(\boldsymbol{\lambda},\boldsymbol{\lambda}) = -nk_1, \tag{5.20}$$

$$\mathbf{N}(\boldsymbol{\mu},\boldsymbol{\mu}) = -nk_2, \tag{5.21}$$

$$\mathbf{N}(\boldsymbol{\lambda},\boldsymbol{\mu}) = \mathbf{N}(\boldsymbol{\mu},\boldsymbol{\lambda}) = -nt_1, \tag{5.22}$$

$$\mathbf{N}(\boldsymbol{\lambda},\boldsymbol{\nu}) = \mathbf{N}(\boldsymbol{\nu},\boldsymbol{\lambda}) = n\gamma_1, \tag{5.23}$$

$$\mathbf{N}(\boldsymbol{\mu},\boldsymbol{\nu}) = \mathbf{N}(\boldsymbol{\nu},\boldsymbol{\mu}) = n\gamma_2, \tag{5.24}$$

$$\mathbf{N}(\boldsymbol{\nu},\boldsymbol{\nu}) = n\gamma_3. \tag{5.25}$$

These equations may be regarded as the abstract versions of (5.14)-(5.19), and strictly speaking they *are not definitions* of the quantities appearing on their right-hand sides, since these quantities have already been defined by the leg calculus and ($\mathcal{B}$). What these equations *do define* are the leg components of the abstract Marussi tensor **N** in the Hotine 3-leg $\{\boldsymbol{\lambda},\boldsymbol{\mu},\boldsymbol{\nu}\}$. These are equivalent to the covariant leg representation of N_{rs} given in (5.11), as well as the contravariant leg representation of **N** which we have not exhibited. The latter is obtained by raising the covariant indices by the usual rule:

$$N^{rs} := g^{rp}g^{sq}N_{pq}, \tag{5.26}$$

but is not ordinarily useful in practice since it involves 'contravariant' derivatives of N whereas covariant derivatives are the commonly occurring derivatives.

VI.6 Bruns Curvature Equation

By virtue of ($\mathcal{B}$) the third vector $\boldsymbol{\nu}$ in the Hotine vectorial 3-leg $\{\boldsymbol{\lambda}_a\}_{a=1}^{3}$ has a privileged rôle in differential geodesy, viz. it is the tangent vector to the normal congruence $\boldsymbol{\Gamma}$ of **S**, $\boldsymbol{\Gamma} \perp$ **S**. A first analytic indication of this is furnished by equation (4.12), and is even more strikingly shown by the *Bruns curvature equation*:

$$\nu^r{}_r = -2H \tag{6.1}$$

where H is the Germain (mean) curvature of **S**. We now proceed to derive this equation.

It is rather remarkable that a result of such elegance is seldom cited in differential geometry texts. Hotine did recognize this equation and made extensive use of it in his treatise. Unfortunately, his proof of it (see [page 41] and [7.19]) is misleading since he cited the wrong equation numbers ! A correct derivation makes use of his [6.19] (see our discussion in SECTION V-7) which is our V-(7.4), viz.

$$b_{\alpha\beta} = -\nu_{rs} x^r_\alpha x^s_\beta. \tag{6.2}$$

Then by contraction with $a^{\alpha\beta}$ and using the definition of H, V-(2.26), we have

$$2H = -\nu_{rs} a^{\alpha\beta} x^r_\alpha x^s_\beta. \tag{6.3}$$

The right-hand side of this expression may be readily evaluated by using the equation

$$a^{\alpha\beta} x^r_\alpha x^s_\beta = g^{rs} - \nu^r \nu^s \tag{6.4}$$

which may be obtained by using the leg representations for $a^{\alpha\beta}$ and x^r_α, i.e. IV-(3.8) and IV-(3.15). Thus, we get (6.1), or a simpler proof may be given by contracting the leg representation of ν_{rs}, i.e. V-(6.3), with g^{rs}.

Equation (6.1) is an intriguing result which prompts one to wonder whether it is possible to also determine an expression for the Gauss (total) curvature K in terms of covariant derivatives of $\boldsymbol{\nu}$. This turns out to be the case, and in PROBLEM VI.4 we will ask the reader to derive such an expression, however, it lacks the elegance and simplicity of (6.1). Thus, in anticipation of this result it is more proper to call (6.1) the *first Bruns curvature equation.*

VI.7 Bruns Equation and Hotine's Theorem

The Marussi tensor N_{rs} as exhibited in (5.11) is of obvious physical interest since (when the leg coefficients k_1, k_2, t_1, γ_1, γ_2, the local gravity n, and the Hotine vectorial 3-leg are known for a given N-surface **S**) it represents the gravity gradients of the geopotential field. However, a more immediate consequence is that since $N_{rs} = N_{,rs}$ contraction of it by g^{rs} yields the 3-dimensional Laplacian of N, viz.

$$\Delta N := g^{rs} N_{rs}. \tag{7.1}$$

Then by the modified Laplace equation (2.3) and computing the indicated contraction using (5.11) we have

$$-2\tilde{\omega}^2 = -2Hn + n_{/3}. \tag{7.2}$$

Following Hotine's usage, see [page 80 or page 148] we will call this result the *Bruns equation.* This equation is noteworthy since it furnishes not only an important relation between the curvature of the N-surface **S** and the local gravity n, but also an expression for the quantity $n_{/3}$:

$$\text{(B)} \qquad n_{/3} = 2\left(Hn - \tilde{\omega}^2\right). \tag{7.3}$$

This is important since although the tangential leg derivatives $n_{/1}$ and $n_{/2}$ have been specified by ($\mathrm{N_{II}}$) and ($\mathrm{N_{III}}$) respectively in SECTION 4, the normal leg derivative $n_{/3}$ requires (2.3) for its specification. Thus, while the former derivatives of n are consequences of ($\mathcal{B}$) and its integrability conditions, the latter derivative is specified by an additional physical requirement, i.e. the modified Laplace equation (2.3)! Indeed, this latter dependence can be made more explicit by proving

<u>Hotine's Theorem</u>

If N satisfies the basic gradient equation ($\mathcal{B}$), then

$$\Delta N = -2\tilde{\omega}^2 \Leftrightarrow n_{/3} - 2Hn = -2\tilde{\omega}^2,$$

viz. the modified Laplace equation is 'equivalent' to the Bruns equation.

The proof is immediate:

$$\begin{aligned} -2\tilde{\omega}^2 &= \Delta N = g^{rs} N_{rs} = N^r{}_r = (n\nu^r)_r \\ &= n_r \nu^r + n\nu^r{}_r = n_{/3} - 2Hn \end{aligned}$$

since these steps are reversible.

This argument was given by Hotine in his treatise (see [page 148] of Chapter 20 and in [20.26] in particular). As he noted, the asserted 'equivalence' requires some care, since $\Delta N = -2\tilde{\omega}^2$ determines N in $\mathcal{R}$ *except* on the N-surface **S**: $N(x^1, x^2, x^3) = c$ while $n_{/3} - 2Hn = -2\tilde{\omega}^2$ is defined on the family of normal congruences of **S**. However, by assumption, the $\boldsymbol{\Gamma}$ are external to **S** and fill the region $\mathcal{R}$ of $\mathbf{E}_3$ so the domains of these equations are 'equivalent' only in this sense. Note that neither of these equations defines the local gravity n on **S**, (see Hotine's discussion on [page 148]).

Finally, by (4.5), (7.2) may be rewritten as an N-equation, i.e.

$$N_{/3/3} - 2HN_{/3} = -2\tilde{\omega}^2, \tag{7.4}$$

where $N_{/3} := N_r \nu^r$. This suggests that if $\boldsymbol{\nu}$ can be aligned in a direction in such a manner that ν^r has only one non-zero component the solution N of (7.4) will depend on a single coordinate. This occurs in the case of spherical polar coordinates $x^r = (\omega, \phi, r)$ (see our example in SECTION IV-10) for which

$$\nu^r = (0, 0, 1)\,. \tag{7.5}$$

Then $\boldsymbol{\nu} = \dfrac{\partial}{\partial r}$ so

$$N_{/3} = \frac{\partial N}{\partial r} = n \tag{7.6}$$

and taking **S** to be a sphere of variable radius r we have

$$H = -\frac{1}{2r}; \tag{7.7}$$

(7.4) reduces to

$$\frac{\partial n}{\partial r}+\frac{2n}{r}=-2\tilde{\omega}^2. \tag{7.8}$$

As Hotine observed [page 149], equation (7.8) may be more conveniently written in the form

$$\frac{\partial}{\partial r}\left(nr^2\right)=-2\tilde{\omega}^2r^2 \tag{7.9}$$

and when $\tilde{\omega}$ = constant, this may be immediately integrated to yield

$$nr^2=-\frac{2}{3}\tilde{\omega}^2r^3+f\left(\omega,\phi\right) \tag{7.10}$$

where f is an arbitrary function of (ω,ϕ) introduced by the integration. By (7.6), we see that (7.10) leads to the following N-equation

$$\frac{\partial N}{\partial r}=-\frac{2}{3}\tilde{\omega}^2r+\frac{f\left(\omega,\phi\right)}{r^2} \tag{7.11}$$

and a further integration gives

$$N=-\frac{1}{3}\tilde{\omega}^2r^2-\frac{f\left(\omega,\phi\right)}{r}+g\left(\omega,\phi\right) \tag{7.12}$$

where again g is an arbitrary function of (ω,ϕ). But for each value of r, N must be a constant on the corresponding sphere, so the functions f and g can at most be constants, say c_1 and c_2. Thus, we have

$$N=-\frac{1}{3}\tilde{\omega}^2r^2-\frac{c_1}{r}+c_2$$

as a very special example of a solution of (7.4). This example is purely of an illustrative nature and there is no suggestion that otherwise it is of any significance. Clearly (7.4) for a tangent vector $\boldsymbol{\nu}$ whose contravariant components ν^r have several non-zero values will be quite complicated and the integration will be difficult if not impossible. We will return to this question in CHAPTER X when an alternate approach to differential geodesy more general than that envisioned by Marussi and Hotine will be proposed.

None of these difficulties occur in the Marussi-Hotine formulation of differential geodesy since a solution of the modified Laplace equation, or equivalently the Bruns equation (by Hotine's Theorem), is assumed to be *known*. In this situation, both (2.3) and (7.2) are identically satisfied by the *given* geopotential function N and the local gravity n. Then (7.3) is a prescription for $n_{/3}$, and this equation plays the rôle of a consistency, or compatibility, equation for n, $n_{/3}$, H (i.e. k_1 and k_2), and the constant $\tilde{\omega}$. In their approach, for a given N the question is then to exhibit the corresponding Hotine 3-leg $\{\boldsymbol{\lambda}, \boldsymbol{\mu}, \boldsymbol{\nu}\}$ and hence determine n and the leg coefficients, i.e. to describe the geometry of the geopotential field.

VI.8 n-Integrability Conditions

By virtue of (7.3), we now have a complete specification of the leg derivatives of n: viz.

$$(\mathrm{N_{II}}) \quad n_{/1} = n\gamma_1, \tag{4.10}$$

$$\{\mathcal{B}'\} \qquad (\mathrm{N_{III}}) \quad n_{/2} = n\gamma_2, \tag{4.11}$$

$$(\mathrm{B}) \quad n_{/3} = 2\left(Hn - \tilde{\omega}^2\right). \tag{7.3}$$

These are a set of leg differential equations and as shown they are meaningful only when their integrability conditions — the *n-integrability conditions* — are identically satisfied. The set $\{\mathcal{B}'\}$ permits us to express the differential of n in Pfaffian form, i.e.

$$dn = n\gamma_1\theta_1 + n\gamma_2\theta_2 + 2\left(Hn - \tilde{\omega}^2\right)\theta_3, \tag{8.1}$$

and the integrability conditions are obtained by exterior differentiation of this expression. Since $d(dn) = d^2n = 0$, this gives

$$\begin{aligned} 0 &= d(n\gamma_1) \wedge \theta_1 + n\gamma_1 d\theta_1 \\ &\quad + d(n\gamma_2) \wedge \theta_2 + n\gamma_2 d\theta_2 \\ &\quad + 2d(Hn - \tilde{\omega}^2) \wedge \theta_3 + 2(Hn - \tilde{\omega}^2)\, d\theta_3. \end{aligned} \tag{8.2}$$

Evaluation of (8.2) requires the product rule:

$$\begin{aligned} d(fg) &= (df)\,g + f\,(dg) \\ &= \left(f_{/1}\theta_1 + f_{/2}\theta_2 + f_{/3}\theta_3\right) g \\ &\quad + f\left(g_{/1}\theta_1 + g_{/2}\theta_2 + g_{/3}\theta_3\right), \end{aligned} \tag{8.3}$$

and use of the expressions for $d\theta_a$ $(a = 1,2,3)$ of SECTION IV-8 when property (T) is in force, i.e.

$$d\theta_1 = \sigma_1\theta_1 \wedge \theta_2 - k_2\theta_3 \wedge \theta_1 + (t_1 - \varepsilon_3)\,\theta_2 \wedge \theta_3, \tag{8.4}$$

$$d\theta_2 = \sigma_2\theta_1 \wedge \theta_2 - (t_1 + \varepsilon_3)\,\theta_3 \wedge \theta_1 + k_2\theta_2 \wedge \theta_3, \tag{8.5}$$

$$d\theta_3 = \gamma_1\theta_3 \wedge \theta_1 - \gamma_2\theta_2 \wedge \theta_3. \tag{8.6}$$

One could replace the γ_1 and γ_2 in (8.6) by the expressions (5.9) involving the leg derivatives of $\log n$, but for our immediate purposes nothing is gained by doing this. Then, upon making these substitutions and equating the coefficients of the resulting exterior products $\theta_1 \wedge \theta_2$, $\theta_3 \wedge \theta_1$ and $\theta_2 \wedge \theta_3$ to zero we have the following n-integrability conditions which we refer to as $\{n\}$:

$$(n_{\mathrm{I}}) \quad \gamma_{1/2} - \gamma_{2/1} = \sigma_1\gamma_1 + \sigma_2\gamma_2, \tag{8.7}$$

$$(n_{\mathrm{II}}) \quad \gamma_{1/3} - 2H_{/1} = \left(4n^{-1}\tilde{\omega}^2 - k_2\right)\gamma_1 + (t_1 + \varepsilon_3)\,\gamma_2, \tag{8.8}$$

$$(n_{\mathrm{III}}) \quad \gamma_{2/3} - 2H_{/2} = (t_1 - \varepsilon_3)\,\gamma_1 + \left(4n^{-1}\tilde{\omega}^2 - k_1\right)\gamma_2. \tag{8.9}$$

As in the case of $\{N\} = \{N_{\mathrm{I}}, N_{\mathrm{II}}, N_{\mathrm{III}}\}$ in his treatise, Hotine did not specifically consider $\{n\} = \{n_{\mathrm{I}}, n_{\mathrm{II}}, n_{\mathrm{III}}\}$ in a systematic manner. However, on [page 148], (n_{II}) and (n_{III}) occur as [20.24] and [20.25] in his (ω, ϕ, N)-system for a constant $\tilde{\omega}$. Condition (n_{I}) does not appear in his analysis.

The condition (n_{I}) is an identity, viz. the Schouten identity $(\mathcal{S}_{\mathrm{III}})$ of SECTION IV-7 with property (T) being assumed; (n_{II}) and (n_{III}) impose physical constraints on the leg coefficients

$$k_1, k_2, t_1, \gamma_1, \gamma_2 \text{ and } \varepsilon_3$$

which relate them to the angular velocity $\tilde{\omega}$ which, for generality, we do not assume to be constant.

VI.9 The Hotine-Marussi Equations

We have now obtained the full system of leg differential equations required for specifying the geopotential field *external* to the N-surface **S**. This system of differential equations will be collectively called the *Hotine-Marussi equations,* denoted by $\{\mathcal{HM}\}$, and the external equations consist of the following sets of equations

$\{\mathcal{B}\} = \{\mathcal{B}_{\mathrm{I}}, \mathcal{B}_{\mathrm{II}}, \mathcal{B}_{\mathrm{III}}\}$: the leg equations which express the basic gradient equation $(\mathcal{B})$, i.e., (4.3), (4.4), (4.5);

$\{N\} = \{N_{\mathrm{I}}, N_{\mathrm{II}}, N_{\mathrm{III}}\}$: the N-integrability equations, i.e., (4.9), (4.10), (4.11);

(B): the Bruns equation, i.e.; (7.3)

$\{n\} = \{n_{\mathrm{I}}, n_{\mathrm{II}}, n_{\mathrm{III}}\}$: the n-integrability equations, i.e., (8.7), (8.8), (8.9);

$\{\mathcal{L}\} = \{\mathcal{L}_{\mathrm{I}}, \cdots, \mathcal{L}_{\mathrm{IX}}\}$: the Lamé equations, i.e., IV-(9.10), ..., IV-(9.18).

The sets $\{\mathcal{B}\}$, $\{N\}$, and $\{n\}$ essentially couple Gaussian differential geometry to the geopotential field: $\{\mathcal{B}\}$ shows that N varies normal to **S** and introduces the local gravity n; $\{N\}$ incorporates property (T) and gives the tangential dependence of n, while (B) specifies the normal dependence of n; $\{n\}$ includes additional conditions on γ_1, γ_2 and tangential conditions on H, i.e. k_1 and k_2, and relates them to the leg coefficients k_1, k_2, t_1, γ_1, γ_2, ε_3 and $\tilde{\omega}$.

If we choose to regard the local gravity n as the basic variable of the theory, then $\{\mathcal{B}\}$ reduces to a definition of this quantity. Then $\{N\}$ and (B) determine the dependence of n, while $\{n\}$ provides new coupling equations on the leg coefficients and $\tilde{\omega}$. Thus, we have the following number of equations

3 equations for n: (N_{II}), (N_{III}) and (B),

3 equations on the leg coefficients: property (T), (n_{II}) and (n_{III}).

The set $\{\mathcal{L}\}$ consists of a set of compatibility conditions which the leg coefficients must satisfy in order that these coefficients be well-defined in $\mathbf{E}_3$. These consist of 6 independent equations which include the Gauss equation ($\mathcal{G}$), i.e. V-(8.22), and the pair of Codazzi equations ($\mathcal{C}_{\text{I}}$) and ($\mathcal{C}_{\text{II}}$), i.e. V-(8.23) and V-(8.24), which insures that **S** is properly defined in $\mathbf{E}_3$. Thus $\{\mathcal{L}\}$ yields three geometric equations; and three leg coefficient compatibility equations.

VI.10 3-Dimensional Laplacian of Local Gravity

In SECTION 7 we saw that the contraction of the Marussi tensor lead to the modified Laplace equation for N. It is natural to wonder whether an analogous calculation for the local gravity n based on a computation of $n_{rs} := n_{,rs}$ leads to a similarly interesting equation. We now investigate this question.

By covariant differentiation of (5.13) we obtain

$$n_{rs} = N_{rts}\nu^r + N_{rt}\nu^r{}_s. \tag{10.1}$$

However, for any vector ξ we have the commutator rule for covariant derivatives

$$\xi_{rst} - \xi_{rts} = R^p{}_{rst}\xi_p, \tag{10.2}$$

and since $R^p{}_{rst} \equiv 0$ in $\mathbf{E}_3$, it follows that

$$\xi_{rst} = \xi_{rts}. \tag{10.3}$$

Using this relationship in (10.1) gives

$$n_{rs} = N_{rst}\nu^t + N_{rt}\nu^t_s, \tag{10.4}$$

and thus

$$\begin{aligned}\Delta n &:= g^{rs}n_{rs}\\ &= g^{rs}N_{rst}\nu^t + g^{rs}N_{rt}\nu^t{}_s\\ &= (g^{rs}N_{rs})_t\,\nu^t + g^{rs}\,(\nu_r n_t + n\nu_{rt})\,\nu^r{}_s,\end{aligned}$$

or

$$\Delta n = n_t \nu^t{}_s \nu^s + n \nu^s{}_t \nu^t{}_s. \tag{10.5}$$

This rather complicated expression may be readily evaluated by using the leg calculus. Then we have

$$\begin{aligned} \Delta n &= n_t \left(\mu_1 \lambda^t + \gamma_2 \mu^t\right) \\ &\quad + n \left(\lambda^r \lambda^s + \mu^r \mu^s + \nu^r \nu^s\right) \nu_{rt} \nu^t{}_s, \end{aligned}$$

and so

$$\begin{aligned} n^{-1} \Delta n &= \gamma_1^2 + \gamma_2^2 \\ &\quad + \left(-k_1 \lambda_t - t_1 \mu_t + \gamma_1 \nu_t\right)\left(-k_1 \lambda^t - t_1 \mu^t\right) \\ &\quad + \left(-t_1 \lambda_t + k_2 \mu_t + \gamma_2 \nu_t\right)\left(-t_1 \lambda^t - k_1 \mu^t\right), \end{aligned}$$

and finally

$$n^{-1} \Delta n = k_1^2 + k_2^2 + 2t_1^2 + \gamma_1^2 + \gamma_2^2. \tag{10.6}$$

Upon introduction of the curvatures K, H of $\mathbf{S}$, and χ of $\boldsymbol{\Gamma}$, this equation reduces to

$$\Delta n = \left(4H^2 - 2K + \chi^2\right) n \tag{10.7}$$

which is the desired 3-dimensional Laplacian of the local gravity. This result was derived by Hotine on [page 81] (see [12.115], but without identifying $\gamma_1^2 + \gamma_2^2$ with χ^2. Equation (10.7) for Δn is not a familiar result in classical theoretical geodesy. We now consider why and prove the following

<u>Theorem</u>

By virtue of the flatness of $\mathbf{E}_3$, all the versions of Δn given above are identities and not new equations for determining n on $\mathcal{R} \subset \mathbf{E}_3$.

To prove this it suffices to consider one of these equations, say (10.7), and recall the expression IV-(8.8) for the 3-dimensional Laplacian of an arbitrary function F:

$$\Delta F = F_{/1/1} + F_{/2/2} + F_{/3/3} + (\sigma_2 - \gamma_1) F_{/1} - (\sigma_1 + \gamma_2) F_{/2} - 2H F_{/3}.$$

Applying this to n, by the set $\{\mathcal{B}'\}$, we have

$$\begin{aligned} n_{/1/1} &= n\gamma_1^2 + n\gamma_{1/1}, \\ n_{/2/2} &= n\gamma_2^2 + n\gamma_{2/2}, \\ n_{/3/3} &= 2nH_{/3} + 4H\left(Hn - \tilde{\omega}^2\right). \end{aligned} \tag{10.8}$$

Hence

$$\begin{aligned} n^{-1}\Delta n &= \gamma_{1/1} + \gamma_{2/2} + \gamma_1^2 + \gamma_2^2 + 2H_{/3} \\ &\quad + \sigma_2\gamma_1 - \sigma_1\gamma_2, \end{aligned} \tag{10.9}$$

and (10.7) requires that

$$\begin{aligned} &\gamma_{1/1} + \gamma_{2/2} + \chi^2 + 2H_{/3} + \sigma_2\gamma_1 - \sigma_1\gamma_2 \\ &= 4H^2 - 2K + \chi^2. \end{aligned} \tag{10.10}$$

However, addition of the Lamé equations $(\mathcal{L}_{\mathrm{V}})$ and $(\mathcal{L}_{\mathrm{IX}})$ yields

$$\gamma_{1/1} + \gamma_{2/2} + 2H_{/3} = k_1^2 + k_2^2 + 2t_1^2 + \chi^2 + \gamma_2\sigma_1 - \gamma_1\sigma_2 \tag{10.11}$$

and substitution of this into (10.10) reveals that (10.7) is an identity.

Hotine did not indicate the significance of this version of (10.7), i.e. [12.115], but merely exhibited it as an another version of his (ω, ϕ, N)-system expression [12.106] (on [page 81]). Without having the full set of Lamé equations $\{\mathcal{L}\}$ in hand, he would have been unlikely to have seen that his version of Δn was an identity.

VI.11 2-Dimensional Laplacian of the Local Gravity

We now derive the 2-*dimensional*, i.e. the *surface, Laplacian* for n. We denote this by $\Delta_2 n$ (notation different from that employed by Hotine [page 81]), and observe that a new derivation is required, viz. the result is not a specialization of the spatial equation, since Δ_2 involves the Gaussian parametrization of **S** whereas Δ is taken with respect to the ambient curvilinear system x^r on $\mathcal{R}$. Hence one must employ

$$\Delta_2 F := a^{\alpha\beta} F_{\alpha\beta} \tag{11.1}$$

where F depends on the Gaussian parameters u^α defined on a neighborhood of **S**. The corresponding Hotine (surface) 2-leg of (11.1) is

$$\begin{aligned}\Delta_2 F &= a^{\alpha\beta}\left(F_\alpha\right)_\beta \\ &= a^{\alpha\beta}\left(F_{/1}\lambda_\alpha + F_{/2}\mu_\alpha\right)_\beta \\ &= a^{\alpha\beta}\{\left(F_{/1/1}\lambda_\beta + F_{/1/2}\mu_\beta\right)\lambda_\alpha + F_{/1}\lambda_{\alpha\beta} \\ &\quad + \left(F_{/2/1}\lambda_\beta + F_{/2/2}\mu_\beta\right)\mu_\alpha + F_{/2}\mu_{\alpha\beta} \\ &= F_{/1/1} + F_{/2/2} + F_{/1}\left(a^{\alpha\beta}\lambda_{\alpha\beta}\right) \\ &\quad + F_{/2}\left(a^{\alpha\beta}\mu_{\alpha\beta}\right).\end{aligned}$$

Then using V-(4.3) and (4.4) we have

$$\Delta_2 F = F_{/1/1} + F_{/2/2} + \sigma_2 F_{/1} - \sigma_1 F_{/2}. \tag{11.2}$$

So consequently for $F = n$, upon using (N_I), (N_{II})

$$\begin{aligned}\Delta_2 n &= n_{/1/1} + n_{/2/2} + \sigma_2 n_{/1} - \sigma_1 n_{/2} \\ &= (n\gamma_1)_{/1} + (n\gamma_2)_{/2} + n\sigma_2\gamma_1 - n\sigma_1\gamma_2\end{aligned}$$

we finally obtain

$$n^{-1}\Delta_2 n = \gamma_{1/1} + \gamma_{2/2} + \chi^2 + \sigma_2\gamma_1 - \sigma_1\gamma_2. \tag{11.3}$$

Addition of ($\mathcal{L}_V$) and ($\mathcal{L}_{IX}$) gives (10.3), which permits (11.3) to be rewritten in the form

$$\Delta_2 n = 2\left(2H^2 - K + \chi^2 - H_{/3}\right)n \tag{11.4}$$

as announced by Hotine — without proof — (see his [12.122] on [page 81]).

Unlike the previous expression (10.7) for the 3-dimensional Laplacian of n, (11.4) is *not* an identity! The meaning of this equation depends on whether one chooses the Marussi-Hotine view of differential geodesy (where a solution of N is assumed known) or a more general view in which the function N is to be determined (this will be further explored in CHAPTER X). In the former case if the known solution is determined by a Neumann problem in which the normal derivative N is prescribed on the surface of the Earth, or a nearby equipotential surface **S**, then this derivative with respect to the Hotine 3-leg is $N_{/3}$, and by (4.5), this is n. Then n is known on **S**, and (11.4) is a *compatibility equation* which the leg coefficients k_1, k_2, γ_1, γ_2 and t_1 must satisfy. On the other hand if, in the more general view, N is unknown then n is also unknown, and (11.4) is a *determining equation* for n on **S**. In either case, (11.4) should be added to the set of Hotine-Marussi equations exhibited in SECTION 9.

PROBLEMS FOR CHAPTER VI

VI.1 Given the set of leg differential equations $\{\mathcal{B}\}$, i.e. (4.3-.5), show that by using the commutators IV-(4.12) one obtains the N-integrability conditions $\{N\}$, viz. (4.9-.11).

VI.2 Read the discussion [pages 72-86] in Hotine's treatise to see how he derives equations (5.14-.19) *without appeal* to the basic equations of the leg calculus, i.e. IV-(6.1-.3). Do you think Hotine knew these equations and chose not to exhibit them, or did not know them and tailored his analysis to avoid them?

VI.3 Show how (6.1) can be expressed in a way which avoids knowing the Christoffel symbols.

Answer: $$\nu^r{}_r = \frac{1}{\sqrt{g}}\left(\sqrt{g}\nu^r\right)_{;r}\,.$$

VI.4 In addition to the first Bruns curvature equation (6.1), there is also a *second Bruns* curvature equation involving the Gauss curvature K. Derive this equation starting from the tensorial expression for K obtained in PROBLEM V.3

Answer: $$2K = \nu_r\Delta_2\nu^r + \left(\nu^r{}_r\right)^2.$$

ZUND (1993)

VI.5 Given the set of leg differential equations $\{\mathcal{B}'\}$, i.e. (4.10-.11) and (7.3), show that by using the commutators as in PROBLEM VI.1 one obtains the n-integrability conditions $\{n\}$, viz. (8.7-.9).

VI.6 The combination of terms $4H^2 - 2K$ appearing on the right hand side of (10.7) suggests the introduction of the *Casorati curvature* C, i.e.

$$2C := 2\left(2H^2 - K\right).$$

Show that equivalent expressions are given by

$$2C = a^{\alpha\beta} c_{\alpha\beta},$$

or in terms of Hotine's curvature parameters by

$$2C = k_1^2 + k_2^2 + 2t_1^2.$$

Can you suggest reasons why the Casorati curvature is in some sense more useful than the Gauss curvature K, and may — loosely speaking — be regarded as a replacement for K?

[Note, this curvature was introduced in CASORATI (1890), and following BIANCHI (1927) (see page 197), it was employed by MARUSSI (1951) (see page 19 of MARUSSI (1985). Both Bianchi and Marussi considered this quantity only when $t_1 = 0$; i.e. for the principal curvatures κ_1 and κ_2, and their expressions differ from ours by a factor of $\frac{1}{2}$.]

VI.7 If $t_{\alpha\beta}$ is an arbitrary symmetric tensor, introduce the quantity

$$t_{\alpha\beta\lambda\mu} := t_{\alpha\lambda} t_{\beta\mu} - t_{\alpha\mu} t_{\beta\lambda}.$$

Then upon successively choosing $t_{\alpha\beta}$ to be the basic tensors $a_{\alpha\beta}$, $b_{\alpha\beta}$ and $c_{\alpha\beta}$ respectively, show that one has the following *canonical forms* for the Riemann tensor:

$$R_{\alpha\beta\lambda\mu} = K a_{\alpha\beta\lambda\mu}, \quad R_{\alpha\beta\lambda\mu} = b_{\alpha\beta\lambda\mu}, \quad R_{\alpha\beta\lambda\mu} = \frac{1}{K} c_{\alpha\beta\lambda\mu},$$

where K is the Gauss curvature.

[Note that formally the non-appearance of the Casorati curvature in the above canonical forms can be used as an argument of why K is more 'fundamental' than C.]

ZUND (1988b)

VI.8 If the angular velocity $\tilde{\omega}$ of the Earth is not a constant, show how a position dependent $\tilde{\omega}^2$ affects the n-integrability conditions $\{n\}$ given in (8.7-.9). Hence, show that such an $\tilde{\omega}^2$ makes a contribution to the expressions (10.7) and (10.9) for the 3-dimensional Laplacian of the local gravity.

VI.9 In HOTINE (1957a), and on [pages 150-151] of his treatise, Hotine gives a discussion of how his curvature parameters are related to the differential geodesy of the Eötvös torsion balance.

Derive the canonical 2-leg representations for the fourth basic tensor $d_{\alpha\beta}$ and show that the difference of the curvature k_1, k_2 (or respectively the principal curvatures κ_1, κ_2), appears in these expressions.

[Note: this suggests that physically our inability to measure these curvature differences may have a geometric origin.]

ZUND (1989)

VI.10 Consider the first basic form

$$\mathrm{I} = \rho_1^2 \cos^2 \phi d\omega^2 + \rho_2^2 d\phi^2,$$

where $u^1 := \omega$, $u^2 := \phi$ are respectively the longitude and latitude on a *spheroid* (an ellipsoid of revolution about the z-axis) having the semi-axes $a > b$, and

$$\rho_1 := \frac{1}{\kappa_1}, \quad \rho_2 := \frac{1}{\kappa_2}$$

with

$$\kappa_1 = -\left(1 - e^2 \sin^2 \phi\right)^{1/2} / a,$$

$$\kappa_2 = -\left(1 - e^2 \sin^2 \phi\right)^{3/2} / a \left(1 - e^2\right),$$

and e is the eccentricity $e^2 := (a^2 - b^2) / a^2$.

Then compute the components of the 2-leg vectors $\boldsymbol{\lambda}$, $\boldsymbol{\mu}$, and exhibit the entries in the matrices $\|a_{\alpha\beta}\|$, $\|b_{\alpha\beta}\|$, $\|c_{\alpha\beta}\|$ and $\|d_{\alpha\beta}\|$ of the four basic tensors.

VII

The Fundamental Theorem of Differential Geodesy

VII.1 Introduction

In this chapter we take the first step towards investigating the global, i.e. large-scale, structure of the geopotential field. We first state the Pizzetti Theorem, which we call the *Fundamental Theorem of Differential Geodesy,* since it essentially specifies the natural domain of validity of differential geodesy. Intuitively, one is accustomed to assuming that the nearby equipotential surfaces of a uniformly rotating Earth are closed surfaces which are locally isometrically imbedded in an Euclidean 3-space $\mathbf{E}_3$, and Pizzetti's result indicates when this is a reasonable assumption. We then examine the *Extension Problem* which, in effect, considers to what extent Gaussian differential geometry can in practice be applied to the study of these equipotential surfaces.

VII.2 Basic Assumptions

We now examine the basic assumptions which underlie all our considerations in this chapter.

PIZZETTI (1901)[1] began his analysis with the assumption that the physical shape of the Earth, denoted by $\mathcal{E}$, is roughly a sphere, or as we will say, $\mathcal{E}$ is *sphere-like.* He then considered the following geometric setup: let O be the center of $\mathcal{E}$, $\{\mathbf{A}, \mathbf{B}, \mathbf{C}\}$ a constant Cartesian 3-leg fixed at O, and suppose that $\mathcal{E}$ is uniformly rotating about the $\mathbf{C}$-axis with (constant) angular

---

[1] **A slightly more detailed version may be found in the elegant monograph PIZZETTI (1913).**

velocity $\tilde{\omega}$. Let $\mathcal{S}_R(O)$ denote a sphere of radius R having its center at O with R being sufficiently large that it encloses the total mass $\mathcal{M}$ of $\mathcal{E}$. This sphere will be called a *toposphere*, and will often be denoted simply by $\mathcal{S}$. Suppose R_m be the mean radius of $\mathcal{E}$, viz. R_m is the radius of a sphere of volume-equal to that of $\mathcal{E}$. Then Pizzetti's first assumption is that

$$(\mathcal{P}_{\mathrm{I}}) \qquad R - R_m < \frac{1}{100} R_m,$$

i.e. R differs from R_m by less than 1%. If R_p and R_e are the respective polar and equatorial radii of $\mathcal{E}$, then $(\mathcal{P}_{\mathrm{I}})$ is clearly satisfied. Indeed, the bound of $\frac{1}{100}$ in $(\mathcal{P}_{\mathrm{I}})$ is a generous one.

Pizzetti's second assumption is that

$$(\mathcal{P}_{\mathrm{II}}) \qquad \frac{\tilde{\omega}^2 R_m^3}{G\mathcal{M}} < \frac{1}{289}$$

where G is the Newtonian gravitational constant. By $(\mathcal{P}_{\mathrm{I}})$ we have

$$R < 1.01 R_m, \tag{2.1}$$

so $(\mathcal{P}_{\mathrm{II}})$ immediately yields the m-inequality:

$$m := \frac{\tilde{\omega}^2 R^3}{G\mathcal{M}} < \frac{1}{280}. \tag{2.2}$$

Note that the left-hand side of this inequality is the ratio of the centrifugal and gravitational accelerations, while the *derived fraction* on the right-hand side is dependent on the bounds chosen in $(\mathcal{P}_{\mathrm{I}})$ and $(\mathcal{P}_{\mathrm{II}})$. The values employed by Pizzetti were of course those available at the close of the nineteenth century. We will retain them in our analysis since they facilitate comparison with his original work and make the arithmetic somewhat simpler. Note the newer values do not affect the choice of the integer which appears in his theorem. In PROBLEMS VII.2, VII.3 the reader will be asked to determine the errors involved when the numerical factors in $(\mathcal{P}_{\mathrm{I}})$ and $(\mathcal{P}_{\mathrm{II}})$ are changed.

VII.3 The Fundamental Theorem of Differential Geodesy

We now state Pizzetti's result as follows:

Fundamental Theorem

Suppose the Earth is rotating with a uniform angular velocity $\tilde{\omega}$ and is sphere-like i.e. it satisfies $(\mathcal{P}_\mathrm{I})$ and $(\mathcal{P}_\mathrm{II})$ so that

$$m < \frac{1}{280}.$$

Then there exists a radius r and a sphere $\mathcal{S}^*$, the *Pizzetti sphere*, such that the set Σ of all equipotential surfaces of $\mathcal{E}$ lying between the radius R of the toposphere $\mathcal{S}$ are compact surfaces whenever

$$0 < h < 5R_m \tag{3.1}$$

where h is the height above $\mathcal{E}$.

Equation (3.1) is called the *Pizzetti Inequality* and the significance of the integer 5 is that bounds the various decimal expressions obtained by slightly changing the values of the fractions on the right hand sides of $(\mathcal{P}_\mathrm{I})$ and $(\mathcal{P}_\mathrm{II})$.

Before passing to a general discussion of the theorem and why we maintain that it should be regarded as a fundamental theorem, let us comment briefly on the logic of its proof. First the reader may perhaps be surprised at the crudity of Pizzetti's argument, which essentially involved only a cunning use of the calculus. However, two things should be kept in mind:

(i) the proof must necessarily involve approximate methods;

(ii) no one has ever claimed that it was a 'best-possible' argument, merely that it worked!

In other words, the integral factor of 5 appearing in (3.1) suffices for sensibly small changes in the fractions appearing in $(\mathcal{P}_\mathrm{I})$ and

($\mathcal{P}_{\mathrm{II}}$), and hence in the m-inequality. Moreover, with the accuracies to which these physical constants are now known only small changes — indeed very small changes — may be envisioned in the future.

Likewise, the choices of bounds in various inequalities are generous and could be *improved.* However, the replacement of our choices by 'better ones' does not change our final result, viz. the choice of the integer 5 in (3.1). In this sense the integer 5 is the best-possible value of the numerical factor appearing on the right-hand side of (3.1).

A more serious defect of Pizzetti's analysis is that he gave no details about what happens as $h \to 5R_m$. Presumably via the sketch in Figure 6, and the obvious rotational symmetry of the situation in the equatorial plane of $\mathcal{E}$, Σ degenerates into a closed curve C which is circular in shape. Moreover, for $h > 5R_m$, Σ unfolds, or *bifurcates* into a 2-sheeted open cylindrical surface. Although this behavior is apparent from a graphical investigation, a precise analysis of the situation has yet to be given. Of course, such an analysis must again be approximate and our suggestion is that a sharpening of the bounds in the proof of the theorem be done as part of an investigation of the bifurcation process.

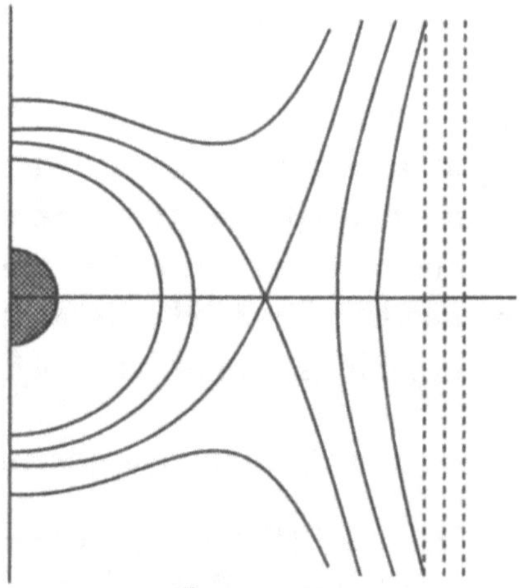

Figure 6

VII.4 Discussion

We now indicate why Pizzetti's theorem may be regarded as the *fundamental theorem of differential geodesy.* There are essentially three reasons for its importance.

i) It is a theorem on differential geometry in the large, viz. global differential geometry, in which a basically simple physical situation leads to an elegant mathematical result. Global results are not easy to obtain in differential geometry and even less common in differential geodesy. It completely answers the question as to when the family $\boldsymbol{\Sigma}$ of equipotential surfaces is closed and bounded — an obvious physical question (!) — and in doing so it provides a precise answer as to what is meant by the adjective 'global'. Here global means the region $\mathbf{E}_3$ defined by (3.1).

ii) The condition that the surfaces of $\boldsymbol{\Sigma}$ are closed and bounded in $\mathbf{E}_3$, means that each of these surfaces are compact, see MUNKRES (1975, 1991). This is an important fact since some of the sharpest results in Gaussian differential geometry apply only to compact surfaces. Hence, by focusing our attention on those surfaces of $\boldsymbol{\Sigma}$ which satisfy (3.1), we are essentially dealing with the part of the geopotential field where differential geodesy can be expected to have its most significant application.

Presumably, as $h \to 5R_m$, the family $\boldsymbol{\Sigma}$ generates to a circular curve C in the equatorial plane of $\mathcal{E}$; and for $h > 5R_m$, $\boldsymbol{\Sigma}$ tends to a cylindrical family having the polar axis of $\mathcal{E}$ as its axis. Cylindrical, or cylinder-like, surfaces are non-compact and have comparatively few deep geometric properties.

iii) By a fundamental property of compact sets, or more generally spaces — again see MUNKRES (1975, 1991) — every covering of the sets, or spaces, by a collection of open sets admits a finite subcollection. Hence, if a surface is compact it possesses a covering consisting of a *finite number* of parametrized coordinate neighborhoods. As is known in the case of a 2-dimensional sphere $\mathbf{S}^2$ in $\mathbf{E}_3$ it is not possible to parametrize the entire surface of $\mathbf{S}^2$ by a single parametrized coordinate neighborhood; at least two such neighborhoods are required to handle the singularities which occur at the North and South poles of $\mathbf{S}^2$. The compactness property then guarantees that each surface of the family $\boldsymbol{\Sigma}$ (of equipotential surfaces of the geopotential field) can be covered by a finite number of parametrized coordinate neighborhoods. In particular, there is no (conceptual) difficulty in evaluating surface integrals on a compact surface.

VII.5 The Extension Problem

The fundamental theorem discussed in the previous four sections immediately leads to the question of to what extent Gaussian differential geometry can be applied to the surfaces of the family $\boldsymbol{\Sigma}$ satisfying the Pizzetti inequality (3.1). For an arbitrarily chosen individual surface of $\boldsymbol{\Sigma}$, there is no difficulty. However, it is not obvious how to relate the analysis for a pair of individual surfaces of $\boldsymbol{\Sigma}$. This is known as the *extension problem*, (**EP**), and it deals with how to 'extend' the geometry on one equipotential surface to another equipotential surface. Before discussing it, let us briefly recall the general situation.

As in CHAPTER VI, the geopotential function N in a coordinate system rotating with the Earth $\mathcal{E}$ is a solution of the modified Laplace equation

$$\Delta N = -2\tilde{\omega}^2, \tag{5.1}$$

where Δ is the 3-dimensional Laplacian on a region $\mathcal{R}$ of $\mathbf{E}_3$ which encloses at least part of $\mathcal{E}$. If the geopotential N is rewritten in terms of Cartesian coordinates

$$\mathbf{x} = (x, y, z)$$

then the equipotential surfaces are given by

$$N(\mathbf{x}) = c \tag{5.2}$$

where $c =$ a constant. As c assumes the successive values

$$c < c_1 < c_2 < \cdots < c_n < \cdots$$

we obtain a family of surfaces

$$\boldsymbol{\Sigma} = \{\mathbf{S}, \mathbf{S}_1, \mathbf{S}_2, \cdots, \mathbf{S}_n, \cdots\} \tag{5.3}$$

where the initial surface $\mathbf{S}$ is called the *base surface.*

Let us suppose that a solution N of (5.1) is known, and the family $\boldsymbol{\Sigma}$ is given by a set of implicitly defined surfaces

$$\mathbf{S} : N(\mathbf{x}) = c, \quad \mathbf{S}_1 : N(\mathbf{x}) = c_1, \quad \mathbf{S}_2 : N(\mathbf{x}) = c_2, \cdots \tag{5.4}$$

which satisfy the Pizzetti inequality (3.1), or $R < r < 6R$. Then the **EP** consists in determining how the Gaussian differential geometry of $\mathbf{S}$ extends to $\mathbf{S}_1, \mathbf{S}_2, \cdots$. A crucial issue in the **EP** is how to relate the various Gaussian parameter systems

$$\begin{aligned} u^\alpha &= (u, v) \text{ on } \mathbf{S}, \\ u_1^\alpha &= (u_1, v_1) \text{ on } \mathbf{S}_1, \\ u_2^\alpha &= (u_2, v_2) \text{ on } \mathbf{S}_2, \end{aligned} \tag{5.5}$$

since the character of N changes as r increases (recall Figure 6). An obvious and convenient *assumption* is that there exists a common Gaussian parametrization on at least *some* of the surfaces of $\boldsymbol{\Sigma}$, e.g. those belonging to one of the types as discussed in the next paragraph. This assumption is the basis of the theory of normal systems given in [Chapter 15] of HOTINE (1969); however, there is no reason to expect that it need be valid for all the surfaces of $\boldsymbol{\Sigma}$ which satisfy the Pizzetti inequality. If one does not make this assumption, then our entire approach to the **EP** is invalid, and in this case the problem appears to be intractable.

As a preliminary to the **EP**, it is useful, and seemingly unavoidable, to delineate *three* particular situations: the *local problem* where the members of $\boldsymbol{\Sigma}$ are neighboring surfaces; the *global problem* where the surfaces need not be nearby; and the *limiting problem* where $r \to 5R_m$. A complete solution of the **EP** requires solving all three of the separate problems in a compatible manner. Our present observation, which will be confirmed by the analysis to be given in the remainder of this chapter, is that it is unrealistic to expect to find a single solution of the **EP** which encompasses *all* of the above situations.

It is well-known that the geometry of surfaces, and in particular of the family $\boldsymbol{\Sigma}$, requires knowing — i.e. specifying — both the first and second basic forms, viz. the basic tensors $a_{\alpha\beta}$ and $b_{\alpha\beta}$, in order for the individual surfaces to be uniquely determined up to position in $\mathbf{E}_3$, see EISENHART (1940), etc. The latter form, and the tensor $b_{\alpha\beta}$ involves the unit normal $\boldsymbol{\nu}$ of the surface; it is convenient to regard these as local tangents to a congruence of curves $\boldsymbol{\Gamma}$ normal to the individual surfaces of $\boldsymbol{\Sigma}$ or at least to pieces of surfaces of $\boldsymbol{\Sigma}$. Thus, at least some of the geometry of $\boldsymbol{\Sigma}$ is directly related to the geometry of $\boldsymbol{\Gamma}$. As we will see, the solution of the **EP** crucially depends on an investigation of *both* $\boldsymbol{\Sigma}$ and $\boldsymbol{\Gamma}$ and two subsidiary problems considered in SECTIONS 8 and 9. In effect these problems specify the general mathematical framework in which a solution of the **EP** — when it exists — must be sought.

Typically a solution of the **EP** should allow one to compare distances, i.e. arc-lengths, involving pairs of points P, Q on $\mathbf{S}$ and P_1, Q_1 on $\mathbf{S}_1$ where the latter are defined by the intersections of $\boldsymbol{\Gamma}$ with $\mathbf{S}_1$ (see Figure 8 for a cross-sectional view of $\boldsymbol{\Sigma}$ and $\boldsymbol{\Gamma}$). For instance, denoting the distance by $|-,-|$ with an attached subscript indicating whether this is a surface or spatial distance, one might wish to compare the distances:

(i) $|P,Q|_{\mathbf{S}}$ and $|P_1,Q_1|_{\mathbf{S}_1}$,

(ii) $|P,Q_1|_{\mathbf{E}_3}$ and $|Q,P_1|_{\mathbf{E}_3}$,

(iii) $|P,P_1|_{\mathbf{E}_3}$ and $|Q,Q_1|_{\mathbf{E}_3}$.

Relative to (i), in general one expects that $|P,Q|_{\mathbf{S}} \neq |P_1,Q_1|_{\mathbf{S}_1}$ with

$$|P,Q|_{\mathbf{S}} < |P_1,Q_1|_{\mathbf{S}_1}$$

indicating a *divergence*, or *expansion*, of Γ, and

$$|P, Q|_{\mathbf{S}} > |P_1, Q_1|_{\mathbf{S}_1}$$

a *convergence*, or *contraction*, of Γ. Likewise one might wish to require the equality of the distances in (ii) or (iii). The former is a 'diagonal property' (see Figure 7) which is of rather surprising importance (known as the *Ivory Property* after Sir James Ivory) while the latter leads to the notion of Σ being a parallel family of surfaces.

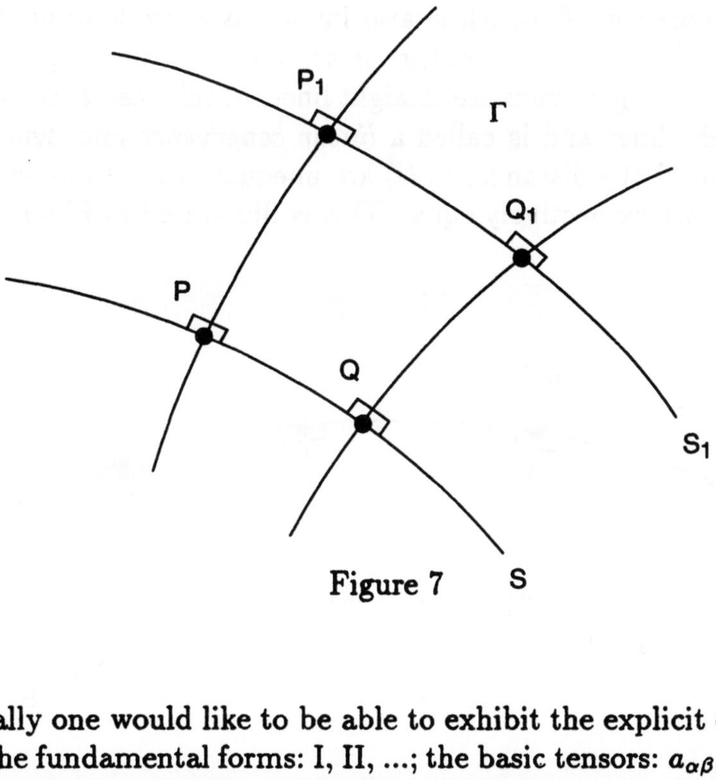

Figure 7

Ideally one would like to be able to exhibit the explicit changes in the fundamental forms: I, II, ...; the basic tensors: $a_{\alpha\beta}$, $b_{\alpha\beta}$, ...; the curvatures K, H; and Hotine's curvature parameters k_1, k_2, t_1, γ_1, γ_2 as one traverses Γ in passing form **S** to $\mathbf{S}_1$, $\mathbf{S}_2$, ... etc.

Before concluding this section we will briefly consider two very special geometric configurations which — at least at first glance — appear *a priori* to be solutions of the **EP**. The first of these is a *triply-orthogonal system* of surfaces which we denote by $\mathcal{O}_3$, or in particular a *separable* $\mathcal{O}_3$ denoted by $\mathcal{O}_3^*$. The adjective 'separable' refers to the separability of the variables in (5.1) which is reflected in the form of the solution $N(\mathbf{x})$ appearing in (5.2). The above properties (i)-(iii) now have an interesting interpretation. In general, the distances (i) and (iii) are unequal; however, as BLASCHKE (1928) has proved, the Ivory Property (ii) is equivalent to the *Stäckel Separation Property*, STÄCKEL (1894), which characterizes $\mathcal{O}_3^*$. The $\mathcal{O}_3^*$ do not exhaust all of $\mathcal{O}_3$, although the most common $\mathcal{O}_3$ are $\mathcal{O}_3^*$ because of their importance in mathematical physics, see MADELUNG (1950) and Figure 8. The second system is that of a *parallel system* of surfaces denoted by $\mathcal{P}$ which is also known as a geodesic parallel system. The latter terminology is somewhat misleading in $\mathbf{E}_3$ since there all geodesics are straight lines. In this case $\boldsymbol{\Gamma}$ consists of straight lines and is called a *linear congruence* and denoted by $\boldsymbol{\Lambda}$. For $\mathcal{P}$ the distances in (i) are unequal, while those in (ii) and (iii) are respectively equal. This is illustrated in Figure 8.

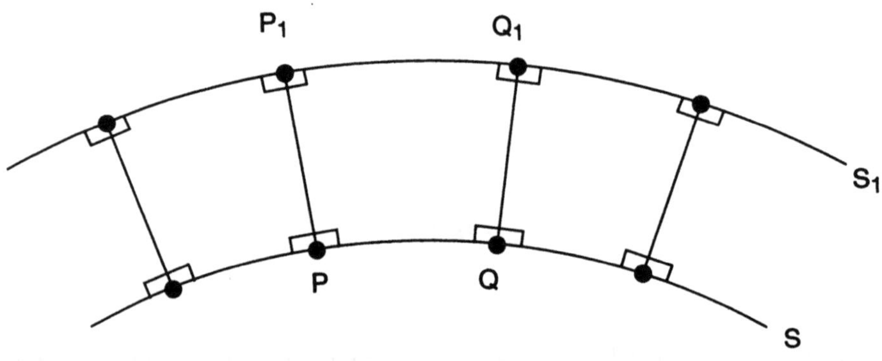

Figure 8

As mentioned previously, this essentially corresponds to the normal systems of Hotine, since he apparently forgot that in $\mathbf{E}_3$ a geodesic congruence $\boldsymbol{\Gamma}$ reduces to a linear congruence $\boldsymbol{\Lambda}$. However, he did correctly observe that if one took his (ω, ϕ), i.e. (longitude, latitude), to be the Gaussian parameters u^α on $\mathbf{S}$, one cannot take $u_1^\alpha = (\omega, \phi)$ on $\mathbf{S}_1$ since $\mathbf{S}$ and $\mathbf{S}_1$ are different types of surfaces, see PROBLEM VII.7. Both of the systems $\mathcal{O}_3^*$ and $\mathcal{P}$ will be further examined in CHAPTER X. The difficulty with both of these systems being solutions of the **EP** is that neither admits the kind of 'evolution', or limiting behavior, exhibited in Figure 6. In other words, as r increases, the type of surface in $\mathcal{O}_3^*$ remains a confocal quadric, i.e. it preserves its type; for $\mathcal{P}$, although the type of the surface changes (see PROBLEMS VII.7 and VII.8), ostensibly, the shape remains roughly the same. The only possible application of $\mathcal{O}_3^*$ and $\mathcal{P}$ to the **EP** is to the local problem when the equipotential surfaces are nearby, and the latter holds only when one may assume that the curvature χ of the plumblines is zero.

VII.6 Two Subsidiary Problems

The **EP** is intimately related to the geometric theory of congruences of curves $\boldsymbol{\Gamma}$ and their associated family of surfaces $\boldsymbol{\Sigma}$ in $\mathbf{E}_3$. The resulting theory has a long and rich history. The geometry of orthogonal trajectories, or curves whose tangents are surface normals, was first systematically studied at the turn of the nineteenth century by G. Monge and his students. Their viewpoint and contributions were summarized in MONGE (1809) and may be regarded as Mongean differential geometry, which is distinct from the Gaussian theory in that it makes no use of a parametrization on the surfaces of $\boldsymbol{\Sigma}$. In particular, Monge's student E.L. Malus applied this theory to the optics of reflected and refracted rays of light. Indeed, MALUS (1808) and (1811) contain a preliminary version of the celebrated Malus-Dupin Theorem of geometric optics, which had also been independently discovered by Sir William Rowan Hamilton while studying MONGE (1809) as an undergraduate. Subsequently, he dealt with the geometric and optical content of the theory in his great memoir, HAMILTON (1828). For a lucid discussion of this memoir, see the account given in Chapter 4 of Hamilton's biography, HANKINS (1980), Note 2 by the editors of his mathematical papers on optics, HAMILTON (1931), and the technical analysis in CAYLEY (1888).

The Cayley reference is noteworthy in that it emphasized that Hamilton had realized that his theory of the systems of rays could be viewed in two ways:

[i] as rays emanating from the points of a surface (or surfaces), and

[ii] given the rays (irrespective of any surface) inquiring when there exists a surface (or surfaces) orthogonal to the ray system.

These two situations are the progenitors of a pair of *subsidiary problems* on which the whole theory of the **EP** depends.

Our analysis consists of translating the physical situations described in [i] and [ii] into the language of partial differential equations and Mongean differential geometry. This will be done in SECTIONS 7 and 8 respectively, and in SECTION 9 we will present our conclusions about the solvability of the **EP** which requires that the mathematical formulations of both [i] and [ii] be satisfied.

VII.7 The Congruence-Forming Property

The situation described in [i] of SECTION 6 is called the *congruence-forming property* of a system of surfaces Σ. In discussing this property it is not necessary to assume that the function defining Σ be a solution of (5.1), merely that it is a known function. Although we will be primarily interested in the case when Σ consists of equipotential surfaces, in order to emphasize the difference between the general situation and the geodetic problem we will *initially assume* that Σ is defined by a smooth function F of the local coordinates x^r defined on some region $\mathcal{R}$ of $\mathbf{E}_3$ which is large enough to include at least part of the region defined by the Pizzetti inequality (3.1). Later we will make the identification of F with the geopotential function N.

Without loss of generality we may take the x^r to be the Cartesian coordinates (x, y, z) so that the F-surfaces $\boldsymbol{\Sigma}$ are defined by

$$F(x, y, z) = c \tag{7.1}$$

where c is an arbitrary parameter which ranges over a set of values. Then since F is a smooth function, its gradient is well-defined and collinear with the field of unit normals of the F-surfaces which we denote — as usual — by $\boldsymbol{\nu}$. Then following Hotine we write

$$F_r = f\nu_r \tag{7.2}$$

where $f \neq 0$ is a proportionality factor that is required to insure that the ν_r are the covariant components of a unit vector, and since F is a scalar function we write $F_r \equiv F_{;r}$ where the semi-colon denotes partial differentiation. Then, since the x^r are Cartesian coordinates in $\mathbf{E}_3$, the covariant and contravariant components of the metric tensor are identical and equal to the Kronecker delta (written covariantly or contravariantly respectively). Hence, we have

$$f^2 = \delta^{rs} F_r F_s \tag{7.3}$$

i.e.

$$f^2 = (F_1)^2 + (F_2)^2 + (F_3)^2 \,. \tag{7.3$'$}$$

Now [i] assumes that one is given a family of F-surfaces $\boldsymbol{\Sigma}$, and asks *when* $\boldsymbol{\Sigma}$ rigorously admits a system of orthogonal trajectories in $\mathbf{E}_3$, viz. a congruence of normal curves $\boldsymbol{\Gamma}$ in $\mathbf{E}_3$. When this is the case, $\boldsymbol{\Sigma}$ is said to be *congruence-forming*, and hence [i] is known as the *congruence-forming property*, viz. property ($\boldsymbol{\Gamma}$). Given a $\boldsymbol{\Sigma}$ defined by (7.1), it is easy to see that the determination of the orthogonal trajectories of $\boldsymbol{\Sigma}$ reduces to the integration of the following system of differential equations:

$$\frac{dx}{F_1} = \frac{dy}{F_2} = \frac{dz}{F_3}. \tag{7.4}$$

The *two* integrals of (7.4) contain a pair of arbitrary constants, and consequently determine a 2-parameter family of orthogonal trajectories which is our desired normal congruence $\boldsymbol{\Gamma}$. There are only two integrals since as in Mongean differential geometry we assume that (7.1) can be solved for one of the variables, say z, as a function of the other pair of variables viz. $z = \varphi(x, y)$.

Suppose now that we have such a $\boldsymbol{\Gamma}$ which is defined by a system of differential equations

$$\frac{dx}{X_1} = \frac{dy}{X_2} = \frac{dz}{X_3} \tag{7.5}$$

where the X_r are given functions of $x^r = (x, y, z)$. If there exists an F-surface $\mathbf{S}$ normal to $\boldsymbol{\Gamma}$, then at each point of $\mathbf{S}$, one must have the following pair of partial differential equations

$$\begin{aligned} p &:= \frac{\partial z}{\partial x} = -\frac{X_1}{X_3} \\ q &:= \frac{\partial z}{\partial y} = -\frac{X_2}{X_3} \end{aligned} \tag{7.6}$$

where as above z is regarded as a function of x and y. Equation (7.5) is an obvious analogue of (7.2) and upon contraction of the latter by dx^r we obtain the Pfaffian expression:

$$\Phi := X_r dx^r = X_1 dx + X_2 dy + X_3 dz, \tag{7.7}$$

which is called *Euler's form.* But, since this must be a perfect differential dF, by (7.1) we are led to seeking a solution of a so called *total differential equation* $\Phi = 0$, i.e.

$$\Phi = X_1 dx + X_2 dy + X_3 dz = 0. \tag{7.8}$$

If $X_3 \neq 0$ with $z = \varphi(x, y)$, then we immediately have

$$dz = -\left(\frac{X_1}{X_3}\right) dx - \left(\frac{X_2}{X_3}\right) dy, \tag{7.9}$$

which yields (7.6). The difficulty with (7.6) is that *both* these equations must be identically satisfied by the *same function* $z =$

φ and — in general — this is not possible. Suppose there exists such a function φ, then clearly we must also have

$$\frac{\partial p}{\partial y} = \frac{\partial q}{\partial x}. \tag{7.10}$$

Upon explicit expression this yields

$$\frac{\partial}{\partial y}\left(\frac{X_1}{X_3}\right) + \frac{\partial}{\partial z}\left(\frac{X_1}{X_3}\right) q = \frac{\partial}{\partial x}\left(\frac{X_2}{X_3}\right) + \frac{\partial}{\partial z}\left(\frac{X_2}{X_3}\right) p, \tag{7.11}$$

which by (7.6) becomes *Euler's condition*

$$X_1 \left(X_{2;3} - X_{3;2}\right) + X_2 \left(X_{3;1} - X_{1;3}\right) + X_3 \left(X_{1;2} - X_{2;1}\right) = 0, \tag{7.12}$$

and is a restriction on the X_r! Equation (7.12) is equivalent to the tensorial expression

$$\varepsilon^{rst} X_r X_{s;t} = 0, \tag{7.13}$$

where the semi-colon may be replaced by a comma indicating covariant differentiation, or more elegantly by the Pfaffian expression

$$\Phi \wedge d\Phi = 0. \tag{7.14}$$

Both (7.13) and (7.14) are valid in *any* local coordinate system on $\mathcal{R}$, and each is equivalent to the complete integrability of (7.8), or more generally $X_r dx^r$, where the x^r need not be Cartesian coordinates. In other words, either $X_r dx^r$ is a perfect differential, i.e. an exact differential 1-form, or can be made into one by the use of an appropriate integrating factor. These two cases correspond to the respective choices:

$$X_r = \nu_r \quad (f = 1)$$

or

$$X_r = f\nu_r \quad (f \neq 1) \tag{7.15}$$

where in (7.2) $f \neq 0$ is the integrating factor. Note that, in general, the former choice does not occur since it would obviously entail a specialization of the vectorial 3-leg, whereas the latter avoids such a requirement. Likewise physically when F is identified with the geopotential N, f becomes the local gravity n and $n = 1$ is a severe restriction.

Thus, the admissibility of the congruence forming property $(\boldsymbol{\Gamma})$ reduces to *three* cases relative to the Euler condition:

Case 1^0 (7.14) is *identically satisfied*, so the pair of equations in (7.6) admit a *common solution* $z = \varphi(x, y)$ which involves *one* arbitrary constant, and $\boldsymbol{\Gamma} \perp \boldsymbol{\Sigma}$;

Case 2^0 (7.14) is not identically satisfied, however, it may be *conditionally satisfied* when the equations in (7.6) admit special solutions depending on particular choices of the constants c appearing in (7.1), and $\boldsymbol{\Gamma} \perp \boldsymbol{\Sigma}_n$ where $\boldsymbol{\Sigma}_n$ is a finite family of F-surfaces;

Case 3^0 neither Case 1^0 nor Case 2^0 holds and hence there does not exist an F-surface $\mathbf{S}$ such that $\boldsymbol{\Gamma} \perp \mathbf{S}$ and thus $\boldsymbol{\Gamma} \not\perp \boldsymbol{\Sigma}$.

Relative to differential geodesy, we must restrict our consideration to Case 1^0 in order to conform to our intuitive picture of the geopotential field being described by a 'space-filling' family of equipotential surfaces and their orthogonal trajectories. Taking $F = N$ and employing an arbitrary ambient coordinate system on a region $\mathcal{R}$ of $\mathbf{E}_3$, (7.2) becomes VI-(3.2) i.e.

$$N_r = n\nu_r,$$

or more conveniently

$$\Phi = dN = n\theta_3, \tag{7.16}$$

where $\theta_3 := \nu_r dx^r$ as in the Hotine leg calculus of CHAPTER IV. We now verify that this Φ identically satisfies (7.14). For the left hand side this is obvious since

$$d\Phi = d(dN) = d^2N = 0$$

by the Poincaré property. The right hand side is less obvious, but

$$d\Phi = dn \wedge \theta_3 + nd\theta_3 \tag{7.17}$$

so

$$\Phi \wedge d\Phi = n^2\theta_3 \wedge d\theta_3 \tag{7.18}$$

which is identically zero by IV-(6.9) and property (T).

In concluding our discussion of [i] let us briefly consider the following argument which is based on that employed in proving the Malus-Dupin theorem in geometric optics (see HERZBERGER (1958)). Let

$$\mathbf{S} : x^r(u,v)$$

be an arbitrary base surface of Σ and

$$\tilde{\mathbf{S}} : x^r(u,v) = x^r(u,v) + \varepsilon\nu^r(u,v) \tag{7.19}$$

be a neighboring surface of **S**. Note that both **S** and $\tilde{\mathbf{S}}$ are assumed to have the same Gaussian parametrization u^α, ε is a parameter labelling members of Σ which *could* depend on the u^α, and ν^r are the components of a unit tangent vector of a congruence Γ — however Γ need not be normal to **S**; SYNGE (1937) calls such a Γ a *skew congruence.* Then covariant differentiation of (7.19) yields

$$\tilde{x}^r_\alpha = x^r_\alpha + \varepsilon_\alpha\nu^r + \varepsilon\nu^r_{,\alpha} \tag{7.20}$$

where $\varepsilon_\alpha := \varepsilon_{;\alpha}$. If we require $\Gamma \perp \tilde{\mathbf{S}}$, then $\tilde{x}^r_\alpha\nu_r = 0$ and (7.20) gives

$$x^r_\alpha\nu_r + \varepsilon_\alpha = 0 \tag{7.21}$$

which is a partial differential equation which ε must satisfy. Clearly (7.21) is identically satisfied when $\Gamma \perp$ **S** and ε is a constant, and the obvious question is whether other solutions are possible. Contraction of the above equation by du^α yields

$$dx^r\nu_r + d\varepsilon = 0$$

viz.

$$d\varepsilon = -\nu_r dx^r = -\theta_3 \tag{7.22}$$

where θ_3 is the spatial Pfaffian form associated with **S**. Since ε is a scalar we must have $d(d\varepsilon) = d^2\varepsilon = 0$ which requires that $d\theta_3 =$

0. Hence, by the leg calculus the direction[2] ν_r associated to θ_3 by (7.22) is that of a congruence $\boldsymbol{\Gamma}$ which is normal to $\mathbf{S}$ and has zero curvature $\chi = 0$; i.e., $\boldsymbol{\Gamma}$ is a normal *linear* congruence $\boldsymbol{\Lambda}$. This derivation suggests that for a normal congruence, the case $\boldsymbol{\Gamma} = \boldsymbol{\Lambda}$ is, in some sense, the most natural choice.

Indeed the choice $\boldsymbol{\Gamma} = \boldsymbol{\Lambda}$ is the one which is of importance in geometric optics which explains the notion of parallel systems $\mathcal{P}$ discussed in SECTION 4. SYNGE (1937) shows how virtually all of this theory can be deduced from Fermat's Principle, and that in a single ordinary optical medium light travels along straight lines, i.e. $\boldsymbol{\Gamma} = \boldsymbol{\Lambda}$.

VII.8 The Surface-Forming Property

The situation described in [ii] is called the *surface-forming property* of a congruence of curves $\boldsymbol{\Gamma}$. It is far more complicated than [i] and — in general — given a $\boldsymbol{\Gamma}$, there need not exist a family of surfaces $\boldsymbol{\Sigma}$ which are orthogonal to the $\boldsymbol{\Gamma}$. The problem is further complicated by the fact that the majority of the discussions of [ii] which occur in the literature concern only the case when $\boldsymbol{\Gamma} = \boldsymbol{\Lambda}$, i.e. the parallel systems $\mathcal{P}$! The general discussion when $\boldsymbol{\Gamma} \neq \boldsymbol{\Lambda}$ is due to BELTRAMI (1864) and does not appear to be well-known. We will present his theory first and then note three variants which occur when $\boldsymbol{\Gamma} = \boldsymbol{\Lambda}$. The latter are of interest to differential geodesy only under very restrictive situations, e.g. in the local version of the **EP** when $\mathcal{P}$ is appropriate; however, it is useful to recognize them explicitly since often the presentations which are given do not make it clear that they apply only when $\boldsymbol{\Gamma} = \boldsymbol{\Lambda}$.

Beltrami's analysis begins by assuming that the local coordinates of points on the congruence are given by equations of the form

$$\boldsymbol{\Gamma} : x^r = x^r\left(u^\alpha, \varepsilon\right) \tag{8.1}$$

where u^α are a pair of parameters — remember $\boldsymbol{\Gamma}$ is a two-

[2] We have denoted this direction by ν, however ν need not be precisely the third vector of the Hotine vectorial 3-leg.

parameter system — and ε is a parameter which defines points on an individual curve **C** of $\boldsymbol{\Gamma}$ (it need not be arc-length!). It is convenient to rewrite the above expression as

$$\boldsymbol{\Gamma} : x^r = x^r\left(u^i\right) \tag{8.2}$$

where $u^i := (u^\alpha, \varepsilon)$ is a generic symbol for the indicated parameters, and the new parameter index i ranges over the values 1, 2, 3 and obeys the summation convention. Then clearly

$$dx^r = \frac{\partial x^r}{\partial u^i} du^i \tag{8.3}$$

and it is convenient to introduce the following special notation:

$$x^r_i := \frac{\partial x^r}{\partial u^i}, \quad x^r_\varepsilon := \frac{\partial x^r}{\partial \varepsilon}. \tag{8.4}$$

Since ε is a parameter along C, we have

$$x^r_i := (x^r_\alpha, \nu^r) \tag{8.5}$$

where x^r_α are normals to $\boldsymbol{\Gamma}$ and ν^r is a tangent to $\boldsymbol{\Gamma}$ (since ε is not necessarily an arc-length ν^r, need not be a unit vector). Thus the equation

$$\nu_r dx^r = 0 \tag{8.6}$$

describes tangential displacements to $\boldsymbol{\Sigma}$.

We now seek to translate (8.6) into an expression involving the u^i parameters and the quantities given in (8.4). Then (8.6) becomes

$$\delta_{rs} x^r_\varepsilon x^s_i du^i = 0 \tag{8.7}$$

and defining

$$\xi_i := \delta_{rs} x^r_\varepsilon x^s_i \tag{8.8}$$

this condition assumes the form

$$\Psi := \xi_i du^i = 0, \tag{8.9}$$

where Ψ is *Beltrami's form*, which is a u^i-parametric analogue of (7.7). The condition that the parametric Pfaffian form Ψ be *completely integrable* is then given by

$$\varepsilon^{ijk} \xi_i \xi_{j;k} = 0, \tag{8.10}$$

or more elegantly by

$$\Psi \wedge d\Psi = 0 \tag{8.11}$$

which is a parametric analogue of (7.14). This condition, which we call the Beltrami condition, is necessary and sufficient for the congruence $\boldsymbol{\Gamma}$ to be surface-forming. By analogy with three cases enumerated previously, BELTRAMI (1864) indicated the following cases:

Case 1⁰ If (8.11) is identically satisfied then (8.9) admits a finite integral containing a single arbitrary constant and hence there is an infinite number of surfaces of $\boldsymbol{\Sigma}$ which are orthogonal to $\boldsymbol{\Gamma}$.

Case 2⁰ If (8.11) is not identically satisfied, it may be possible to establish a relationship between the u^i which satisfy (8.9) but is not a particular integral of (8.11), and there exists only a single surface **S** of $\boldsymbol{\Sigma}$ which is orthogonal to $\boldsymbol{\Gamma}$.

Case 3⁰ If neither Cases 1⁰ nor 2⁰ are satisfied, there exists no surface orthogonal to $\boldsymbol{\Gamma}$ and the surface-forming property is not satisfied.

Note that (7.14) and (8.11) are formally similar but the latter is more stringent and hence less likely to be satisfied in practice.

Heuristically we may contrast the difference between (7.14) and (8.11) as follows:

(i) the Euler condition $\Phi \wedge d\Phi = 0$ is an integrability condition on the coordinate dependent functions of $\boldsymbol{\Sigma}$ which permit it to define a normal congruence $\boldsymbol{\Gamma}$;

(ii) the Beltrami condition $\Psi \wedge d\Psi = 0$ is an integrability condition on the parameter dependent functions on $\boldsymbol{\Gamma}$ which permit it to admit an orthogonal family $\boldsymbol{\Sigma}$ of surfaces.

i) *Kummer's (Linear) Ray Systems*

This was originally given in KUMMER (1860), however, we derive it — in two different versions — via the Beltrami theory.

The first step is to take ε to be the arc-length along $\boldsymbol{\Lambda}$ and choose

$$\xi_3 = 1.$$

Then (8.10) reduces to

$$\varepsilon^{\alpha\beta}\left(\xi_{\alpha;\beta} + \xi_\alpha \xi_{\beta\varepsilon}\right) = 0 \tag{8.12}$$

where

$$\xi_{\beta\varepsilon} := \frac{\partial \xi_\beta}{\partial \varepsilon}. \tag{8.13}$$

The next step is to *assume* that $\tilde{x}^r(u^\alpha; \varepsilon)$ has the form

$$\tilde{x}^r(u, v) = x^r(u, v) + \varepsilon \nu^r(u, v) \tag{8.14}$$

where ε is a constant parameter. Then (8.8) yields

$$\xi_\alpha := \delta_{rs} \nu^r x^s_\alpha \tag{8.15}$$

with

$$\xi_{\alpha\varepsilon} = 0. \tag{8.16}$$

Now (8.12) in this situation reduces to

$$\varepsilon^{\alpha\beta} \xi_{\alpha;\beta} = 0, \tag{8.17}$$

viz.

$$\xi_{\alpha;\beta} - \xi_{\beta;\alpha} = 0, \tag{8.18}$$

or explicitly we obtain

$$\delta_{rs} \nu^r{}_{,\alpha} x^s_\beta = \delta_{rs} \nu^r{}_{,\beta} x^s_\alpha \tag{8.19}$$

which is known as *Kummer's equation* and is a parametrized version of (8.10) or (8.11).

Suppose ν^r is specified in terms of x^r, then (8.10) becomes

$$\begin{aligned}(\nu_{2;3} - \nu_{3;2}) \frac{\partial(y, z)}{\partial(u, v)} &+ (\nu_{3;1} - \nu_{1;3}) \frac{\partial(z, x)}{\partial(u, v)} \\ &+ (\nu_{1;2} - \nu_{2;1}) \frac{\partial(x, y)}{\partial(u, v)} = 0\end{aligned} \tag{8.20}$$

which upon introducing

$$\zeta^r := \varepsilon^{rmn} \nu_{m;n} \tag{8.21}$$

permits (8.20) to be written more concisely as

$$\zeta^r \frac{\partial (x^s, x^t)}{\partial (u, v)} \varepsilon_{rst} = 0. \tag{8.22}$$

If **S** is given implicitly by (7.1), then by (7.2) we have

$$F_r x^r_\alpha = 0. \tag{8.23}$$

Hence, since

$$F_r = \rho \varepsilon_{rst} \frac{\partial (x^s, x^t)}{\partial (u, v)}, \tag{8.24}$$

where ρ is a non-zero proportionality factor, (8.22) becomes

$$\zeta^r F_r = 0 \tag{8.25}$$

or

$$\varepsilon^{rst} \nu_{r,s} F_t = 0 \tag{8.26}$$

and by (7.2) we recognize this as being equivalent to (7.13) with the identification (7.15). In particular, by contraction of (8.15) with du^α we obtain

$$\xi_\alpha du^\alpha = \delta_{rs} \nu^r x^s_\alpha du^\alpha = \delta_{rs} \nu^r dx^s = \nu_r dx^r, \tag{8.27}$$

which shows that — in this special situation — the property of a congruence being surface forming implies that a system of surfaces is also congruence forming. In other words

$$\text{[ii]} \Rightarrow \text{[i] but [i]} \not\Rightarrow \text{[ii]}$$

when $\boldsymbol{\Gamma} = \boldsymbol{\Lambda}$.

ii) *Line geometry approach*

Tentatively we will ascribe this to KOENIGS (1895), however it is possible that it is due to PLÜCKER (1868-69) and in an optical context to MALUS (1808, 1811) and HAMILTON (1828).

The theory begins with the observation that the lines of a linear congruence $\boldsymbol{\Lambda}$ can be assembled into a one-parameter family of developable surfaces in two different ways. Furthermore, the lines of $\boldsymbol{\Lambda}$ are common tangents of a pair of surfaces which are known as the *focal surfaces* of $\boldsymbol{\Lambda}$. The pair of points at which a line of $\boldsymbol{\Lambda}$ touches the focal surfaces of $\boldsymbol{\Lambda}$ are called the *focal points* of the line, and the tangent planes of the focal surfaces at the focal points of a line are said to be the *focal planes* of the line. These loci have a number of elegant properties, e.g. the focal surface is the locus of the edges of regression of the developable surfaces which can be formed from the lines of $\boldsymbol{\Lambda}$.

All of this is essentially pure line geometry, and the connection between the linear congruence $\boldsymbol{\Lambda}$ and the normal congruences of Gaussian differential geometry is given by the following result:

A necessary and sufficient condition that the lines of a given (linear) congruence $\boldsymbol{\Lambda}$ be the normals of some surfaces $\mathbf{S}$ of $\boldsymbol{\Sigma}$ is that the focal planes through every line of $\boldsymbol{\Lambda}$ be orthogonal to each other.

iii) *Darboux's approach*

This is given in volume 2 of the treatise DARBOUX (1915), however, it could well have been common knowledge among French geometers several decades earlier, viz. KOENIGS (1895) and VESSIOT (1906). The result is nice in that it involves the notion of a Malus cone $\mathbf{C}_M$, and having determined the equation of $\mathbf{C}_M$ one can readily rederive the Euler condition (7.14). Thus, Darboux's result may be stated as follows:

The necessary and sufficient condition that the lines of $\boldsymbol{\Lambda}$ admit an orthogonal family of surfaces $\boldsymbol{\Sigma}$ is that the *Malus cone* $\mathbf{C}_M$ relative to each point of $\mathbf{E}_3$ be equilateral.

We first derive the equation of $\mathbf{C}_M$. Upon differentiation of

$$\tilde{x}^r = x^r + \varepsilon\nu^r \tag{8.28}$$

we obtain

$$0 = dx^r + \varepsilon d\nu^r + \nu^r d\varepsilon, \tag{8.29}$$

and elimination of ε and $d\varepsilon$ from these equations yields

$$\varepsilon_{rmn}\nu^r dx^m d\nu^n = 0. \tag{8.30}$$

Now

$$d\nu^n = \nu^n_{;t} dx^t,$$

and substitution of this into (8.30) and defining

$$M_{rs} := \varepsilon_{pqr} \nu^p \nu^q_{;s} \tag{8.31}$$

the equation of $\mathbf{C}_M$ becomes

$$M_{rs} dx^r dx^s = 0. \tag{8.32}$$

Then we that

$$\delta^{rs} M_{rs} = 0 \tag{8.33}$$

is equivalent to the Euler condition written in the form (7.13).

VII.9 Conclusions

An excellent historical study of the geometry of line system is given in the thesis of ATZEMA (1993).

The discussion given in the previous sections strongly suggests that the prospects of obtaining a *general solution* of the **EP** are not encouraging. The first two of these sections essentially disposed of the obvious candidates, i.e. the systems $\mathcal{O}_3$ or $\mathcal{O}_3^*$ and $\mathcal{P}$, as solutions except under very restrictive and idealized conditions; while the latter section indicated why it is unlikely — in general — to expect that both integrability conditions, viz.

$$\Psi \wedge d\Phi = 0, \tag{9.1}$$

$$\Psi \wedge d\Psi = 0, \tag{9.2}$$

of the subsidiary problems will be identically satisfied. Indeed, as noted in SECTION 5, (9.1) poses no particular difficulty provided that the surfaces of Σ are sufficiently smooth. However, (9.2) is highly non-trivial and when it is not identically, but only *conditionally satisfied*, there may exist only a *finite number* of solutions, i.e. a Σ_n, *not* a 1-parameter family of surfaces (5.2) or (7.1) as one would naively expect.

On the other hand, while (9.1) and (9.2) are demanding — as integrability conditions — they insure only the existence of solutions, viz. the compatibility/consistency of the basic geometric equations, but not the equations which exhibit the functional form of the *developmental equations* $x^r(u^\alpha;\varepsilon)$ subject to the crucial assumption that $u^\alpha = u_1^\alpha = u_\alpha^\alpha = \cdots$. This is clear since in differential geometry the choice of u^α on **S** is arbitrary, and likewise ε need not always be the arc-length along $\boldsymbol{\Gamma}$. Indeed, Beltrami recognized that the latter was an *additional* assumption and indicated it apart from his general considerations. Likewise, Kummer, notwithstanding the elegance of his analysis, is obliged to *assume* that $\tilde{x}^r$, i.e. our $\tilde{x}^r(u^\alpha;\varepsilon)$, has the functional form

$$\tilde{x}^r(u^\alpha) = x^r(u^\alpha) + \varepsilon\nu^r(u^\alpha). \tag{9.3}$$

In other words, the system (9.1), (9.2) leaves *unspecified* both the identification of the parameters $u^i \equiv (u^\alpha;\varepsilon)$ on $\boldsymbol{\Sigma}$ and $\boldsymbol{\Gamma}$, as well as the explicit functional form of the $x^r(u^\alpha;\varepsilon)$. Hence, while the conditions (9.1), (9.2) are inadequate to completely solve the **EP**, their failure renders the problem totally intractable.[3] This may strike the reader as a very unsatisfactory situation; however we believe — after due consideration — that this is precisely as it should be! It is our contention that the choice of *both* the parameters u^i *and* the form of the developmental equations are boundary/initial conditions which must be selected by taking into account the geodetic/geometric context of the geophysical situation. Conversely, if this were not the case then there would be some privileged choice of these quantities which would *a fortiori* be independent of these geodetic/geometric conditions.

Our view also has the advantage that it relieves us of the requirement of seeking a common choice of these quantities for the general **EP**. In particular, it suggests the naturalness of making these choices *separately* for the respective local, global, and limiting problems. Thus, the hope of a general solution of the **EP** — in the sense that it would automatically accommodate

[3]**We have in mind here that (9.1) is satisfiable, however (9.2) cannot be even conditionally satisfied.**

all three of the afore-mentioned problems — is a chimerical pipe dream. This viewpoint — far from being discouraging or negative — frees us from trying to solve an undetermined problem and encourages us to seek particular solutions of the local, global and limiting problems.

PROBLEMS FOR CHAPTER VII

VII.1 (Research Problem) Give a new proof of the Fundamental Theorem which makes Pizzetti's argument more rigorous. [Pizzetti's proof is essentially an intuitive one — which although fundamentally sound — makes no contact with mathematical rigor at the level of advanced calculus.]

VII.2 Redo the analysis given in PIZZETTI (1901), or the references cited in PROBLEM VII.4, by changing the numerical factor on the right-hand side of $(\mathcal{P}_{\mathrm{I}})$. Show that the resulting Pizzetti inequality is not sensibly changed, viz. (3.1) still holds with the integer 5, for either the m-inequality (2.2) or the one suggested in PROBLEM VII.3 when one replaces the numerical factor of $\frac{1}{100}$ in $(\mathcal{P}_{\mathrm{I}})$ by an ε

$$\frac{1}{1000} < \varepsilon < \frac{1}{100}.$$

[Such choices of ε are messier than $\frac{1}{100}$ and as indicated, do *not significantly improve* the accuracy. Hence $\frac{1}{100}$ is a nice choice!]

VII.3 Redo the analysis by replacing the numerical factor on the right-hand side of $(\mathcal{P}_{\mathrm{II}})$ by a modern value, i.e.

$$m < \frac{1}{290},$$

see LAMBECK (1988).

VII.4 The Pizzetti theorem despite its basic importance is seldom cited in the literature. Read the brief discussion of it in PIZZETTI (1906), pages 123-130; HOPFNER (1949), pages 152-154; and the pictures given — without reference to the Pizzetti inequality (3.1) — on page 537 of JUNG (1956). The derivation in HOPFNER (1933), pages 306-310, essentially reproduces Pizzetti's argument with little alteration. Finally, a delightful exposition of the result, with a reference to Laplace and nice pictures, may be found in LAMBERT (1921). Note that Lambert makes no reference to Pizzetti or his work!

VII.5 Within the context of the restricted 3-body problem in celestial mechanics, i.e. the sun, a planet, and a satellite, deduce the following notions of a gravitation sphere:

(i) a *sphere of activity* in which a planet is regarded as the central body and the sun is the perturbing body;

(ii) a *gravitational sphere of a planet* in which the planetary attraction is greater than the solar attraction;

(iii) a *Hill gravitational sphere* whose surface can be regarded as the theoretical boundary of the system of satellites of a given planet.

Compute these values for the moon and sun. [The notion of a sphere of activity is due to Laplace (1805) and can be found in his *Traité de méchanique céleste*, Vol. IV, Book IX, Chap. 2 i.e. LAPLACE (1839).]

CHEBOTAREV (1964)

VII.6 Read BODE and GRAFAREND (1982).

VII.7 Given an ellipse $C : x^2/a^2 + y^2/b^2 = 1$, which is a curve of order 2, show that the curve C_1 *parallel* to C which is defined as the locus of unit normals to C is a curve of degree 8. This is the plane analogue of a parallel system of surfaces.

CAYLEY (1860a), SALMON (1879)

VII.8 Given an ellipsoid $\mathbf{S} : x^2/a^2 + y^2/b^2 + z^2/c^2 = 1$ which is a surface of order 2, show that the parallel surface $\mathbf{S}_1$ to $\mathbf{S}$ which is defined by the locus of unit normals to $\mathbf{S}$ is a surface of order 12 when $a \neq b \neq c$ and a surface of order 8 when $a = b \neq c$.

CAYLEY (1860b), ROBERTS (1872)

VII.9 Given a surface $\mathbf{S}$ in $\mathbf{E}_3$ of order n, prove that its parallel surface $\mathbf{S}_1$ has order $2\,(n^3 - n^2 + n)$.

ROBERTS (1873)

VII.10 The theory of parallel surfaces in $\mathbf{E}_3$ for the linear case $\boldsymbol{\Gamma}=\boldsymbol{\Lambda}$ is reasonably well-known, but often very inadequately discussed in the differential-geometric literature (the restriction to $\boldsymbol{\Lambda}$ being usual implicit rather than explicitly stated). For a reasonably complete discussion, read the treatment in Section 24, pages 102-107, of THOMAS (1965).

[Note, that parallel surfaces are discussed in [Chapter 17, see in particular his pages 118-119] of Hotine's treatise as an example of one of his coordinate systems. The exposition given in THOMAS (1965) is a nice companion to Hotine's treatment.]

VIII

Algebraic Theory of the Marussi Tensor

VIII.1 Introduction

The *Marussi tensor*, or the *tensor of gravity gradients*, or simply the *gravity tensor*, has long been regarded as the centerpiece of differential geodesy. It was introduced in MARUSSI (1949) primarily by using the homographic calculus and in MARUSSI (1951a) he called it the Eötvös tensor since, as he noted, "... it was this geodesist who gave the first systematic consideration to the application of the second derivatives of the potential to geodetic and geophysical problems." In his formulation the theory of the tensor is essentially that of the Eötvös and generalized Burali-Forti homographies. As he noted, up to second order this tensor systematically synthesizes all of the dynamical and geometric properties of the Earth's gravity field.

Hotine (1957a, 1969) gave the most detailed purely tensorial study of this tensor, and called it the Marussi tensor. Unfortunately, Hotine's analysis in the later reference was scattered over several chapters and sections of his book, viz. Chapters 12 and 20, and this investigation made extensive use of his general coordinate system (ω, ϕ, N) where ω is the longitude, ϕ the latitude, and N is the geopotential. Hence, many general results and properties occur in HOTINE (1957a, 1969) *only* in this coordinate system. Thus, it is desirable to see how much of the theory of the Marussi tensor can be carried out without making a choice of a particular local coordinate system in $\mathbf{E}_3$. Indeed, the vector 3-leg calculus offers a natural framework in which to develop the theory of the Marussi tensor.

VIII.2 Eigenstructure of Symmetric Tensors

In this section, as a preliminary to the investigation of the eigenstructure of the Marussi tensor, we will consider the algebraic structure of an arbitrary symmetric tensor. This involves not only consideration of the *eigenvectors/eigenvalues* of this tensor, but also the *basic system* of its *algebraic invariants.* Such study is well-known mathematically; physically, it plays an important rôle in classical elasticity theory, where the tensor in question may be taken to be either the stress or strain tensor. There the eigenvectors correspond to the principal directions of stress or strain, while the eigenvalues give the respective lengths of these directions. In particular, we will be interested in the geometric realization of this situation where the tensor defines a quadric surface in Euclidean space and the principal directions correspond to the principal axes of this surface.

Let $\mathbf{T}$ be a symmetric second order tensor which has covariant components T_{rs} where $r, s = 1, 2, 3$. Then the algebraic invariants of $\mathbf{T}$ are determined by the third order determinant

$$\det \|T_{rs} - \tau g_{rs}\| = 0 \tag{2.1}$$

where τ is an unknown scalar, and g_{rs} is the metric tensor in $\mathbf{E}_3$. Equation (2.1) is often called the *secular equation*, and when the components T_{rs} and g_{rs} are known, it gives a cubic equation for the determination of τ.

There are two methods for studying this algebraic problem: the *matrix approach* and the *tensorial approach.* We will follow the former approach; the key steps in the latter approach will be outlined in the Problems at the end of this chapter. The matrix approach begins by rewriting (2.1) as

$$\det(\mathbf{T} - \tau \mathbf{g}) = 0, \tag{2.2}$$

where

$$\mathbf{T} := \|T_{rs}\|, \quad \mathbf{g} := \|g_{rs}\|. \tag{2.3}$$

In particular, relative to an orthonormal 3-leg such as the Hotine 3-leg of CHAPTER VI, we may choose

$$\mathbf{g} = \|\delta_{rs}\| = \mathbf{I} \tag{2.4}$$

where $\mathbf{I}$ is the 3×3 identity matrix. Then upon expansion (2.2) yields

$$J_3 - J_2\tau + J_1\tau^2 - \tau^3 = 0 \tag{2.5}$$

where J_1, J_2, J_3 are the basic algebraic invariants of the matrix $\mathbf{T}$. We will refer to them as 'J-invariants' in order to clearly distinguish them from the 'I-invariants' which are studied in PROBLEMS VIII.3 and VIII.4. Matrically these invariants may be expressed as follows:

$$\begin{aligned} J_1 &:= \operatorname{tr} \mathbf{T}, \\ J_2 &:= \operatorname{tr} \operatorname{adj} \mathbf{T}, \\ J_3 &:= \det \mathbf{T}, \end{aligned} \tag{2.6}$$

where 'tr' denotes the *trace*, and 'adj' the *adjoint* of the indicated matrix. The subscripts on the J-invariants denote the degree of these invariants in terms of the τ-roots of (2.2). Indeed, if these τ-roots are known and denoted by τ_1, τ_2, τ_3, respectively, then when $\mathbf{T}$ is reduced to its canonical (diagonal) form, then it is easy to check that

$$\begin{aligned} J_1 &= \tau_1 + \tau_2 + \tau_3, \\ J_2 &= \tau_1\tau_2 + \tau_2\tau_3 + \tau_3\tau_1, \\ J_3 &= \tau_1\tau_2\tau_3. \end{aligned} \tag{2.7}$$

Let ξ^r denote an arbitrary set of three variables, technically called *indeterminates,* which have <u>no relation</u> to the local coordinates x^r that occur as variables in the components T_{rs} of the tensor $\mathbf{T}$. Then in the Euclidean space $\mathbf{E}_3$ having ξ^r as variables, the *central quadric* Ω associated with $\mathbf{I}$ has the form

$$T_{rs}\xi^r\xi^s = \pm\epsilon^2 \tag{2.8}$$

where ϵ is a constant. Matrically (2.8) can be written as

$$\xi'\mathbf{T}\xi = \pm\epsilon^2 \tag{2.9}$$

where

$$\xi = \left\|\xi^1\xi^2\xi^3\right\|' \tag{2.10}$$

and the dash denotes the transpose operation, i.e. ξ is a 3×1 column matrix. When the τ-roots are determined then Ω is reduced to canonical (diagonal) form

$$\tau_1 \left(\xi^1\right)^2 + \tau_2 \left(\xi^2\right)^2 + \tau_3 \left(\xi^3\right)^2 = \pm\epsilon^2, \tag{2.11}$$

and the specific geometric character of Ω, and its existence as a *proper*, i.e. *non-degenerate*, *quadric* depends on the values of the J-invariants, or ultimately on the τ-roots.

A vector $\mathbf{v}$ is said to be an *eigenvector* of $\mathbf{T}$ whenever

$$\mathbf{T}\mathbf{v} = \tau\mathbf{v} \tag{2.12}$$

and τ is the *eigenvalue* corresponding to this eigenvector. Note that given $\mathbf{T}$ and τ with $\mathbf{v}$ unknown, (2.12) is a set of three linear equations for the components of $\mathbf{v}$, and upon rewriting this equation as

$$(\mathbf{T} - \tau\mathbf{I})\,\mathbf{v} = 0 \tag{2.13}$$

we see that (2.2) with $\mathbf{g} = \mathbf{I}$ is the necessary and sufficient condition for the existence of a non-trivial (solution) vector $\mathbf{v}$. Since there are three τ-roots of the secular equation, the inequality of these roots insures the existence of three eigenvectors, which we denote by $\mathbf{v}_{(1)}$, $\mathbf{v}_{(2)}$ and $\mathbf{v}_{(3)}$, respectively. If any of the τ-roots vanishes, then by the last equation in (2.6) and (2.7), respectively, we have

$$J_3 = \det \mathbf{T} = 0 \tag{2.14}$$

and Ω is not a proper quadric. In other words, we are primarily interested in the case when there exist three distinct non-zero τ-roots and this amounts to requiring that

$$J_3 = \det \mathbf{T} \neq 0. \tag{2.15}$$

The classification of central quadrics in $\mathbf{E}_3$ has long been known, and an exhaustive treatment is readily available in the literature. Including the possibilities of complex or pure imaginary τ-roots, *seventeen* types of quadratic loci occur. However, only *four* of these are *real* and satisfy (2.15). For $\epsilon \neq 0$ these are

(i) an ellipsoid,

(ii) a hyperboloid of one-sheet,

(iii) a hyperboloid of two-sheets;

and when $\epsilon = 0$

(iv) a quadric cone.

The cases (i)-(iii) occur in elasticity theory, and seem to be the prime candidates for having a physical interpretation in differential geodesy.

On the other hand, the tensorial approach offers the simplest method for determining the eigenstructure of **T** when a canonical leg representation is available. For example, in such a case (2.12), or equivalently

$$T_{rs}\mathbf{v}^s = \tau \mathbf{v}_r \tag{2.16}$$

is very simple when **T** is expressed in a 3-leg representation, e.g. the Hotine 3-leg $\{\boldsymbol{\lambda}, \boldsymbol{\mu}, \boldsymbol{\nu}\}$. One merely expresses **v** as a linear combination of the 3-leg vectors, and examines

$$T_{rs}\lambda^s, \quad T_{rs}\mu^s, \quad T_{rs}\nu^s$$

etc. Then (2.16) is obtained by forming the linear combinations.

We now note that having determined the values of the J-invariants finding the eigenvalues requires solving this cubic equation. Depending on the values of the J-invariants which are the coefficients in this cubic equation, this may be a complicated process. Some insight into it may be obtained by considering the so-called *discriminant* of (2.5). In doing so it is convenient to denote the *negative* of this equation by $f(\tau)$ and write

$$f(\tau) := a_0\tau^3 + a_1\tau^2 + a_2\tau + a_3 = 0, \tag{2.17}$$

in our case,

$$a_0 = 1, \quad a_1 = -J_1, \quad a_2 = J_2, \quad a_3 = -J_3. \tag{2.18}$$

The discriminant D of $f(\tau)$ is given by VAN DER WAERDEN (1966) as

$$D = a_1^2 a_2^2 - 4a_0 a_2^3 - 4a_1^3 a_3 - 27a_0^2 a_3^2 + 18a_0 a_1 a_2 a_3, \qquad (2.19)$$

which in our case reduces to

$$D = J_1^2 J_2^2 - 4J_2^3 - 4J_1^3 J_3 - 27J_3^2 + 18J_1 J_2 J_3. \qquad (2.20)$$

Then from the theory of algebraic equations, *loc. cit supra*,

$$D = a_0^4 (\tau_1 - \tau_2)^2 (\tau_2 - \tau_3)^2 (\tau_3 - \tau_1)^2 \qquad (2.21)$$

where τ_1, τ_2, τ_3 are now the τ-roots of (2.17). Consequently, we have the following results:

[i] $D = 0$ if and only if $f(\tau) = 0$ has at least *two* equal roots;

however, if all the roots of $f(\tau) = 0$ are unequal and D is real, then

[ii] $D > 0$ whenever *all* the roots are real;

[iii] $D < 0$ whenever at least *one* of the roots is complex.

Needless to say, for moderately complicated J-invariants, it may not be apparent which of these cases occurs. This indicates the difficulty of solving (2.5) and determining the eigenstructure of **T**.

Finally, we return to the general classification of proper quadrics and indicate how these are related to the J-invariants. Upon taking the minus sign on the right-hand side of (2.8), or (2.9), and defining

$$\delta := \epsilon^2 J_3 \qquad (2.22)$$

we have the following classification scheme for Ω as given in GROTEMEYER (1969):

(i) an *ellipsoid*:

$$J_2 > 0, \quad J_1 J_3 > 0, \quad \delta > 0;$$

(ii) a *hyperboloid of one-sheet*:

1) $J_2 > 0,\ J_1J_3 > 0,\ \delta > 0;$

2) $J_2 < 0, \begin{cases} J_1J_3 \geq 0 \\ \text{or} \\ J_1J_3 < 0 \end{cases}, \delta > 0;$

(iii) a *hyperboloid of two-sheets*:

1) $J_2 \geq 0,\ J_1J_3 < 0,\ \delta < 0;$

2) $J_2 < 0, \begin{cases} J_1J_3 \geq 0 \\ \text{or} \\ J_1J_3 < 0 \end{cases}, \delta < 0;$

(iv) a *quadric cone*:

1) $J_2 \geq 0,\ J_1J_3 < 0,\ \delta = 0;$

2) $J_2 < 0, \begin{cases} J_1J_3 \geq 0 \\ \text{or} \\ J_1J_3 < 0 \end{cases}, \delta = 0.$

VIII.3 Eigenstructure of the Marussi Tensor

We now apply the theory of SECTION 2 to the Marussi tensor with the goal of exhibiting the basic invariants and the eigenvalues/eigenvectors of this tensor.

First, we note that by virtue of the canonical leg representations II-(2.13),

$$g_{rs} = \lambda_r\lambda_s + \mu_r\mu_s + \nu_r\nu_s, \tag{3.1}$$

and VI-(5.11),

$$\begin{aligned} N_{rs} &= -nk_1\lambda_r\lambda_s - nt_1\left(\lambda_r\mu_s + \mu_r\lambda_s\right) \\ &\quad -nk_2\mu_r\mu_s + n\gamma_1\left(\lambda_r\nu_s + \nu_r\lambda_s\right) \\ &\quad +n_{/3}\nu_r\nu_s + n\gamma_2\left(\mu_r\nu_s + \nu_r\mu_s\right). \end{aligned} \tag{3.2}$$

Upon displaying the coefficients of the products of the 3-leg vectors in an obvious manner (corresponding to the usual enumeration) we have the following matrices:

$$\mathbf{g} = \mathbf{I} = \begin{Vmatrix} 1 & 0 & 0 \\ 0 & 1 & 0 \\ 0 & 0 & 1 \end{Vmatrix} \tag{3.3}$$

$$\mathbf{N} = \begin{Vmatrix} -nk_1, & -nt_1, & n\gamma_1 \\ -nt_1, & -nk_2, & n\gamma_2 \\ n\gamma_1, & n\gamma_2, & n_{/3} \end{Vmatrix} \tag{3.4}$$

relative to the Hotine 3-leg of CHAPTER VI. Evaluation of $\det(\mathbf{N} - \tau\mathbf{I})$, or equivalently,

$$\det \begin{Vmatrix} nk_1 + \tau & nt_1 & -n\gamma_1 \\ nt_1 & nk_2 + \tau & -n\gamma_2 \\ n\gamma_1 & n\gamma_2 & n_{/3} - \tau \end{Vmatrix},$$

leads to the following values of the J-invariants

$$\begin{aligned} J_1 &= -2\tilde{\omega}^2, \\ J_2 &= 4Hn\tilde{\omega}^2 - 4\{4H^2 - K + \chi^2\}n^2, \\ J_3 &= -2Kn^2\tilde{\omega}^2 + [2HK + (k_1\gamma_2^2 + k_2\gamma_1^2 - 2t_1\gamma_1\gamma_2)]\,n^3 \end{aligned} \tag{3.5}$$

where we have employed the curvatures H and K of the N-surface and the principal curvature χ of its normal congruence Γ.

In seeking to evaluate these quantities for particular N-surfaces, we note first that $H = 0$ and $K = 0$ respectively characterize *distinct* classes of equipotential surfaces. Moreover, for a general N-surface we expect that both H and K are non-zero, and of course physically we must have $n \neq 0$ and $\tilde{\omega} \neq 0$. Hence, we have

(i) $J_1 \neq 0$ always,

(ii) $J_2 \neq 0$ in general, but $J_2 = 0$ for a *particular value* of $\tilde{\omega}^2$,

(iii) $J_3 \neq 0$ for a proper quadric Ω (recall 2.15)).

We call the quadric Ω defined by (2.8) and (2.9), with $\mathbf{T}$ replaced by Marussi tensor $\mathbf{N}$, the *Marussi quadric.* By analogy with elasticity theory we will restrict our analysis to proper central quadrics. Then (i) and (iii) are general results. The particular value of $\tilde{\omega}^2$ which makes J_2 vanish is determined by

$$H\tilde{\omega}^2 = \left(4H^2 - K + \chi^2\right) n. \tag{3.6}$$

For a sphere-like (ellipsoidal) equipotential surface, by virtue of Hotine's conventions $n > 0$, with $k_1 < 0$, $k_2 < 0$. Thus, $H < 0$, and indeed for a sphere of radius r_0,

$$H = -\frac{1}{r_0}, \quad K = \frac{1}{r_0^2}.$$

Hence (3.6) is sensible only when the following *K-inequality* is satisfied:

$$K > 4H^2 + \chi^2, \tag{3.7}$$

and then, subject to the restriction imposed by this inequality when $H \neq 0$, the resulting real value of $\tilde{\omega}$ is given by

$$\tilde{\omega} = \sqrt{(4H^2 + \chi^2 - K)\, n/H}. \tag{3.8}$$

Clearly if (3.7) is not satisfied, then $\tilde{\omega}$ will be pure imaginary. Note that if $H = 0$, by replacing K and χ by their values in terms of the curvature parameters, the right-hand side of (3.6) reduces to

$$\left(k_1^2 + t_1^2 + \gamma_1^2 + \gamma_2^2\right) n = 0. \tag{3.9}$$

Then, since $n \neq 0$ and $k_2 = -k_1$, this equation requires that all the curvature parameters k_1, k_2, t_1, γ_1, γ_2 vanish which means that the equipotential surface is a plane.

By using (3.2) the eigenstructure of the Marussi tensor is very simply obtained. One merely contracts the canonical leg representation of N_{rs} successively with the leg vectors $\boldsymbol{\lambda}$, $\boldsymbol{\mu}$ and $\boldsymbol{\nu}$, which yields

$$\begin{aligned} N_{rs}\lambda^s &= -nk_1\lambda_r - nt_1\mu_r + n\gamma_1\nu_r, \\ N_{rs}\mu^s &= -nt_1\lambda_r - nk_2\mu_r + n\gamma_2\nu_r, \\ N_{rs}\nu^s &= n\gamma_1\lambda_r + n\gamma_2\mu_r + 2\left(Hn - \tilde{\omega}^2\right)\nu_r. \end{aligned} \tag{3.10}$$

These equations show that *none* of the leg vectors, or simple combinations of them with constant coefficients, are automatically eigenvectors of the Marussi tensor. The leg vectors are eigenvectors only in the presence of rather severe geometric conditions. Indeed, upon inspection of (3.10), we have the

Eigenstructure Theorem

The Marussi tensor $\mathbf{N}$ admits the following leg vectors as eigenvectors

[i] $\boldsymbol{\lambda}$ with eigenvalue $-nk_1$ if and only if $t_1 = 0$ and $\gamma_1 = 0$;

[ii] $\boldsymbol{\mu}$ with eigenvalue $-nk_2$ if and only if $t_1 = 0$ and $\gamma_2 = 0$;

[iii] $\boldsymbol{\nu}$ with eigenvalue $2(Hn - \tilde{\omega}^2)$, i.e. $n_{/3}$, if and only if $\chi = 0$, i.e. $\gamma_1 = 0$ and $\gamma_2 = 0$.

The only result of apparent interest is the observation that when both $\boldsymbol{\lambda}$ and $\boldsymbol{\mu}$ are eigenvectors, then $\boldsymbol{\nu}$ is also an eigenvector. The converse holds only when $t_1 = 0$. It is also interesting to note that none of the above restrictions involve the normal curvatures k_1 and k_2 of $\mathbf{S}$ in the $\boldsymbol{\lambda}$ and $\boldsymbol{\mu}$ directions while the eigenvalues do.

VIII.4 The Singularity Condition

We now consider the case when condition (iii) of SECTION 3 is not satisfied, i.e.

$$J_3 = 0 \tag{4.1}$$

which makes the quadric Ω improper. This is called the *singularity condition*, and by the third equation in (3.5), reduces to

$$2K\tilde{\omega}^2 = \left[2HK + \left(k_1\gamma_2^2 + k_2\gamma_1^2 - 2t_1\gamma_1\gamma_2\right)\right] n. \tag{4.2}$$

But as previously mentioned in SECTION 3 for a sphere-like equipotential surface, we expect that $K > 0$, $H < 0$, $k_1 < 0$,

$k_2 < 0$; so in order for the left-hand side of (4.2) to be positive, we must have

$$2HK + \left(k_1\gamma_2^2 + k_2\gamma_1^2 - 2t_1\gamma_1\gamma_2\right) n > 0, \tag{4.3}$$

or

$$2HK + \alpha\gamma_1 + \beta\gamma_2 > 0, \tag{4.4}$$

where

$$\begin{aligned} \alpha &:= k_2\gamma_1 - t_1\gamma_2, \\ \beta &:= k_1\gamma_2 - t_1\gamma_1. \end{aligned} \tag{4.5}$$

Assuming that (4.3) holds, then the resulting real value of $\tilde{\omega}$ is given by

$$\tilde{\omega} = \sqrt{\frac{\{2HK + (k_1\gamma_2^2 + k_1\gamma_1^2 - 2t_1\gamma_1\gamma_2)\}\, n}{2K}}. \tag{4.6}$$

For quadrics, unlike conics, the distinction between proper and improper loci is not so pronounced. Indeed, a quadric Ω for which (4.1) holds is said to be of *paraboloidal type*, and the surface is no longer a central quadric. In this case (2.5) reduces to

$$\tau^2 - J_1\tau + J_2 = 0, \tag{4.7}$$

which admits a pair of equal τ-roots if and only if

$$J_1^2 = 4J_2. \tag{4.8}$$

By the last equation in (2.7), when (4.1) holds, at least one τ-root is zero, and then the above equation shows that (2.11) represents a surface of revolution. It is easy to see that the *axis of the paraboloid* corresponds to an eigenvector of $\mathbf{N}$ which has zero eigenvalue. Let $\mathbf{v}$ denote this eigenvector and in terms of the Hotine vectorial 3-leg $\{\boldsymbol{\lambda}, \boldsymbol{\mu}, \boldsymbol{\nu}\}$ we have

$$\mathbf{v} = a\boldsymbol{\lambda} + b\boldsymbol{\mu} + c\boldsymbol{\nu} \tag{4.9}$$

where a, b and c are scalars. By virtue of (3.10) it follows that

$$\mathbf{N}\mathbf{v} = 0 \tag{4.10}$$

is equivalent to

$$\left\| \begin{matrix} -nk_1, & -nt_1, & n\gamma_1 \\ -nt_1, & -nk_2, & n\gamma_2 \\ n\gamma_1, & n\gamma_1, & n_{/3} \end{matrix} \right\| \cdot \left\| \begin{matrix} a \\ b \\ c \end{matrix} \right\| = \left\| \begin{matrix} 0 \\ 0 \\ 0 \end{matrix} \right\|$$

which can be written as

$$\begin{aligned} ak_1 + bt_1 - c\gamma_1 &= 0, \\ at_1 + bk_2 - c\gamma_2 &= 0, \\ a\gamma_1 + b\gamma_2 + c(\log n)_{/3} &= 0. \end{aligned} \tag{4.11}$$

The first two of these equations gives

$$\left\| \begin{matrix} k_1, & t_1 \\ t_1, & k_2 \end{matrix} \right\| \cdot \left\| \begin{matrix} a \\ b \end{matrix} \right\| = c \left\| \begin{matrix} \gamma_1 \\ \gamma_2 \end{matrix} \right\|, \tag{4.12}$$

and since $K = k_1 k_2 - t_1^2 \neq 0$, we have the solutions:

$$\begin{aligned} a &= \frac{c}{K} \left| \begin{matrix} \gamma_1, & t_1 \\ \gamma_2, & k_2 \end{matrix} \right| = \frac{c}{K}\alpha, \\ b &= \frac{c}{K} \left| \begin{matrix} k_1, & \gamma_1 \\ t_1, & \gamma_2 \end{matrix} \right| = \frac{c}{K}\beta, \end{aligned} \tag{4.13}$$

where α, β were defined in (4.5). Substitution of these values into the third equation of (4.11), gives an equation for $n_{/3}$, viz.

$$n_{/3} = -(\alpha\gamma_1 + \beta\gamma_2)\, n/K,$$

and by using VI-(7.3) we obtain

$$(2HK + \alpha\gamma_1 + \beta\gamma_2)\, n = 2\tilde{\omega}^2 K \tag{4.14}$$

which can be used to confirm that (4.4) is satisfied. The resulting components of **v** relative to the Hotine 3-leg are given by

$$\mathbf{v} = c(\alpha/K, \beta/K, 1)\,. \tag{4.15}$$

We have in (4.14) an expression for $\tilde{\omega}^2$ i.e.

$$\tilde{\omega}^2 = (2HK + \alpha\gamma_1 + \beta\gamma_2)\, n/2K \tag{4.16}$$

which is an alternate version of (4.6). However, (4.8) yields the following quartic equation

$$\tilde{\omega}^4 - 4Hn\tilde{\omega}^2 + 4\left[4H^2 - K + \chi^2\right]n^2 = 0 \qquad (4.17)$$

which can be solved for $\tilde{\omega}^2$, i.e.

$$\tilde{\omega}^2 = \left\{2H \pm 2\sqrt{K - 3H^2 - \chi^2}\right\}n, \qquad (4.18)$$

and again we have a *K-inequality*:

$$K > 3H^2 + \chi^2 \qquad (4.19)$$

which is reminiscent of (3.7).

The preceding discussion has been purely geometrical and has essentially been concerned with the question of whether the Marussi quadric Ω is proper ($J_3 \neq 0$), or improper ($J_3 = 0$). The singularity problem, is due to the fact that it has been suggested that the condition

$$\det \mathbf{N} = 0 \qquad (4.20)$$

is a physical one and in some sense characterizes a *singularity* of the external gravity field of the Earth. Hence, the corresponding analysis is known as the *geodetic singularity problem* in the literature (see the references cited in PROBLEM VIII.9). The resulting theory is very delicate and elegant in that (4.20) is related to the question of when anholonomic, i.e. Pfaffian, expressions can be replaced by ordinary, i.e. holonomic, differentials. In this respect, our discussion has considered the critical condition (4.20) without reference to any particular local coordinate systems, has exhibited the corresponding inequalities on the angular velocity $\tilde{\omega}$, and has shown that geometrically the singularity problem can be visualized in terms of whether the Marussi quadric Ω is improper or not.

Finally, a general caveat on the nature and notion of a singularity. First, the term has a clearcut definition and meaning *only* within the context of algebraic and differential topology (see MILNOR (1965) or GUILLEMIN and POLLACK (1974)) where one speaks of *singularities of a vector field* defined on a surface of $\mathbf{E}_3$. This notion is "coordinate-free" and since the singularity should occur in the components of the vector field, or perhaps in the local basis, i.e. 3-leg vectors in which it is represented, it is natural to expect that ultimately such a vectorial singularity should be manifested in terms of the local coordinates x^r on the surface. Second, there is no available procedure indicating *how* the above scheme can be carried out. In particular, there is no known process to test whether a given singularity is a 'true' one, i.e. of a geometric/topological character, or a 'fictitious' one which is due to the choice of a particular local coordinate system.

While the available work on the Geodetic Singularity Problem is ingenious, it is susceptible to the difficulties indicated in the previous paragraph. It is by no means clear that it really characterizes a singularity in the *true* sense, and much of it is explicitly coordinate dependent. It represents an intriguing and suggestive *first step*; however clearly, much remains to be done before the final words are written on singularities in differential geodesy.

VIII.5 Conclusions

The material in the previous two sections represents — to the best of our knowledge — a new approach to the algebraic theory of the Marussi tensor. It differs from the usual methodology in the fact that it is "coordinate-free" in the sense that it makes no appeal to a particular local coordinate system, and hence everything is related to the local gravity n and Hotine's five curvature parameters: k_1, k_2, t_1, γ_1, γ_2. Marussi's original approach via the Eötvös homography was also "coordinate-free", but failed to produce a well-defined computational formalism rivaling that constructed by Hotine in his treatise.

There are essentially *five* new aspects to our approach:

1) the introduction and evaluation of the J-invariants in (3.5);

2) recognition of the situation when $J_2 = 0$, the resulting special value of $\tilde{\omega}^2$ given in (3.6), and the K-inequality (3.7);

3) the eigenstructure theorem of SECTION 3;

4) the singularity problem, recognition of the situation when $J_3 = 0$, the resulting special value of $\tilde{\omega}^2$ given in (4.2), the K-inequality (4.14), and the paraboloid case when (4.8) holds;

5) the interpretation of these algebraic results in terms of the geometry of the Marussi quadric Ω.

PROBLEMS FOR CHAPTER VIII

VIII.1 Show that (2.1), or (2.2), is equivalent to the tensorial expression

$$\varepsilon^{mpr}\varepsilon^{nqs}\left(T_{mn} - \tau g_{mn}\right)\left(T_{pq} - \tau g_{pq}\right)\left(T_{rs} - \tau g_{rs}\right) = 0.$$

VIII.2 Evaluation of the expression in PROBLEM VIII.1 requires the following identities which the reader should derive

(i) $\varepsilon^{mnr}\varepsilon_{mn}{}^{s} = 2g^{rs}$,

(ii) $\varepsilon^{pmn}\varepsilon_{p}{}^{rs} = \begin{vmatrix} g^{mr} & g^{ms} \\ g^{nr} & g^{ns} \end{vmatrix}$,

(iii) $\varepsilon^{mpr}\varepsilon^{nqs} = \begin{vmatrix} g^{mn} & g^{mq} & g^{ms} \\ g^{pn} & g^{pq} & g^{ps} \\ g^{rn} & g^{rq} & g^{rs} \end{vmatrix}$.

VIII.3 Using the expressions in PROBLEMS VIII.1 and VIII.2 show that the J-invariants of $\mathbf{T}$ are given by

$$\begin{aligned} J_1 &:= \varepsilon^{mnr}\varepsilon_{mn}{}^{s}T_{rs}/2! \\ J_2 &:= \varepsilon^{pmn}\varepsilon_{p}{}^{rs}T_{mr}T_{ns}/2! \\ J_3 &:= \varepsilon^{mpr}\varepsilon^{nqs}T_{mn}T_{pq}T_{rs}/3! \end{aligned}$$

VIII.4 Introducing the I-invariants of $\mathbf{T}$ defined by

$$I_1 := g^{rs}T_{rs}$$
$$I_2 := T^{rs}T_{rs}$$
$$I_3 := T_{rs}T^{rp}T_p{}^s$$

express the J-invariants in terms of these I-invariants.

Answer: $J_1 = I_1,\ J_2 = \{I_1^2 - I_2\}/2,$
$J_3 = \{I_1^3 + 2I_3 - 3I_1I_2\}/3!$

VIII.5 Evaluate the expressions for the I-invariants in PROBLEM VIII.4 for the Marussi tensor $\mathbf{N}$.

Answers: $I_1 = \Delta N = -2\tilde{\omega}^2$
$I_2 = n_s n^s + n^2 \nu_{rs}\nu^{rs}$
$I_3 = \nu^p n_p n_q n^q + n\nu_{rs}n^r n^s$
$+n^2\nu_{rs}\nu^s\nu^{rt}n_t + n^3\nu_{rs}\nu^{rp}\nu_p{}^s$

VIII.6 Use the canonical leg representations of CHAPTER VI to evaluate the I-invariants of PROBLEM VIII.5 in terms of the five curvature parameters and the curvatures H, K and χ.

VIII.7 Prove the following two results on the τ-roots of a general symmetric tensor $\mathbf{T}$:

[i] the number of non-vanishing τ-roots is equal to the rank of the matrix $\mathbf{T}$,

[ii] a pair of τ-roots are equal if and only if the quadric Ω associated with the matrix $\mathbf{T}$ is a surface of revolution.

OLMSTED (1947)

VIII.8 Note that equations (4.16) and (4.18) give two equations for $\tilde{\omega}^2$; use these to obtain an expression involving α, β, γ_1, γ_2, K and H.

Answer: $(\alpha\gamma_1 + \beta\gamma_2)^2 - 4HK(\alpha\gamma_1 + \beta\gamma_2) + 4H^2K^2$
$= 16K^2(K - 3H^2 - \chi^2).$

VIII.9 Examine the discussions in the literature, e.g. LIVIERATOS (1976), BOCCHIO (1981, 1982) and GROSSMAN (1978), on the geodetic singularity problem.

VIII.10 Read the papers RUMMEL (1985, 1986) and RUMMEL and VAN GELDEREN (1992) on the use of the Marussi tensor in satellite/airborne gradiometry. See also P. HOLOTA (1989) for connections with boundary value problems.

VIII.11 Show that the general eigenvalue problem $\det(\mathbf{T} - \lambda\mathbf{G}) = 0$ is equivalent to the special eigenvalue problem

$$\det\left(\mathbf{T}\mathbf{G}^{-1} - \lambda\mathbf{I}\right) = 0$$

where $\mathbf{I}$ is the unit matrix. Accordingly the J-invariants are

$$J_1 := tr\mathbf{T}\mathbf{G}^{-1}, \qquad J_2 := tr\,\mathrm{adj}\,\mathbf{T}\mathbf{G}^{-1},$$

and

$$J_3 := \det\mathbf{T}\mathbf{G}^{-1}.$$

GRAFAREND (1994)

VIII.12 For a detailed version of the eigenvalue problem read GRAFAREND (1994).

IX

Conformal Differential Geodesy

IX.1 Introduction

The use of conformal transformations has a long history in the geodetic sciences. Apparently, the topic was initially taken up in 1768 by Euler in a study of transformations of orthogonal trajectories. In the next decade it was successively applied to cartography by Lambert, Euler, Lagrange, and Schubert. Indeed, in 1788, Schubert in an analysis of the stereographic projection of a spheroid (an ellipsoid of revolution) onto a plane used the term *projection conformis*, however, the use of the terminology conformal transformation or mapping, viz. *conforme Abbildung*, dates back to Gauss in 1822.

The purely geodetic use of conformal transformations was proposed by MARUSSI (1951c) for mappings between surfaces, and in MARUSSI (1953) for the study of paths of rays of light in a continuous isotropic refringent media. Strictly speaking neither of these applications was new: the former is a classical problem in the differential geometry of surfaces, while the latter is known in optics as the geometrical theory of optical images. Nevertheless, Marussi's work introduced the topic to geodesists and stimulated numerous papers by his coworkers: Bocchio, Caputo, Bozzi Zadro, Carminelli, Taucer, *et al.* He personally wrote over a half dozen papers on the subject and five of them are reprinted in his monograph MARUSSI (1985). Likewise, Hotine had long been interested in conformal geometry, and indeed it was the subject of one of his most elaborate mathematical investigations, *vide* HOTINE (1946/47), before taking up differential geodesy. He wrote a basic paper on the use of 3-dimensional conformal transformations in geodesy, i.e. HOTINE (1969), and he devoted two chapters to them — Chapters 10 and 24 — in his treatise.

Our investigation has two goals. First, to give a unified theory of conformal transformations of space and surfaces by using exterior differential forms. This includes a conformal approach to the Marussi tensor which we believe is new. Our second goal is to apply the method of *generalized conformal transformations.* This was a proposal of Hotine's which was considered by CHOVITZ (1972). Our discussion essentially follows the notation and terminology of Hotine's treatise which in turn was heavily influenced by that of LEVI-CIVITA (1926). Hotine's presentation in his Chapter 10, although it is brief and highly selective, has some novel features not found in the standard expositions in textbooks on differential geometry. Indeed, Marussi was of the opinion — see ZUND (1990a) — that Hotine's most important contribution to theoretical geodesy was his application of conformal techniques.

IX.2 Fundamental Notions of Conformal Geometry

We consider transformations such that

$$\bar{g}_{rs} = m^2 g_{rs} \tag{2.1}$$

where $m = f(x^r)$ is a smooth function. Then in $\mathbf{E}_3$

$$\bar{g}^{rs} = m^{-2} g^{rs} \tag{2.2}$$

since

$$\bar{g} = m^6 g \tag{2.3}$$

where $g := \det \|g_{rs}\|$ etc. Equation (2.1) is said to be a *conformal transformation*, or *mapping*, $\mathbf{E}_3 \to \mathbf{E}_3$; (2.2) is the *inverse transformation*, and m is known as the *conformal function.*

By a direct calculation in which the quantities appearing in the definition of Γ^r_{st} are replaced by barred quantities, (see

PROBLEM IX.1) we have

$$\overline{\Gamma}^r_{st} = \Gamma^r_{st} + \mathcal{M}^r_{st} \tag{2.4}$$

where

$$\mathcal{M}^r_{st} := \delta^r_s M_t + \delta^r_t M_s - g_{st} g^{rp} M_p \tag{2.5}$$

for

$$M := \log m \tag{2.6}$$

$m > 0$, and we have written

$$M_r = (\log m)_{,r}\,. \tag{2.7}$$

We denote covariant differentiation with respect to

Γ^r_{st} by a comma, i.e. "," ,
$\overline{\Gamma}^r_{st}$ by a period, i.e. "." ,

and partial differentiation by a semi-colon, i.e. ";" . The presence of two different kinds of covariant derivatives makes Hotine's habit of employing no notation for covariant differentiation impractical. However, as in (2.7) and (2.5), we may omit the semicolon for partial differentiation of scalars, i.e. first derivatives, without ambiguity.

Note that (2.1) assumes that under a conformal transformation the local coordinates x^r on $\mathbf{E}_3$ remain unchanged, i.e.

$$\overline{x}^r = x^r, \tag{2.8}$$

and hence $d\overline{x}^r = dx^r$ so that the line elements on $\mathbf{E}_3$ and $\overline{\mathbf{E}}_3$ are related by

$$d\overline{s}^2 = m^2 ds^2. \tag{2.9}$$

Thus, we are led to the

Fundamental Principle of Conformal Geometry

Algebraic expressions are converted into their conformally related expressions by 'barring', while differential expressions involving covariant derivatives require replacing "," by "." and (2.4).

An important example, which the reader is asked to work through in PROBLEM IX.2, is that

$$\overline{g}_{rs.t} = m^2 g_{rs,t} \tag{2.10}$$

and since $g_{rs,t} = 0$ by Ricci's lemma (see SECTION III-8) we have

$$\overline{g}_{rs.t} = 0. \tag{2.11}$$

It is important to note that *by definition*, a conformal transformation $\mathbf{E}_3 \to \mathbf{E}_3$ involves only the metric tensor and quantities whose definitions depend on this tensor.

We now establish a useful

Lemma

If under a conformal transformation an arbitrary smooth vector $\mathbf{v}$ having covariant components v_r undergoes the change

$$v_r \mapsto \overline{v}_r = m v_r \tag{2.12}$$

then

$$\overline{v}_{r.s} = m v_{r,s} - m_r v_s + g_{rs} v^t m_t. \tag{2.13}$$

The proof is straightforward, and we include only the basic steps:

$$\begin{aligned} \overline{v}_{r.s} &= \overline{v}_{r;s} - \overline{\Gamma}^p_{rs} \overline{v}_p \\ &= (m v_r)_{;s} - m \overline{\Gamma}^p_{rs} v_p \\ &= m v_{r,s} + v_r m_s - m \mathcal{M}^p_{rs} v_p \end{aligned}$$

and (2.13) then follows by using (2.5).

The realization of (2.1) in the leg calculus is particularly interesting. The obvious way to do this is to employ (2.12) of the above lemma and to take

$$\begin{aligned} \lambda_r &\mapsto \overline{\lambda}_r = m\lambda_r, \\ \mu_r &\mapsto \overline{\mu}_r = m\mu_r, \\ \nu_r &\mapsto \overline{\nu}_r = m\nu_r, \end{aligned} \tag{2.14}$$

under a conformal transformation, with the contravariant leg components undergoing the inverse transformations:

$$\begin{aligned}\lambda^r &\mapsto \overline{\lambda}^r = m^{-1}\lambda^r,\\ \mu^r &\mapsto \overline{\mu}^r = m^{-1}\mu^r,\\ \nu^r &\mapsto \overline{\nu}^r = m^{-1}\nu^r.\end{aligned} \tag{2.15}$$

Then it is clear that the canonical leg representations of g_{rs}, i.e. II-(2.13), and g^{rs}, i.e. II-(2.14) reproduce (2.1) and (2.2) as required. Furthermore, the leg re-scalings (2.14) and (2.15) have the effect of preserving the magnitudes of all vectors under a conformal mapping (2.1).

Hence, as a consequence of these results, an arbitrary vector $\mathbf{v}$ which has a leg representation

$$\mathbf{v} = a\boldsymbol{\lambda} + b\boldsymbol{\mu} + c\boldsymbol{\nu} \tag{2.16}$$

where a, b, c are scalars, i.e. invariants, transforms as follows under a conformal transformation $v_r \to \overline{v}_r$ as in (2.12) and

$$v^r \mapsto \overline{v}^r = m^{-1}v^r. \tag{2.17}$$

An important example of (2.12) occurs when $\mathbf{v}$ is a gradient — in which case the v_r may be written as

$$v_r = F_r \tag{2.18}$$

where $F_r := F_{;r}$.

$$v_r = F_{/1}\lambda_r + F_{/2}\mu_r + F_{/3}\nu_r \tag{2.19}$$

Then, by (2.14) we have

$$\overline{v}_r = \overline{F}_r = mF_r. \tag{2.20}$$

Upon passing to the contravariant components of (2.17) we have

$$\overline{v}^r := \overline{g}^{rs}\overline{F}_s = m^{-1}F^r. \tag{2.21}$$

These considerations have an important application when F is chosen to be the geopotential N. Then the basic gradient equation ($\mathcal{B}$), i.e. VI-(3.2), upon barring becomes

$$\overline{N}_r = \overline{n}\,\overline{\nu}_r. \tag{2.22}$$

Then by using the last expression in (2.14) and $\overline{N} = N$ we have

$$mN_r = \overline{n} m \nu_r \tag{2.23}$$

and upon recalling ($\mathcal{B}$) we have established the following conformal behavior of the local gravity:

$$\overline{n} = n. \tag{2.24}$$

The same result is readily confirmed by using VI-(3.3), which upon barring yields

$$\overline{n}^2 = \overline{g}^{rs}\overline{N}_r\overline{N}_s. \tag{2.25}$$

Then if $\overline{N} = N$, so $\overline{N}_r = N_r$, and using (2.3) we get

$$\overline{n}^2 = n^2.$$

IX.3 Conformal Structural Equations

We now investigate how the notion of conformal transformations carries over to the formalism of the Cartan calculus. This will lead not only to the conformal analogues of the structural equations, but also yield expressions exhibiting the conformal behavior of the full set of leg coefficients

$$k_1, k_2, t_1, \gamma_1, \gamma_2, \sigma_1, \sigma_2, \varepsilon_3.$$

First, by virtue of the definitions of the basic Pfaffian forms, $\{\theta_a\}_{a=1}^3$, see III-(5.3) or VI-(3.9), and the transformation rules (2.14), it is clear that, under a conformal transformation,

$$\overline{\theta}_a = m\theta_a. \tag{3.1}$$

Then the first structural equation III-(6.1), i.e.

$$d\theta_a + \omega_{ab} \wedge \theta_b = 0,$$

upon barring in accord with the fundamental principle of SECTION 2, yields

$$d\overline{\theta}_a + \overline{\omega}_{ab} \wedge \overline{\theta}_b = 0. \tag{3.2}$$

The individual terms in this equation are

$$\begin{aligned} d\bar{\theta}_a &= dm \wedge \theta_a + m d\theta_a \\ &= m_{/b}\theta_b \wedge \theta_a + m\left(-\omega_{ab} \wedge \theta_b\right) \\ &= m_{/b}\theta_b \wedge \theta_a - m\omega_{ab} \wedge \theta_b, \end{aligned}$$

and

$$\begin{aligned} \bar{\omega}_{ab} \wedge \bar{\theta}_b &= \bar{\omega}_{ab} \wedge (m\theta_b) \\ &= m\bar{\omega}_{ab} \wedge \theta_b, \end{aligned}$$

where the expressions for $\bar{\omega}_{ab}$ are to be determined. Then (3.2) becomes

$$m_{/b}\theta_b \wedge \theta_a + m\left(\bar{\omega}_{ab} - \omega_{ab}\right) \wedge \theta_b = 0. \tag{3.3}$$

Upon expansion and writing

$$M_{/a} := (\log m)_{/a} = m^{-1} m_{/a}, \tag{3.4}$$

(3.3) gives the following system of equations:

$$\begin{aligned} \left(\bar{\omega}_{31} - \omega_{31} + M_{/3}\theta_1\right) \wedge \theta_3 &= \left(\bar{\omega}_{12} - \omega_{12} - M_{/2}\theta_1\right) \wedge \theta_2, \\ \left(\bar{\omega}_{12} - \omega_{12} + M_{/1}\theta_2\right) \wedge \theta_1 &= \left(\bar{\omega}_{23} - \omega_{23} - M_{/3}\theta_2\right) \wedge \theta_3, \\ \left(\bar{\omega}_{23} - \omega_{23} + M_{/2}\theta_3\right) \wedge \theta_2 &= \left(\bar{\omega}_{31} - \omega_{31} - M_{/1}\theta_3\right) \wedge \theta_1, \end{aligned} \tag{3.5}$$

which we will solve for the $\bar{\omega}_{ab}$. We solve these equations by reducing each of them to the trivial equation $0 = 0$! We illustrate this by requiring that

$$\bar{\omega}_{12} - \omega_{12} - M_{/2}\theta_1 = B\theta_2$$

on the right-hand side of the first equation of (3.5) and

$$\bar{\omega}_{12} - \omega_{12} + M_{/1}\theta_2 = A\theta_1$$

from the left-hand side of the second equation of (3.5). In these two expressions the factors A and B are scalars, and the pair is consistent if and only if

$$B = -M_{/1} \qquad \text{and} \qquad A = M_{/2},$$

viz. the pair of equations reduces to

$$\bar{\omega}_{12} - \omega_{12} - M_{/2}\theta_1 = -M_{/1}\theta_2.$$

Hence, the solutions for $\overline{\omega}_{ab}$ are given by

$$\begin{aligned}\overline{\omega}_{12} &= \omega_{12} + M_{/2}\theta_1 - M_{/1}\theta_2,\\ \overline{\omega}_{31} &= \omega_{31} + M_{/1}\theta_3 - M_{/3}\theta_1,\\ \overline{\omega}_{23} &= \omega_{23} + M_{/3}\theta_2 - M_{/2}\theta_3,\end{aligned} \tag{3.6}$$

where the pattern exhibited in the subscripts is the same for each ω_{ab}. These solutions for $\overline{\omega}_{ab}$ identically satisfy (3.5), and this method of determining the conformally related connection 1-forms is the Cartan analogue of solving for the $\overline{\Gamma}^r_{\ st}$ in SECTION 2.

The second structural equation III-(6.2), i.e.

$$\Omega_{ab} = d\omega_{ab} + \omega_{ac} \wedge \omega_{cb}$$

gives an expression for the curvature 2-form Ω_{ab}, (recall III-(6.16)). However, in $\mathbf{E}_3$ we know that these curvature forms are zero. Indeed, from CHAPTER V we know that these give the Gauss equation: $\Omega_{12} = 0$, and the Codazzi equations: $\Omega_{31} = 0$ and $\Omega_{23} = 0$. Hence, barring the second structural equation will lead to the conformal versions of the Gauss and Codazzi equations. We will obtain these equations in SECTION 6.

In CHAPTER IV we exhibited expressions IV-(6.2) for the connection 1-forms in terms of the leg coefficients, and using property (T) these are:

$$\begin{aligned}\omega_{12} &= -\sigma_1\theta_1 - \sigma_2\theta_2 - \varepsilon_3\theta_3,\\ \omega_{31} &= k_1\theta_1 + t_1\theta_2 - \gamma_1\theta_3,\\ \omega_{23} &= -t_1\theta_1 - k_2\theta_2 + \gamma_2\theta_2.\end{aligned}$$

Upon barring these we have

$$\begin{aligned}\overline{\omega}_{12} &= m\left(-\overline{\sigma}_1\theta_1 - \overline{\sigma}_2\theta_2 - \overline{\varepsilon}_3\theta_3\right),\\ \overline{\omega}_{31} &= m\left(\overline{k}_1\theta_1 + \overline{t}_1\theta_2 - \overline{\gamma}_1\theta_3\right),\\ \overline{\omega}_{23} &= m\left(-\overline{t}_1\theta_1 - \overline{k}_2\theta_2 + \overline{\gamma}_2\theta_3\right),\end{aligned} \tag{3.7}$$

where the barred leg coefficients are to be determined. However, by (3.6) we have

$$\begin{aligned}\omega_{12} + M_{/2}\theta_1 - M_{/2}\theta_2 &= m\left(-\overline{\sigma}_1\theta_1 - \overline{\sigma}_2\theta_2 - \overline{\varepsilon}_3\theta_3\right),\\ \omega_{31} + M_{/1}\theta_3 - M_{/3}\theta_1 &= m\left(\overline{k}_1\theta_1 + \overline{t}_1\theta_2 - \overline{\gamma}_1\theta_3\right),\\ \omega_{23} + M_{/3}\theta_2 - M_{/2}\theta_3 &= m\left(-\overline{t}_1\theta_1 - \overline{k}_2\theta_2 + \overline{\gamma}_2\theta_3\right).\end{aligned} \tag{3.8}$$

Finally upon employing the above equations for the connection 1-forms we obtain:

$$\begin{aligned}
&\left(-\sigma_1 + M_{/2}\right)\theta_1 + \left(-\sigma_2 - M_{/1}\right)\theta_2 - \varepsilon_3\theta_3 \\
&\quad = m\left(-\overline{\sigma}_1\theta_1 - \overline{\sigma}_2\theta_2 - \overline{\varepsilon}_3\theta_3\right) \\
&\left(k_1 - M_{/3}\right)\theta_1 + t_1\theta_2 - \left(\gamma_1 - M_{/1}\right)\theta_3 \\
&\quad = m\left(\overline{k}_1\theta_1 + \overline{t}_1\theta_2 - \overline{\gamma}_1\theta_3\right) \\
&-t_1\theta_1 - \left(k_2 - M_{/3}\right)\theta_2 + \left(\gamma_2 - M_{/2}\right)\theta_3 \\
&\quad = m\left(-\overline{t}_1\theta_1 - \overline{k}_2\theta_2 + \overline{\gamma}_2\theta_3\right)
\end{aligned}$$

which upon equating coefficients — and listing the leg coefficients in their usual order — gives

$$\overline{k}_1 = m^{-1}\left(k_1 - M_{/3}\right), \tag{3.9}$$

$$\overline{k}_2 = m^{-1}\left(k_2 - M_{/3}\right), \tag{3.10}$$

$$\overline{t}_1 = m^{-1}t_1, \tag{3.11}$$

$$\overline{\gamma}_1 = m^{-1}\left(\gamma_1 - M_{/1}\right), \tag{3.12}$$

$$\overline{\gamma}_2 = m^{-1}\left(\gamma_2 - M_{/2}\right), \tag{3.13}$$

$$\overline{\sigma}_1 = m^{-1}\left(\sigma_1 - M_{/2}\right), \tag{3.14}$$

$$\overline{\sigma}_2 = m^{-1}\left(\sigma_2 + M_{/1}\right), \tag{3.15}$$

$$\overline{\varepsilon}_3 = m^{-1}\varepsilon_3. \tag{3.16}$$

The above equations lead to the following interesting properties which we enumerate into cases:

(i) one can make $\overline{t}_1 = 0$ and $\overline{\varepsilon}_3 = 0$ *only* by choosing $t_1 = 0$ and $\varepsilon_3 = 0$ respectively;

(ii) since both $\overline{k}_1$ and $\overline{k}_2$ involve the same right-hand side, if $k_1 \neq k_2$ then one can make only *one* of the barred quantities zero by choosing k_1 or k_2 equal to $M_{/3}$;

(iii) one can make $\overline{\gamma}_1$ and $\overline{\gamma}_2$ *both* equal to zero by choosing

$$\gamma_1 = M_{/1}, \quad \gamma_2 = M_{/2};$$

respectively;

(iv) one can make $\overline{\sigma}_1$ and $\overline{\sigma}_2$ *both* equal to zero by choosing

$$\sigma_1 = M_{/2}, \quad \sigma_2 = -M_{/2}$$

respectively.

Note that cases (ii)-(iv) impose leg differential conditions on M, so if such conditions are imposed, then their integrability conditions must be considered. Moreover, (iii) and (iv) both involve $M_{/1}$, $M_{/2}$. Taking either (iii) or (iv) entails a specialization of the other, viz. the leg coefficient restrictions

$$\gamma_1 = -\sigma_2, \quad \gamma_2 = \sigma_1.$$

The significance of these cases will be considered in the following sections.

Finally, we observe that upon barring the familiar definitions of the Gaussian and Germain curvatures, viz.

$$\begin{aligned} \overline{K} &= \overline{k}_1\overline{k}_2 - \overline{t}_1^2, \\ \overline{H} &= \left(\overline{k}_1 + \overline{k}_2\right)/2, \end{aligned}$$

we obtain the following conformal curvatures:

$$\overline{K} = m^{-2}\left(K - 2HM_{/3} + \left(M_{/3}\right)^2\right), \tag{3.17}$$

$$\overline{H} = m^{-1}\left(H - M_{/3}\right). \tag{3.18}$$

We will investigate and re-derive these results using our conformal version of the Cartan calculus when we consider the conformal aspects of classical surface theory in SECTION 5.

IX.4 Conformal Ricci Coefficients

The conformal behavior of the leg coefficients exhibited in (3.9)-(3.16) is very important. Indeed, it is *la raison d'être* for conformal differential geodesy in the sense that the use of a conformal transformation offers the *possibility* of simplifying the geometry for an appropriate choice of M in cases (ii)-(iv).

We now present two additional derivations of these leg coefficient formulas. These show that the same results can be obtained without appeal to the Cartan calculus, and also provide a useful check on the consistency of many of our previous equations.

First we give a purely Ricci calculus derivation. The Ricci coefficients were defined in II-(4.1) and upon barring we obtain

$$\overline{\gamma}_{abc} = \overline{\lambda}_{ar.s}\overline{\lambda}_{b}{}^{r}\overline{\lambda}_{c}{}^{s}.$$

Then by (2.13) and (2.15) we have

$$\begin{aligned}\overline{\lambda}_{ar.s} &= m\lambda_{ar,s} - m_r\lambda_{as} + g_{rs}\lambda_a^p m_p \\ &= m\lambda_{ar,s} - m_r\lambda_{as} + g_{rs}m_{/a}\end{aligned} \tag{4.1}$$

and

$$\overline{\lambda}_{b}{}^{r} = m^{-1}\lambda_{b}{}^{r}. \tag{4.2}$$

Thus,

$$\overline{\gamma}_{abc} = m^{-1}\left(\lambda_{ar,s} - M_r\lambda_{as} + g_{rs}M_{/a}\right)\lambda_{b}{}^{r}\lambda_{c}{}^{s}$$

which upon simplification yields

$$\overline{\gamma}_{abc} = m^{-1}\left(\gamma_{abc} + M_{/a}\delta_{bc} - M_{/b}\delta_{ac}\right). \tag{4.3}$$

Then upon expansion and employing the canonical identification IV-(3.1), we get

$$\overline{\gamma}_{121} := \overline{\sigma}_1 = m^{-1}\left(\sigma_1 - M_{/2}\right), \tag{4.4}$$

$$\overline{\gamma}_{122} := \overline{\sigma}_2 = m^{-1}\left(\sigma_2 + M_{/1}\right), \tag{4.5}$$

$$\overline{\gamma}_{123} := \overline{\varepsilon}_3 = m^{-1}\varepsilon_3, \tag{4.6}$$

$$\overline{\gamma}_{231} := \overline{t}_1 = m^{-1} t_1, \tag{4.7}$$

$$\overline{\gamma}_{131} := \overline{k}_1 = m^{-1} \left(k_1 - M_{/3}\right), \tag{4.8}$$

$$\overline{\gamma}_{232} := \overline{k}_2 = m^{-1} \left(k_2 - M_{/3}\right), \tag{4.9}$$

$$\overline{\gamma}_{313} := \overline{\gamma}_1 = m^{-1} \left(\gamma_1 - M_{/1}\right), \tag{4.10}$$

$$\overline{\gamma}_{323} := \overline{\gamma}_2 = m^{-1} \left(\gamma_2 - M_{/2}\right), \tag{4.11}$$

$$\overline{\gamma}_{312} := \overline{t}_2 = m^{-1} t_2, \tag{4.12}$$

and by property (T), (4.12) reproduces (4.7). Apart from their ordering, these equations reproduce the content of our previous set (3.9)-(3.16).

A direct leg calculation derivation of these equations can be obtained by barring the basic leg differential equations IV-(6.1), (6.2) and (6.3) with property (T) being in force. We illustrate this only for the

$$\overline{\lambda}_{r.s} = m^2 \left\{\overline{\sigma}_1 \mu_r \lambda_s + \overline{\sigma}_2 \mu_r \mu_s + \overline{\varepsilon}_3 \mu_r \nu_s + \overline{k}_1 \nu_r \lambda_s + \overline{t}_1 \nu_r \mu_s - \overline{\gamma}_1 \nu_r \nu_s\right\}$$

while by applying (2.13) to λ_r we have

$$\overline{\lambda}_{r.s} = m\lambda_{r,s} + m_{/1} \left(\mu_r \mu_s + \nu_r \nu_s\right) - m_{/2} \mu_r \lambda_s - m_{/3} \nu_r \lambda_s$$

(see PROBLEM IX.6). Equating these two expressions and the corresponding products of the leg vectors we get the transformation formulas for *six* of the leg coefficients, i.e.

$$\overline{\sigma}_1, \overline{\sigma}_2, \overline{\varepsilon}_3, \overline{k}_1, \overline{t}_1, \overline{\gamma}_1.$$

To obtain the transformation formulas for $\overline{k}_2$ and $\overline{\gamma}_2$, one need only examine the conformal analogue of $\mu_{r,s}$. Likewise, by computing the conformal behavior of $\nu_{r,s}$ we obtain the transformation rules for the *five* curvature parameters. The reader is invited to check these assertions in PROBLEM IX.7.

IX.5 Conformal Surface Theory

Before considering the effect of a conformal transformation in classical surface theory, let us recall some of the basic equations which govern the situation. By V-(1.2) we have

$$\theta_3 = 0,$$

and subject to this specialization the first set of structural equations III-(6.22) reduces to

$$d\theta_1 = -\omega_{12} \wedge \theta_2,$$

$$d\theta_2 = \omega_{12} \wedge \theta_1,$$

$$0 = \omega_{31} \wedge \theta_1 - \omega_{23} \wedge \theta_2,$$

where the connection 1-forms are given by V-(7.5), (7.6), and (7.7) (with property (T) being in force):

$$\begin{aligned} \omega_{12} &= -\sigma_1\theta_1 - \sigma_2\theta_2 \\ \omega_{31} &= k_1\theta_1 + t_1\theta_2 \\ \omega_{23} &= -t_1\theta_1 - k_2\theta_2. \end{aligned}$$

Note that by virtue of $\theta_3 = 0$ (see IV-(6.1), (6.2) and (6.3)) the leg coefficients γ_1, γ_2 and ε_3 do not appear in the above expressions for the connection 1-forms. In other words, $\theta_3 = 0$ is an intrinsic condition which characterizes the intrinsic geometry of surfaces, viz. γ_1, γ_2 and ε_3 are not purely intrinsic quantities. Actually γ_1 and γ_2 are purely extrinsic parameters, whereas ε_3 has a mixed intrinsic and extrinsic character. Consequently, we are concerned with only the following *five* leg coefficients and their conformal behavior:

$$\overline{k}_1 = m^{-1}\left(k_1 - M_{/3}\right),$$

$$\overline{k}_2 = m^{-1}\left(k_2 - M_{/3}\right),$$

$$\overline{t}_1 = m^{-1}t_1,$$

$$\overline{\sigma}_1 = m^{-1}\left(\sigma_1 - M_{/2}\right),$$

$$\overline{\sigma}_2 = m^{-1}\left(\sigma_2 + M_{/1}\right),$$

which tentatively we carry over from the set of equations (3.9)-(3.16).

Having recalled these preliminaries, we now proceed to consider what they mean in the conformal analogue of Gaussian differential geometry. Then the conformal transformation is not between $\mathbf{E}_3$ and $\overline{\mathbf{E}}_3$, but between a pair of surfaces $\mathbf{S}$ and $\overline{\mathbf{S}}$. Hence (2.1) is replaced by the following transformation law on the first basic tensor of $\mathbf{S}$, i.e.

$$\overline{a}_{\alpha\beta} = m^2 a_{\alpha\beta} \tag{5.1}$$

where $m = f(u^\alpha)$. Note that while m does not depend on a third spatial parameter, that $M_{/3}$ is not identically zero, viz. recall that

$$\begin{aligned} M_{/3} &:= \nu^r M_r = \nu^1 M_1 + \nu^2 M_2 + \nu^3 M_3 \\ &= \nu^1 (\log m)_1 + \nu^2 (\log m)_2 \neq 0. \end{aligned}$$

It is convenient in the following to write

$$\mathcal{M} := M_{/3} \tag{5.2}$$

to disguise the reference to the third leg derivative which seems out of place in our 2-dimensional picture of Gaussian differential geometry.

The above specialization of our previous 3-dimensional equations — as listed in the unnumbered equations exhibited at the beginning of this section — is completely justified by the discussion given in SECTION V-3, and indeed the fundamental equation V-(2.7), i.e.

$$a_{\alpha\beta} = g_{rs} x^r_\alpha x^s_\beta,$$

shows that if m is chosen as the conformal function in (2.1), then the same notation should be employed in (6.1). Of course, in these equations, these functions depend on different variables: in the former they are the (spatial) local coordinates x^r, and in the latter they are the local Gaussian parameters u^α on **S**. Indeed, upon barring the above equation we obtain

$$\overline{a}_{\alpha\beta} = \overline{g}_{rs}\overline{x}^r_\alpha \overline{x}^s_\beta, \tag{5.3}$$

however

$$\overline{x}^r_\alpha = x^r_\alpha, \tag{5.4}$$

(see PROBLEM IX.7) so (2.1) yields (5.1). Thus, we have the following conformal expressions

$$\overline{\theta}_1 = m\theta_1, \quad \overline{\theta}_2 = m\theta_2 \tag{5.5}$$

and

$$\begin{aligned} \overline{\omega}_{12} &= \omega_{12} + M_{/2}\theta_1 - M_1\theta_2, \\ \overline{\omega}_{31} &= \omega_{31} - \mathcal{M}\theta_1, \\ \overline{\omega}_{23} &= \omega_{23} + \mathcal{M}\theta_2. \end{aligned} \tag{5.6}$$

Upon barring the respective definitions of the basic forms I, II, and III of **S** (recall V-(3.14), (4.36) and (5.9)), viz.

$$\mathrm{I} = (\theta_1)^2 + (\theta_2)^2,$$

$$\mathrm{II} = \theta_1\omega_{31} - \theta_2\omega_{23},$$

$$\mathrm{III} = (\omega_{31})^2 + (\omega_{23})^2,$$

we get

$$\overline{\mathrm{I}} = m^2\mathrm{I}, \tag{5.7}$$

$$\overline{\mathrm{II}} = m(\mathrm{II} - \mathcal{M}\mathrm{I}), \tag{5.8}$$

$$\overline{\mathrm{III}} = \mathrm{III} - 2\mathcal{M}\mathrm{II} + \mathcal{M}^2\mathrm{I}. \tag{5.9}$$

We now verify our previous equations (3.17) and (3.18) for the conformal curvatures $\overline{K}$ and $\overline{H}$ by using the Cartan formalism. The latter is simpler so we consider it first. Upon barring V-(4.47) we have

$$2\overline{H}\,\overline{\theta}_1 \wedge \overline{\theta}_2 = \overline{\omega}_{31} \wedge \overline{\theta}_2 + \overline{\omega}_{23} \wedge \overline{\theta}_1 \tag{5.10}$$

so by using (6.5) and (6.6) we have

$$2m^2\overline{H}\theta_1 \wedge \theta_2 = m(\omega_{31} - \mathcal{M}\theta_1) \wedge \theta_2 + m(\omega_{23} + \mathcal{M}\theta_2) \wedge \theta_1$$

which yields (3.18) which we now write as

$$\overline{H} = m^{-1}(H - \mathcal{M}). \tag{5.11}$$

The derivation of the expression for $\overline{K}$ is messier, since barring V-(4.48) gives

$$d\overline{\omega}_{12} = \overline{K}\,\overline{\theta}_1 \wedge \overline{\theta}_2 \tag{5.12}$$

whose left-hand side requires exterior differentiation of the first equation of (5.5). This gives

$$d\overline{\omega}_{12} = d\omega_{12} + dM_{/2} \wedge \theta_1 + M_{/2}d\theta_1 - dM_{/1} \wedge \theta_2 - M_{/1}d\theta_2$$

and upon recalling that IV-(2.5) holds for any F, in particular $F = M_{/2}$ and $F = M_{/1}$, we have

$$\begin{aligned} d\overline{\omega}_{12} &= d\omega_{12} - \left(M_{/1/1} + M_{/2/2}\right)\theta_1 \wedge \theta_2 \\ &\quad + M_{/2}\left(-\omega_{12} \wedge \theta_2\right) - M_{/1}\left(\omega_{12} \wedge \theta_1\right). \end{aligned}$$

Upon using our previous expression for ω_{12} this gives

$$d\overline{\omega}_{12} = d\omega_{12} - \left(M_{/1/1} + M_{/2/2} - \sigma_1 M_{/2} + \sigma_2 M_{/1}\right)\theta_1 \wedge \theta_2,$$

and hence

$$m^2\overline{K} = K - \left(M_{/1/1} + M_{/2/2} - \sigma_1 M_{/2} + \sigma_2 M_{/1}\right). \tag{5.13}$$

However, recognizing that by VI-(11.2) the expression in the parenthesis is the surface Laplacian of M, i.e. $\Delta_2 M$, we ultimately have

$$\overline{K} = m^{-2}(K - \Delta_2 M) \tag{5.14}$$

which is known as *Souslow's equation.* This differs from our previous equation (3.17) for $\overline{K}$, but they are equivalent when

$$\Delta_2 M = 2H\mathcal{M} - \mathcal{M}^2. \tag{5.15}$$

As noted in SECTION III, the conformal analogues of the second structural equations must be considered. Thus, we must examine the conformal behavior of the *three* curvature 2-forms:

$$\overline{\Omega}_{ab} = 0. \tag{5.16}$$

From V-(8.18), (8.19) and (8.20) we know that

$$\Omega_{12} = 0 \Leftrightarrow d\omega_{12} = K\theta_1 \wedge \theta_2,$$

$$\Omega_{31} = 0 \Leftrightarrow d\omega_{31} = \omega_{12} \wedge \omega_{23},$$

$$\Omega_{23} = 0 \Leftrightarrow d\omega_{23} = -\omega_{12} \wedge \omega_{31}.$$

We have just examined the first of these equations, so it remains only to consider the other two which lead to the pair of Codazzi equations $(\mathcal{C}_{\mathrm{I}})$ and $(\mathcal{C}_{\mathrm{II}})$ of SECTION V.8. To obtain the conformal analogue of $(\mathcal{C}_{\mathrm{I}})$ we must compute the consequences of the barred equation $\overline{\Omega}_{31} = 0$, i.e.

$$d\overline{\omega}_{31} = \overline{\omega}_{12} \wedge \overline{\omega}_{23}. \tag{5.17}$$

The computation is straightforward and makes rather obvious use of (6.6), i.e.

$$\begin{aligned} d\overline{\omega}_{31} &= d\omega_{31} - d\mathcal{M} \wedge \theta_1 - \mathcal{M} d\theta_1 \\ &= d\omega_{31} + \left(\mathcal{M}_{/2} - \sigma_1 \mathcal{M}\right) \theta_1 \wedge \theta_2, \end{aligned}$$

$$\begin{aligned} \overline{\omega}_{12} \wedge \overline{\omega}_{23} &= \left(\omega_{12} + M_{/2}\theta_1 - M_{/2}\theta_2\right) \wedge \left(\omega_{23} + \mathcal{M}\theta_2\right) \\ &= \omega_{12} \wedge \omega_{23} + \left\{-k_2 M_{/2} - t_1 M_{/1} + \left(M_{/2} - \sigma_1\right) \mathcal{M}\right\} \\ &\qquad \theta_1 \wedge \theta_2. \end{aligned}$$

Substitution of these into (5.17) then gives

$$\begin{aligned} d\omega_{31} = \omega_{12} \wedge \omega_{23} + \{&-k_2 M_{/2} - t_1 M_{/1} + \sigma_1 \mathcal{M} - \mathcal{M}_{/2} \\ &+ \left(M_{/2} - \sigma_1\right) \mathcal{M}\} \theta_1 \wedge \theta_2 \end{aligned}$$

and upon writing out $(\mathcal{C}_{\mathrm{I}})$ we obtain

$$k_{1/2} - t_{1/1} = \sigma_1 \left(k_1 - k_2\right) + 2\sigma_1 t_1 + \left\{M_{/1} t_1 + M_{/2} \left(k_2 - \mathcal{M}\right) + \mathcal{M}_{/2}\right\}. \tag{5.18}$$

This shows that the effect of a conformal transformation on $(\mathcal{C}_{\mathrm{I}})$ is a purely additive term, viz. the terms in the parentheses to the Codazzi equation.

The conformal analogue of $(\mathcal{C}_{\mathrm{II}})$ is quite similar and involves considering $\overline{\Omega}_{23} = 0$, viz.

$$d\overline{\omega}_{23} = -\overline{\omega}_{12} \wedge \overline{\omega}_{31}. \tag{5.19}$$

The evaluation of this expression is left as PROBLEM IX.10, and the result is

$$\begin{aligned} k_{2/1} - t_{1/2} &= \sigma_2 (k_1 - k_2) - 2\sigma_1 t_1 \\ &\quad + \left\{ M_{/1} (k_1 - \mathcal{M}) + M_{/2} t_1 + \mathcal{M}_{/1} \right\}. \end{aligned} \tag{5.20}$$

IX.6 Conformal Curve Theory

In this section we consider to what extent a conformal transformation may be used to simplify the geometry of a curve. We consider three topics:

(i) the conformal Frenet equations in $\overline{\mathbf{E}}_3$,

(ii) the behavior of the equation of a geodesic under $\mathbf{E}_3 \to \overline{\mathbf{E}}_3$,

(iii) the effect of a conformal transformation on a surface curve, i.e. under $\mathbf{S} \to \overline{\mathbf{S}}$.

For (i) we consider the vectorial Frenet 3-leg $\{\boldsymbol{\ell}, \mathbf{m}, \mathbf{n}\}$ of SECTION V.9 for which

$$g_{rs} = \ell_r \ell_s + m_r m_s + n_r n_s \tag{6.1}$$

so that each leg vector undergoes the obvious transformations (2.12) and (2.16). Then the barring of the Frenet equations V-(9.2), (9.3) and (9.4) yields

$$\overline{\ell}_{r.s} \overline{\ell}^s = \overline{\chi}\, \overline{m}_r, \tag{6.2}$$

$$\overline{m}_{r,s} \overline{\ell}^s = -\overline{\chi} \overline{\ell}_r + \overline{\tau}\, \overline{n}_r, \tag{6.3}$$

$$\overline{n}_{r.s} \overline{\ell}^s = -\overline{\tau}\, \overline{m}_r, \tag{6.4}$$

and our goal is to determine the transformation behavior of χ and τ, i.e. the explicit forms of $\overline{\chi}$ and $\overline{\tau}$. By using (2.13) of our lemma, and taking care not to confuse the gradient of m with m_r, (6.2) becomes

$$\chi m_r - M_r + \ell_r \ell^t M_t = \overline{\chi} m m_r. \tag{6.5}$$

Upon resolving M_r along the vectorial 3-leg $\{\boldsymbol{\ell}, \mathbf{m}, \mathbf{n}\}$, i.e.

$$M_r = M_{/1}\ell_r + M_{/2}m_r + M_{/3}n_r, \tag{6.6}$$

we obtain

$$\chi m_r - \left(M_{/1}\ell_r + M_{/2}m_r + M_{/3}n_r\right) + \ell_r M_{/1} = \overline{\chi} m m_r$$

which immediately yields

$$\overline{\chi} = m^{-1}\left(\chi - M_{/2}\right) \tag{6.7}$$

$$M_{/3} = 0. \tag{6.8}$$

Likewise barring (6.4) gives

$$n_{r,s}\ell^s - M_r\left(n_s\ell^s\right) + \ell_r\left(M_t n^t\right) = -\overline{\tau} m m_r;$$

employing (6.6) yields (6.8) and

$$\overline{\tau} = m^{-1}\tau. \tag{6.9}$$

Although we have determined the required results, i.e. (6.7)-(6.9), it is interesting to check (6.3). It gives

$$-\chi\ell_r + \tau n_r + \ell_r M_{/2} = -\overline{\chi} m\ell_r + \overline{\tau} m n_r$$

which reproduces (6.7) and (6.9), but curiously not (6.8).

The above results are of limited interest: all they conclusively show is that $\overline{\tau}$ cannot be made to vanish unless $\tau = 0$, viz. if originally the curve $\mathcal{C}$ was not a plane curve, then its image $\overline{\mathcal{C}}$ under a conformal transformation cannot be made planar by a conformal transformation $\mathbf{E}_3 \to \overline{\mathbf{E}}_3$. In theory one could make $\overline{\chi}$ vanish when $\chi \neq$ o by choosing

$$\chi = M_{/2}, \tag{6.10}$$

which makes $\overline{\mathcal{C}}$ into a straight line in $\overline{\mathbf{E}}_3$. However, the integrability conditions, viz. the M-commutators, must be satisfied in order for (6.10) to be meaningful. These commutators are not known to us — the commutators given in CHAPTER IV are those of the vectorial 3-leg $\{\boldsymbol{\lambda}, \boldsymbol{\mu}, \boldsymbol{\nu}\}$ and we have not determined how this 3-leg is related to the Frenet 3-leg $\{\boldsymbol{\ell}, \mathbf{m}, \mathbf{n}\}$.

A more useful approach is to consider (ii) and a particular congruence, i.e. $\boldsymbol{\Gamma}=\boldsymbol{\Gamma}_3$ having $\boldsymbol{\nu}$ as its unit tangent vector. Then the conformal transformation of IV-(6.24)

$$\nu_{r,s}\nu^s = \gamma_1\lambda_r + \gamma_2\mu_r$$

is

$$\begin{aligned} \overline{\nu}_{r.s}\overline{\nu}^s &= \gamma_1\lambda_r\gamma_2\mu_r - M_r + \mathcal{M}\nu_r \\ &= \left(\gamma_1 - M_{/1}\right)\lambda_r + \left(\gamma_2 - M_{/2}\right)\mu_r. \end{aligned} \tag{6.11}$$

If $\boldsymbol{\Gamma}$ is not a geodesic, viz. a straight line in $\mathbf{E}_3$, then $\gamma_1 \neq 0$ and $\gamma_2 \neq 0$, i.e. $\chi \neq 0$. However, (6.11) clearly shows that $\overline{\boldsymbol{\Gamma}}$ will be a geodesic whenever $\overline{\gamma}_1 = 0$ and $\overline{\gamma}_2 = 0$; i.e.,

$$\begin{aligned} \overline{\gamma}_1 = 0 \Leftrightarrow M_{/1} = \gamma_1 \\ \overline{\gamma}_2 = 0 \Leftrightarrow M_{/2} = \gamma_2. \end{aligned} \tag{6.12}$$

By IV-(4.12) with property (T) being in force, the integrability conditions of (6.12) are given by the following M-commutators:

$$\begin{aligned} (M_{\mathrm{I}}): &\quad \gamma_{1/2} - \gamma_{2/1} = \sigma_1\gamma_1 + \sigma_2\gamma_2 \\ (M_{\mathrm{II}}): &\quad \mathcal{M}_{/1} - \gamma_1\mathcal{M} = \gamma_{1/3} - k_1\gamma_1 - (t_1 + \varepsilon_3)\gamma_2 \\ (M_{\mathrm{III}}): &\quad \mathcal{M}_{/2} - \gamma_2\mathcal{M} = \gamma_{2/3} - (t_1 - \varepsilon_3)\gamma_1 - k_2\gamma_2. \end{aligned}$$

We note that these conditions are closely related to the n-integrability conditions $\{n\}$ of differential geodesy, viz. VI-(8.7), (8.8) and (8.9). Indeed, (M_{I}) is identically (n_{I}) i.e. $(\mathcal{S}_{\mathrm{III}})$ while (M_{II}) and (M_{III}) are similar to (n_{II}) V-(8.8) and (n_{III}). In fact, $\mathcal{M}$ must then satisfy

$$\begin{aligned} \mathcal{M}_{/1} - \gamma_1\mathcal{M} = \left(4n^{-1}\tilde{\omega}^2 - 2H\right)\gamma_1 + 2H_{/1} \\ \mathcal{M}_{/2} - \gamma_2\mathcal{M} = \left(4n^{-1}\tilde{\omega}^2 - 2H\right)\gamma_2 + 2H_{/2}. \end{aligned} \tag{6.13}$$

Likewise, let us examine the conformal images of the congruences $\boldsymbol{\Gamma}_1$ and $\boldsymbol{\Gamma}_2$ having the respective unit tangent vectors $\boldsymbol{\lambda}$ and $\boldsymbol{\mu}$. Then by IV-(6.22) and IV-(6.23):

$$\begin{aligned} \lambda_{r,s}\lambda^s &= \sigma_1\mu_r + k_1\nu_r, \\ \mu_{r,s}\mu^s &= -\sigma_2\lambda_r + k_2\nu_r, \end{aligned}$$

and hence we will have

$$\begin{aligned}\overline{\lambda}_{r.s}\overline{\lambda}^s = 0 &\Leftrightarrow \left(\sigma_1 - M_{/2}\right)\mu_r + (k_1 - \mathcal{M})\,\nu_r = 0\\ \overline{\mu}_{r.s}\overline{\mu}^s = 0 &\Leftrightarrow -\left(\sigma_2 + M_{/1}\right)\lambda_r + (k_2 - \mathcal{M})\,\nu_r = 0.\end{aligned} \tag{6.14}$$

So the conditions for $\overline{\boldsymbol{\Gamma}}_1$ and $\overline{\boldsymbol{\Gamma}}_2$ respectively to be geodesics, viz. straight lines in $\overline{\mathbf{E}}_3$, are given by

$$M_{/2} = \sigma_1, \quad \mathcal{M} = k_1; \tag{6.15}$$

$$M_{/1} = -\sigma_2, \quad \mathcal{M} = k_2, \tag{6.16}$$

respectively. Taking (6.15) with $f = M_{/1}$ being an unknown arbitrary function, the M-commutators are

$$\begin{aligned}(M_{\mathrm{I}}):&\quad f_{/2} - \sigma_1 f = \sigma_{1/1} + \sigma_1\sigma_2,\\ (M_{\mathrm{II}}):&\quad f_{/3} - k_1 f = k_{1/1} - \gamma_1 k_1 + (t_1 + \varepsilon_3)\,\sigma_1,\\ (M_{\mathrm{III}}):&\quad \sigma_{1/3} - k_{1/2} + \gamma_2 k_1 - \sigma_2 k_2 = (t_1 - \varepsilon_3)\, f;\end{aligned}$$

and likewise for (7.16) with $g = M_{/2}$ being an unknown arbitrary function, we have:

$$\begin{aligned}(M_{\mathrm{I}}):&\quad g_{/1} + \sigma_2 g = \sigma_{2/2} - \sigma_1\sigma_2,\\ (M_{\mathrm{II}}):&\quad k_{2/1} + \sigma_{2/3} - \sigma_2 k_1 - \gamma_1 k_2 = -(t_1 + \varepsilon_3)\, g,\\ (M_{\mathrm{III}}):&\quad g_{/3} - k_2 g = k_{2/2} - \gamma_2 k_2 + (\varepsilon_3 - t_1)\,\sigma_2.\end{aligned}$$

None of the integrability conditions are identically satisfied, however some of the combination of terms resembles that in the Schouten identities ($\mathcal{S}_{\mathrm{I}}$) and ($\mathcal{S}_{\mathrm{II}}$), see V-(8.6) and (8.7).

Finally, we consider the situation (iii). As in SECTION V-4, let $\boldsymbol{\Gamma}'_1$ be the surface congruence having $\boldsymbol{\lambda}$ as tangent vector, and $\boldsymbol{\Gamma}'_2$ the analogous congruence with tangent vector $\boldsymbol{\mu}$. Then the conformal version of V-(4.9) is

$$\overline{\lambda}_{\alpha.\beta}\overline{\lambda}^{\beta} = \lambda_{\alpha,\beta}\lambda^{\beta} - M_{\alpha} + M_{/1}\lambda_{\alpha} \tag{6.17}$$

which gives

$$\overline{\lambda}_{\alpha.\beta}\overline{\lambda}^{\beta} = \left(\sigma_1 - M_{/2}\right)\mu_{\alpha} \text{ or } = \overline{\sigma}_1\overline{\mu}_{\alpha} \tag{6.18}$$

by using (3.14). Hence, $\boldsymbol{\Gamma}_1'$ will be a geodesic congruence on $\overline{\mathbf{S}}$ whenever

$$M_{/1} = f, \quad M_{/2} = \sigma_2 \tag{6.19}$$

where f is an unknown arbitrary function. The required M-commutator is

$$f_{/2} - \sigma_1 f = \sigma_{1/1} + \sigma_1\sigma_2. \tag{6.20}$$

Likewise the conformal version of V-(4.10) is

$$\overline{\mu}_{\alpha.\beta}\overline{\mu}^{\beta} = \mu_{\alpha,\beta}\mu^{\beta} - M_\alpha + M_{/2}\mu_\alpha \tag{6.21}$$

which gives

$$\overline{\mu}_{\alpha.\beta}\overline{\mu}^{\beta} = -\left(\sigma_2 + M_{/1}\right)\lambda_\alpha \ = -\overline{\sigma}_2\overline{\lambda}_\alpha \tag{6.22}$$

by using (3.15). Hence $\boldsymbol{\Gamma}_2'$ will be a geodesic congruence on $\overline{\mathbf{S}}$ whenever

$$M_{/1} = -\sigma_2, \quad M_{/2} = g \tag{6.23}$$

where g is an unknown arbitrary function. The required M-commutator is

$$g_{/1} + \sigma_2 g = -\sigma_{2/2} - \sigma_1\sigma_2. \tag{6.24}$$

Note that neither of the integrability conditions (6.19) or (6.23) are identically satisfied. On the other hand, if both $\boldsymbol{\Gamma}_1'$ and $\boldsymbol{\Gamma}_2'$ are required to be geodesic congruences on $\overline{\mathbf{S}}$, then instead of (6.18) and (6.22) we have

$$M_{/1} = -\sigma_2, \quad M_{/2} = \sigma_1 \tag{6.25}$$

which involves *no* unknown arbitrary functions. The M-commutator then reduces to

$$\sigma_{1/1} + \sigma_{2/2} = 0 \tag{6.26}$$

which is a non-trivial restriction on the geodesic curvatures of $\boldsymbol{\Gamma}_1'$ and $\boldsymbol{\Gamma}_2'$.

IX.7 Generalized Conformal Transformations

In a letter of August 12, 1968 to B. Chovitz, Hotine suggested he investigate "more general deformations of space than conformal transformations. Instead of a single scale factor m you might introduce three: m, p, q in the coordinate directions so that in some coordinate systems for both spaces, you have

$$ds^2 = m^2 g_{11}\left(dx^1\right)^2 + \cdots 2mpg_{12}dx^1dx^2 + \cdots . \tag{7.1}$$

CHOVITZ (1972) subsequently considered this possibility and interpreted (7.1) as being a special case of the transformation:

$$\overline{g}_{rs} = m_r^p m_s^q g_{pq}, \tag{7.2}$$

where the matrix $\|m_r^p\|$ has diagonal form

$$\|m_r^p\| = \begin{Vmatrix} m & 0 & 0 \\ 0 & p & 0 \\ 0 & 0 & q \end{Vmatrix}, \tag{7.3}$$

and of course when

$$\|m_r^p\| = \|m\delta_r^p\| \tag{7.4}$$

(7.2) reduces to (2.1).

The leg calculus readily admits a nice method of analytically handling (7.3). Instead of (2.4) we consider a *generalized conformal transformation* $\mathbf{E}_3 \to \overline{\mathbf{E}}_3$ in which each of the 3-leg vectors undergoes a different scale change, viz.

$$\begin{aligned} \lambda_r \mapsto \overline{\lambda}_r &= m\lambda_r, \\ \mu_r \mapsto \overline{\mu}_r &= p\mu_r, \\ \nu_r \mapsto \overline{\nu}_r &= q\nu_r; \end{aligned} \tag{7.5}$$

with the inverse transformation:

$$\begin{aligned} \lambda^r \mapsto \overline{\lambda}^r &= m^{-1}\lambda^r, \\ \mu^r \mapsto \overline{\mu}^r &= p^{-1}\mu^r, \\ \nu^r \mapsto \overline{\nu}^r &= q^{-1}\nu^r. \end{aligned} \tag{7.6}$$

Thus, under a generalized conformal transformation the canonical leg representations of g_{rs}, i.e. II-(2.13), and g^{rs}, i.e. II-(2.14) are replaced by

$$\overline{g}_{rs} = m^2\lambda_r\lambda_s + p^2\mu_r\mu_s + q^2\nu_r\nu_s \tag{7.7}$$

and

$$\overline{g}^{rs} = m^{-2}\lambda^r\lambda^s + p^{-2}\mu^r\mu^s + q^{-2}\nu^r\nu^s \tag{7.8}$$

respectively. Since g_{pq} is a positive definite metric, there is no loss of generality in replacing (7.2) by (7.7) because the latter essentially involves reducing $\|m_r^p\|$ to diagonal form as in (7.3). We will see that (7.7) transformation rules are easier to work with than (7.2) with $\|m_r^p\|$ given by (7.3). Indeed, the Pfaffian 3-leg rules analogous to (7.5) are given by:

$$\overline{\theta}_1 = m\theta_1, \quad \overline{\theta}_2 = p\theta_2, \quad \overline{\theta}_3 = q\theta_3. \tag{7.9}$$

Before proceeding with our analysis of these generalized transformations, an obvious question is whether when $m \neq p \neq q$ they deserve the adjective 'conformal'? We recall that under a conformal transformation the angle ϑ between a pair of non-zero vectors given by

$$\cos\vartheta = \frac{g_{rs}\xi^r\eta_s}{\sqrt{g_{mn}\xi^m\xi^n}\cdot\sqrt{g_{pq}\eta^p\eta^q}}, \tag{7.10}$$

is preserved, i.e.

$$\cos\vartheta \mapsto \cos\overline{\vartheta} = \cos\vartheta \tag{7.11}$$

under a conformal transformation. In the ordinary tensor calculus, the numerators and denominators of the right-hand side of (8.10) yield

$$\frac{m^2\cdots}{\sqrt{m^2\cdots}\cdot\sqrt{m^2\cdots}}$$

and hence cancel. On the other hand, in the leg-calculus when we have

$$\begin{aligned} \xi^r &= X\lambda^r + Y\mu^r + Z\nu^r, \\ \eta^r &= X'\lambda^r + Y'\mu^r + Z'\nu^r, \end{aligned} \tag{7.12}$$

where X, Y, Z, X', Y', Z' are scalars, then under (2.15) we have

$$\begin{aligned} \xi^r &\mapsto \overline{\xi}^r = m^{-1}\xi^r \\ \eta^r &\mapsto \overline{\eta}^r = m^{-1}\eta^r \end{aligned} \tag{7.13}$$

and the factors of m cancel in each product before forming the quotient in (7.10). Now consider the effect of a generalized transformation on (7.10). Then it is easy to check that (7.12) must be replaced by

$$\begin{aligned} \overline{\xi}^r &= Xm^{-1}\lambda^r + Yp^{-1}\mu^r + Zq^{-1}\nu^r \\ \overline{\eta}^r &= X'm^{-1}\lambda^r + Y'p^{-1}\mu^r + Z'q^{-1}\nu^r \end{aligned} \tag{7.14}$$

so

$$\begin{aligned} \overline{g}^{rs}\overline{\xi}^r\overline{\eta}^s &= (m^2\lambda_r\lambda_s + p^2\mu_r\mu_s + q^2\nu_r\nu_s) \\ &\quad (Xm^{-1}\lambda^r + Yp^{-1}\mu^r + Zq^{-1}\nu^r) \\ &\quad (X'm^{-1}\lambda^s + Y'p^{-1}\mu^s + Z'q^{-1}\nu^s) \\ &= (m^2\lambda_r\lambda_s + p^2\mu_r\mu_s + q^2\nu_r\nu_s) \\ &\quad (XX'm^{-2}\lambda^r\lambda^s + YY'p^{-2}\mu^r\mu^s \\ &\quad +ZZ'q^{-2}\nu^r\nu^s + \text{ cross terms}). \end{aligned}$$

However, the cross terms contribute nothing by virtue of the orthogonality properties of the vectorial 3-leg. Hence, we have

$$\begin{aligned} \overline{g}_{rs}\overline{\xi}^r\overline{\eta}^s &= XX' + YY' + ZZ' \\ \overline{g}_{rs}\overline{\xi}^r\overline{\xi}^s &= X^2 + Y^2 + Z^2, \\ \overline{g}_{rs}\overline{\eta}^r\overline{\eta}^s &= (X')^2 + (Y')^2 + (Z')^2, \end{aligned} \tag{7.15}$$

under a generalized transformation and (7.12) holds. This justifies our retention of the adjective 'conformal' for the rules (7.5)-(7.8).

We now proceed to investigate such transformations by determining the barred Christoffel symbols or equivalently — and simpler by virtue of (7.9) — the barred connection 1-forms. This involves determining the explicit form of the first structural equation III-(6.1), upon expansion, this consists of the following set of barred equations:

$$\begin{aligned} d\overline{\theta}_1 &= -\overline{\omega}_{12} \wedge \overline{\theta}_2 + \overline{\omega}_{31} \wedge \overline{\theta}_3, \\ d\overline{\theta}_2 &= \overline{\omega}_{12} \wedge \overline{\theta}_1 - \overline{\omega}_{23} \wedge \overline{\theta}_3, \\ d\overline{\theta}_3 &= -\overline{\omega}_{31} \wedge \overline{\theta}_1 + \overline{\omega}_{23} \wedge \overline{\theta}_2, \end{aligned} \tag{7.16}$$

which are obtained by using (7.9).

By direct calculation, the left-hand sides of these are given by

$$\begin{aligned} d\bar{\theta}_1 &= -m_{/2}\theta_1 \wedge \theta_2 + m_{/3}\theta_3 \wedge \theta_1 \\ &\quad + m\left(-\omega_{12} \wedge \theta_2 + \omega_{31} \wedge \theta_3\right), \\ d\bar{\theta}_2 &= p_{/1}\theta_1 \wedge \theta_2 - p_{/3}\theta_2 \wedge \theta_3 \\ &\quad + p\left(-\omega_{23} \wedge \theta_3 + \omega_{12} \wedge \theta_1\right), \\ d\bar{\theta}_3 &= -q_{/1}\theta_3 \wedge \theta_1 + q_{/2}\theta_2 \wedge \theta_3 \\ &\quad + q\left(-\omega_{31} \wedge \theta_1 + \omega_{23} \wedge \theta_2\right). \end{aligned} \tag{7.17}$$

while the right-hand sides of (7.16) are respectively

$$\begin{aligned} d\bar{\theta}_1 &= -p\bar{\omega}_{12} \wedge \theta_2 + q\bar{\omega}_{31} \wedge \theta_3, \\ d\bar{\theta}_2 &= -q\bar{\omega}_{23} \wedge \theta_3 + m\bar{\omega}_{12} \wedge \theta_1, \\ d\bar{\theta}_3 &= -m\bar{\omega}_{31} \wedge \theta_1 + p\bar{\omega}_{23} \wedge \theta_2, \end{aligned} \tag{7.18}$$

where the barred connection 1-forms are to be determined. Hence, combining the above two sets of equations we obtain

$$\begin{aligned} -p\bar{\omega}_{12} \wedge \theta_2 + q\bar{\omega}_{31} \wedge \theta_3 &= -m_{/2}\theta_1 \wedge \theta_2 + m_{/3}\theta_3 \wedge \theta_1 \\ &\quad + m\left(-\omega_{12} \wedge \theta_2 + \omega_{31} \wedge \theta_3\right), \\ -q\bar{\omega}_{23} \wedge \theta_3 + m\bar{\omega}_{12} \wedge \theta_1 &= p_{/1}\theta_1 \wedge \theta_2 - p0_{/3}\theta_2 \wedge \theta_3 \\ &\quad + p\left(-\omega_{23} \wedge \theta_3 + \omega_{12} \wedge \theta_1\right), \\ -m\bar{\omega}_{31} \wedge \theta_1 + p\bar{\omega}_{23} \wedge \theta_2 &= -q_{/1}\theta_3 \wedge \theta_1 + q_{/2}\theta_2 \wedge \theta_3 \\ &\quad + q\left(-\omega_{31} \wedge \theta_1 + \omega_{23} \wedge \theta_2\right); \end{aligned}$$

which we rewrite in a form reminiscent of (3.5) as

$$\begin{aligned} \left(q\bar{\omega}_{31} - m\omega_{31} + m_{/3}\theta_1\right) \wedge \theta_3 &= \left(p\bar{\omega}_{12} - m\omega_{12} - m_{/2}\theta_1\right) \wedge \theta_2, \\ \left(m\bar{\omega}_{12} - p\omega_{12} + p_{/1}\theta_2\right) \wedge \theta_1 &= \left(q\bar{\omega}_{23} - p\omega_{23} - p_{/3}\theta_2\right) \wedge \theta_3, \\ \left(p\bar{\omega}_{23} - q\omega_{23} + q_{/2}\theta_3\right) \wedge \theta_2 &= \left(m\bar{\omega}_{31} - q\omega_{31} - q_{/1}\theta_3\right) \wedge \theta_1. \end{aligned} \tag{7.19}$$

By attempting to follow the procedure employed in SECTION 3, let us examine the equations involving $\bar{\omega}_{12}$. The right-hand side of the first equation (7.19) is identically zero if we put

$$p\bar{\omega}_{12} - m\omega_{12} - m_{/2}\theta_1 = B\theta_2,$$

while the left-hand side of the second equation is similarly

$$m\overline{\omega}_{12} - p\omega_{12} - p_{/1}\theta_2 = A\theta_1$$

where A, B are scalars. The method of solving the 'analogous' equations in SECTION 3 was essentially one of inspection, and clearly is not applicable in the present situation. However, in a personal communication, Dr. Wilkes has suggested a more direct approach to their solution. Under a general rescaling of the 3-leg vectors defined by (7.5) and (7.6), the connection 1-forms transform as follows:

$$\begin{aligned}\overline{\omega}_{12} &= -\overline{\sigma}_1 m\theta_1 - \overline{\sigma}_2 p\theta_2 - \overline{\varepsilon}_3 q\theta_3,\\ \overline{\omega}_{23} &= -\overline{t}_1 m\theta_1 - \overline{k}_2 p\theta_2 + \overline{\gamma}_2 q\theta_3,\\ \overline{\omega}_{31} &= \overline{k}_1 m\theta_1 - \overline{t}_2 p\theta_2 - \overline{\gamma}_1 q\theta_3.\end{aligned} \tag{7.20}$$

Upon substitution of these expressions, together with their unbarred counterparts, and introducing the quantities

$$P := \log p, \quad Q := \log q, \tag{7.21}$$

(which are analogous to M defined in (2.6)), into (7.19) eventually yields the following set of transformed leg coefficients:

$$\begin{aligned}\overline{k}_1 &= q^{-1}\left(k_1 - M_{/3}\right),\\ \overline{k}_2 &= q^{-1}\left(k_2 - P_{/3}\right),\\ \overline{\gamma}_1 &= m^{-1}\left(\gamma_1 - Q_{/1}\right),\\ \overline{\gamma}_2 &= m^{-1}\left(\gamma_2 - Q_{/2}\right),\\ \overline{\sigma}_1 &= p^{-1}\left(\sigma_1 - M_{/2}\right),\\ \overline{\sigma}_2 &= p^{-1}\left(\sigma_2 + P_{/1}\right),\\ 2\overline{t}_1 &= \left[(q/mp)(t_1+t_2) - (m/qp)(t_2+\varepsilon_3) + (p/mq)(t_1+\varepsilon_3)\right],\\ 2\overline{t}_2 &= \left[(q/mp)(t_1+t_2) + (m/qp)(t_2+\varepsilon_3) - (p/mq)(t_1+\varepsilon_3)\right],\\ 2\overline{\varepsilon}_3 &= \left[-(q/mp)(t_1+t_2) + (m/qp)(t_2+\varepsilon_3) + (p/mq)(t_1+\varepsilon_3)\right].\end{aligned} \tag{7.22}$$

The last three of these equations require the solution of a simple system of three equations involving three unknowns. When property (T) is imposed, the last three equations reduce to merely two equations:

$$\begin{aligned}2\overline{t}_1 &= (mpq)^{-1}\left[(m^2+p^2)\,t_1 - (m^2-p^2)\,\varepsilon_3\right],\\ 2\overline{\varepsilon}_3 &= (mpq)^{-1}\left[(m^2+p^2)\,\varepsilon_3 - (m^2-p^2)\,t_1\right].\end{aligned} \tag{7.23}$$

The details of the derivation and their analysis will be presented in a forthcoming paper.

IX.8 Conclusions

Conformal geometry, and hence conformal differential geodesy, has a bittersweet, almost melancholy aspect to it. Few who have ever seriously studied it have failed to be intrigued and mystified by its mathematical charm and elegance. Certainly both Marussi and Hotine both found it fascinating. Unfortunately, for such a beautiful mathematical theory its physical applicability has — so far — provided depressingly meager results.

In 2-dimensions, the theory as exemplified by conformal mapping techniques in complex function theory, or mathematical cartography, is indisputably useful and of major importance and interest. In 3-dimensions, the applications significantly decrease and one is left more with hints of possibilities rather than concrete applications. In 4-dimensions, within relativity theory, several puzzling things occur: the source-free Maxwell equations of electromagnetism are conformal-invariant (a result which holds in both flat and curved space-time!), however the source-free Einstein equations of gravitation are *not* conformal-invariant. Since their discovery (almost seventy-five years ago) what — if anything — these results mean is still a matter of contention and debate. Nevertheless, conformal geometry remains a popular topic in physical circles and true believers have no doubt that it is an area full of promise.

Our purpose in this chapter has been to give an orderly and systematic exposition of conformal geometry by using the leg calculus and the Cartan version of it in particular. Our principal results are the following:

1) a formulation of conformal geometry via the vectorial 3-leg and Pfaffian 3-leg formalisms;

2) determination of the transformation rules for the full set of leg coefficients under a conformal transformation $\mathbf{E}_3 \rightarrow \overline{\mathbf{E}}_3$;

3) geometric properties attainable by use of the rules given in 2);

4) the meaning of the generalized conformal transformations proposed by Hotine and subsequently investigated by Chovitz.

PROBLEMS FOR CHAPTER IX

IX.1 Using the usual definitions of the Christoffel symbols

$$\Gamma_{stq} = \frac{1}{2}\left(g_{tq;s} + g_{qs;t} - g_{st;q}\right)$$
$$\Gamma^{r}_{st} := g^{rq}\Gamma_{stq}$$

and (2.1), (2.2), by 'barring' show that

$$\overline{\Gamma}^{r}_{st} = \Gamma^{r}_{st} + \delta^{r}_{s}(\log m)_{t} + \delta^{r}_{t}(\log m)_{s} - g_{st}g^{rp}(\log m)_{p}\,.$$

By (2.5), (2.6) and (2.7) this expression becomes (2.4).

IX.2 Defining the conformal covariant derivative of an arbitrary second order covariant tensor T_{rs} as follows

$$T_{rs.t} := T_{rs;t} - \overline{\Gamma}^{p}_{rt}T_{ps} - \overline{\Gamma}^{p}_{st}T_{rp},$$

derive (2.10). Hint: replace T_{rs} by $\overline{g}_{rs}$ to obtain

$$\overline{g}_{rs.t} = 2mg_{rs}m_{t} + m^{2}g_{rs,t} - m^{2}\mathcal{M}^{p}_{rt}g_{ps} - m^{2}\mathcal{M}^{p}_{st}g_{rp}$$

which upon using (2.5) yields (2.10).

IX.3 Show that (2.11) is equivalent to

$$\overline{g}_{rs;t} = \overline{\Gamma}^{p}_{rt}\overline{g}_{ps} + \overline{\Gamma}^{p}_{st}\overline{g}_{rp},$$

and that this system of equations can be solved for the barred Christoffel symbols. In this problem assume that $\overline{g}_{rs}$ is unknown, but that it admits an inverse $\overline{g}^{rs}$ such that $\overline{g}^{rt}\overline{g}_{st} = \delta^{r}_{s}$.

IX.4 Rederive (2.21) by using the contravariant version of (2.18).

IX.5 Give an example of a (partial) derivative F_r such that

$$\overline{F}_r \neq mF_r.$$

Answer: Consider $F_r = \Gamma^t_{tr}$ and bar!

IX.6 Apply the Lemma of SECTION 2 and equation (2.13) to the Hotine 3-leg vectors to get the following results:

$$\overline{\lambda}_{r.s} = m\lambda_{r,s} + m_{/1}\left(\mu_r\mu_s + \nu_r\nu_s\right) - \left(m_{/2}\mu_r + m_{/3}\nu_r\right)\lambda_s$$

$$\overline{\mu}_{r.s} = m\mu_{r,s} + m_{/2}\left(\nu_r\nu_s + \lambda_r\lambda_s\right) - \left(m_{/3}\nu_r + m_{/1}\lambda_r\right)\mu_s$$

$$\overline{\nu}_{r.s} = m\nu_{r,s} + m_{/3}\left(\lambda_r\lambda_s + \mu_r\mu_s\right) - \left(m_{/1}\lambda_r + m_{/2}\mu_r\right)\nu_s.$$

N.B. The cyclic nature of these expressions.

IX.7 Complete the second derivation of the conformal transformation rules for the leg coefficients given in SECTION 4. This involves examining not only the conformal analogues of $\lambda_{r,s}$, but also those of $\mu_{r,s}$ and $\nu_{r,s}$.

IX.8 Give *two* proofs of the key equation (5.4): one should be general, i.e. independent of the leg calculus, while the other may be leg-oriented!

IX.9 Show that by choosing the conformal function such that $\overline{\sigma}_1 = 0$ and $\overline{\sigma}_2 = 0$, a surface **S** having $K \neq 0$ can be transformed into a surface $\overline{\mathbf{S}}$ with $\overline{K} = 0$.

IX.10 Derive (5.20), viz. the conformal analogue of (C_{II}), by computing (5.19).

X

Coordinates in Differential Geodesy

X.1 Introduction

In the preceding nine chapters we have attempted to present a unified approach to the analytical foundations of the Marussi-Hotine formulation of differential geodesy. We believe that our formulation, which is based on the leg calculus and unites the classical Ricci congruence theory with the Cartan calculus of differential forms, permits the Marussi-Hotine theory to be given in an especially transparent form. Indeed, it exhibits their theory in what Marussi would have called an *absolute form*, in that it makes no use of a particular local coordinate system, (see ZUND (1991) for an exposition of Marussi's ideas). Actually Marussi, since he was so devoted to the methodology of the homographic calculus of Burali-Forti and Marcolongo, may not have recognized that the rudimentary form of the leg calculus which was developed and employed by Hotine in his treatise and various papers was in fact an *absolute formalism*. Such an oversight would have been quite natural since almost *ab initio*, Hotine particularized his leg theory to a special coordinate system — usually the (ω, ϕ, N)-system which had been invented by Marussi! We completely agree with Marussi's view that abstractly, theories should be set out in absolute form, and that they should assume a *relative form* in terms of local coordinates only in their application to various physical situations. However, we differ from him, at least with regard to his earlier work — his final views being unknown — in that we claim that the leg calculus is actually an absolute formalism, and that his *intrinsic coordinates* (of which the (ω, ϕ, N)-system is a prime example) may not be the most natural realization of the theory in its relative form.

Thus, our goal in this chapter is two-fold. First, to ascertain what is the status of the known local coordinate systems which were employed by Marussi/Hotine in their formulation of differential geodesy; and second, to suggest an alternate interpretation of their approach which we call the *generalized Marussi-Hotine* theory. The latter offers an entirely new approach to the issue of coordinate systems, i.e. what they mean and how they are selected.

X.2 Marussi Hypothesis and the Hotine Problem

The conceptual basis of the Marussi-Hotine approach to differential geodesy has been considered briefly in CHAPTER I -*vide* the Marussi Conditions of SECTION 1-2. We now consider the rôle of local coordinates in their theory. A reference to the absolutist and relativist viewpoints has already been mentioned in SECTION 1 and CHAPTER I. The issue is now that of the existence of intrinsic coordinates, i.e. those having a geometric/physical meaning, viz. local coordinates well-suited for geodetic problems/situations, and whether an adequate supply of them actually exists. For example, if there were only *one* 'universal' local curvilinear coordinate system, then the formalism of the tensor calculus etc. would be unnecessary. There is of course one such 'universal' — actually — global system, i.e. the Cartesian system

$$x^r = (x, y, z),$$

which is the obvious ambient coordinate system in $\mathbf{E}_3$. It is suitable for the purposes of differential geodesy only to the extent that Mongean differential geometry is an acceptable substitute for Gaussian differential geometry, i.e. only in rather limited circumstances.

We call the assertion that such coordinates exist, and that an adequate supply of them is available the *Marussi Hypothesis*. For purposes of discussion it is useful to delineate *two* extreme forms of this hypothesis:

> *The Strong Form: All geodetic problems can be posed in terms of intrinsic coordinates;*
>
> *The Weak Form: Some geodetic problems can be posed in terms of intrinsic coordinates.*

The former asserts that coordinates are the appropriate manner of formulating mathematical geodesy, while the latter states that coordinates may not always be available for all physical situations. The Weak Form then poses the question of determining under what circumstances appropriate coordinates do exist. The truth of the Strong Form would imply that the relativist form of the tensor calculus is the natural mathematical formalism for differential geodesy, whereas the truth of the Weak Form would lead us to the Cartan calculus or more generally the leg calculus.

Only partial results relative to the general validity of the Marussi Hypothesis are known. Although it must have been of concern to him, it seems that Marussi never formally discussed it in his publications. Indeed, for the most part the only intrinsic coordinate system that he employed in his work was the local astronomical system, which is equivalent to Hotine's system $\mathfrak{H}_0$, i.e. his (ω, ϕ, N)-system, to be discussed in the next section.

The discovery of the utility of non-holonomic coordinate/ reference, or leg, systems by GRAFAREND (1971) and GROSSMAN (1974) shows that the Strong Form of the Marussi Hypothesis is untenable. Indeed, these systems are not merely useful, they occur in situations which cannot be described by ordinary (holonomic) coordinates. Grafarend discussed his discovery with Marussi and was amazed at how quickly he understood and accepted this rather unexpected classical result. Subsequently, in MARUSSI (1979), he employed a non-holonomic reference system in an investigation of tidal forces. Nevertheless, the failure of what we have termed the Strong Form must have been a shock to Marussi for it is clear that in the early formulations of his intrinsic geodesy, the Marussi Conditions were always formulated in terms of coordinates rather than a reference system.

Furthermore, while the adjective 'intrinsic' is appropriate with respect to local coordinate systems it is less than obvious that it can be applied meaningfully to non-holonomic reference systems. This ambiguity is why we prefer to refer to the theory as differential, rather than intrinsic, geodesy. Likewise we have employed 'leg system' without prejudging whether it is a coordinate or a more general reference system. Finally, as discussed in SCHOUTEN (1951, 1954) the terminology non-holonomic (anholonomic) *coordinate system* is a misnomer, since it really means a *reference* — not a coordinate — *system*!

Actually the corresponding cases of the Strong and Weak Forms of Marussi's Hypothesis occur in classical dynamics in the distinction between holonomic and non-holonomic systems. For example, the freedom implicit in the choice of generalized coordinates and the simplicity of the ensuing equations of motion in Lagrangian dynamics makes it very easy to believe in the 'Strong Form' as exemplified by holonomic systems, i.e. those mechanical systems having integrable constraints. The 'universal applicability' of the Lagrangian formulation went unchallenged for almost eighty-five years until 1873 when N.M. Ferrers indicated its failure in the case of a heavy uniform circular disc rolling on a perfectly rough horizontal plane! The complete adoption of the Lagrangian formalism to non-holonomic systems, resulting in the replacement of Lagrange's equations by Appell's equations, was attained only at the turn of the century, but unfortunately is usually omitted or very poorly presented in contemporary textbooks. For example, in a beautiful and extremely lucid text, LANCZOS (1949), Appell's work is accorded two sentences and a footnote! Hence, it is hardly surprising or unexpected that most people are unfamiliar with the subtleties encountered in non-holonomic dynamics. Moreover, who can doubt that in the real world non-holonomic systems are just as prevalent — perhaps more so — as the *idealized* holonomic systems of the classroom or textbook?

Relative to the Weak Form of the Marussi Hypothesis one also must ask whether in real geodetic situations one should expect to have intrinsic coordinates available. In other words, do intrinsic coordinates usually exist, or are they *scarce* in the sense that they occur only in very specialized and ideal situations? This question is non-trivial in that it requires an amalgamation of both mathematical and physical ideas. Mathematically, coordinates must exist not only at a point, but on a domain (i.e. an open connected subset) of $\mathbf{E}_3$ which is large enough to be useful, while physically they must be susceptible of measurement. Moreover, these requirements must be consistent with each other in that excludes the possibility of the other being satisfied. Borrowing the language used by Hadamard in his profound study of the Cauchy Problem, we may say that the Marussi Hypothesis requires that our coordinate/leg systems be *bien posé*, i.e. 'well-posed', both mathematically and physically. Since one can imagine different physical situations occurring in a given coordinate/leg system, only the mathematical part of the requirement is unambiguous: if the coordinate/leg system does not exist mathematically, then there is no possibility of utilizing it in a physical situation. Actually, although by 'existence' in mathematical terms one usually understands a complete specification of the context within which a quantity is well-defined and meaningful, in our usage we ask for less: namely that the set of equations defining the quantity be consistent! We are led to this weaker notion of existence — one is tempted to call it *weak existence* — since the other conditions which the quantity must satisfy need not be known and are likely to depend on the physical situation.

For example, suppose that we are required to determine a quantity, in our case a function $F(x^1, x^2, x^3)$, which satisfies the partial differential equations

$$F_{;r} = f_r \tag{2.1}$$

for a given set of functions f_r on a domain $\mathcal{R}$ of $\mathbf{E}_3$. Then in order for F to exist in the weak sense, it must possess *at least* continuous second derivatives and satisfy the commutation rule:

$$F_{;r;s} - F_{;s;r} = 0 \tag{2.2}$$

which requires the functions f_r to be continuously differentiable and satisfy

$$f_{r;s} - f_{s;r} = 0. \tag{2.3}$$

In other words, (2.3) is required for the consistency of (2.1), but we have specified only the *minimal* smoothness conditions on F and said *nothing* about the boundary conditions which F may be required to satisfy on $\mathcal{R}$ or its boundary. The conditions (2.3) are the *integrability conditions* of (2.1), and if they fail to be satisfied then there exists *no* F satisfying (2.1). Thus, while (2.3) by no means uniquely determine F, they must be satisfied in order for us to meaningfully speak of it.

Finally, we call the problem of constructing local coordinate/leg systems which satisfy the requirements of the Marussi Conditions, or an abridged version of them, the *Hotine Problem.* The first systematic attempt to solve this problem in the *weak sense*, i.e. by making no specific reference to the Marussi Conditions *per se*, was attempted by Hotine and will be considered in the next section. In SECTION 3 we will consider Hotine's bold attempt to solve the Hotine Problem, and then in SECTION 4 we give a critical analysis of the Marussi-Hotine approach to differential geodesy.

X.3 Hotine's Hierarchy of Local Coordinate Systems

In Part II of his treatise, Hotine made a bold and ingenious attempt to solve what we have called the *Hotine Problem*, **HP**. We now give a survey of the status of his proposed local coordinate systems which it is convenient to collectively call Hotine's hierarchy $\mathfrak{H}$, with its individual members labelled as follows,

$\mathfrak{H}_0$: (ω, ϕ, N)-system of [Chapter 12];

$\mathfrak{H}_1$: normal-system of [Chapter 15];

$\mathfrak{H}_2$: triply orthogonal-system of [Chapter 16];

$\mathfrak{H}_3$: (ω, ϕ, h)-system of [Chapter 17];

$\mathfrak{H}_4$: symmetric (ω, ϕ, h)-system of [Chapter 18].

Before beginning our discussion of the individual members of $\mathfrak{H}$, it is necessary to consider it as a whole and try to understand how Hotine regarded it and intended it to be understood. Unfortunately, Hotine gave us little guidance on this matter and much of what he said or inferred is somewhat ambiguous. For example, in the Preface of his treatise [page xi] he said (with italics being added for emphasis):

> "Part II deals with coordinate systems of special interest in geodesy. In Chapter 12 the properties of a *general class* of three-dimensional systems are developed from a single-valued, continuous and differentiable scalar N which serves as one coordinate, while the other two coordinates are defined by the direction of the gradient of N. In Chapters 15 through 18, the scalar N is restricted to provide *simpler systems whose properties can then be derived at once from the general results* of Chapter 12."

The obvious implication is that $\mathfrak{H}_0$ represents a general class of 3-dimensional local coordinate systems, and *somehow* the $\mathfrak{H}_i$ $(i = 1, 2, 3, 4)$ can be regarded as derived from $\mathfrak{H}_0$ by appropriate specializations. Note that nowhere in Part II does he explicitly indicate the implied interdependence or even state it as an ironclad fact. Clearly, $\mathfrak{H}_3$ and $\mathfrak{H}_4$ can be logically considered as subcases of $\mathfrak{H}_0$, but as we will see later the situations for $\mathfrak{H}_1$ and $\mathfrak{H}_2$ do *not* permit such an identification.

The above discussion can be put in more formal terms by introducing the inclusion relation "$\subset$" which we intend to be read as indicating a subcase, viz. as above we would have $\mathfrak{H}_3 \subset \mathfrak{H}_0$, $\mathfrak{H}_4 \subset \mathfrak{H}_0$ with $\mathfrak{H}_1 \not\subset \mathfrak{H}_0$, $\mathfrak{H}_2 \not\subset \mathfrak{H}_0$. The question is then whether Hotine actually believed in the strict specializations of $\mathfrak{H}_0$ according to the two scenarios:

$$(*) \qquad \mathfrak{H}_4 \subset \mathfrak{H}_3 \subset \mathfrak{H}_2 \subset \mathfrak{H}_1 \subset \mathfrak{H}_0 \tag{3.1}$$

or

$$(**) \qquad \mathfrak{H}_i \subset \mathfrak{H}_0 \text{ for } (i = 1, 2, 3, 4)\,, \tag{3.2}$$

or even intend such precision? The evidence is inconclusive, but we would suggest that he probably did, or at least harbor this as a fond hope. First, *none* of the $\mathfrak{H}_i$ ($i = 1, 2, 3, 4$) is worked out in his treatise to the extent that $\mathfrak{H}_0$ was done and he could well have regarded it as a model, or general case, which he expected to ultimately be emulated in the individual $\mathfrak{H}_i$. In this view, even if the individual $\mathfrak{H}_i$ did not follow (∗) or (∗∗), the analysis for $\mathfrak{H}_0$ would be indicative of what could, or should, be done for each of the $\mathfrak{H}_i$. He could reasonably have expected that the precise interrelationships would become apparent *only* when all the details were available. It seems likely to us that both of these suggestions may incorporate at least some of his thoughts on the matter.

However, there is another aspect to the issue which must be taken into account, and probably is the decisive factor to keep in mind. Hotine's treatise was not written in sequential order, either with respect to its Parts or Chapters within a given Part[1]. It has not been possible to reconstruct an exact chronology, but even a cursory examination shows that some chapters are clearly more polished than others. Likewise, one may also establish a correlation between some chapters and the papers/reports collected in the monograph HOTINE (1991). Our suspicion is that it is probable that Part II was among the last written, and indeed *no mention* of $\mathfrak{H}_1$, $\mathfrak{H}_3$ on $\mathfrak{H}_4$ occurs in his other work. Thus, we suggest that having worked out $\mathfrak{H}_0$ to the extent that he did, Hotine may well have been merely proposing the $\mathfrak{H}_i$ as possibilities — perhaps generalizations (?) of $\mathfrak{H}_0$ — *without* paying much attention to (∗) or (∗∗) or worrying about the logical interdependencies involved in $\mathfrak{H}$. Since his treatise was completed (in manuscript) only months before his death, this last possibility is quite understandable and has some credence.

Actually, as our analysis will show, the above mentioned possibility is mathematically really the case, viz. $\mathfrak{H}_0$ is a valid local coordinate system while the remaining $\mathfrak{H}_i$ are invalid. The failure

[1] Private communication from B. Chovitz who served as an official reviewer of Hotine's treatise.

of these $\mathfrak{H}_i$ ($i = 1, 2, 3, 4$) is hardly a trivial matter, i.e. the result of a computational omission or slip, but rather a structural one in which apparently Hotine did not grasp the complexity of the underlying geometric situation. In other words, the difficulties which one encounters in the $\mathfrak{H}_i$ ($i = 1, 2, 3, 4$) cannot be resolved by a more careful/detailed analysis.

Finally, although in the following we are very critical of what Hotine attempted to do in Part II, we are the first to applaud and praise him for his ingenuity and vision. As we noted in SECTION 3, the **HP** is a difficult one, and even with its flaws Hotine's solution, i.e. the hierarchy $\mathfrak{H}$, was a noble — even heroic — attempt to solve it. Moreover, over twenty years later *no one* has succeeded in replacing $\mathfrak{H}$ with a new hierarchy.

In the following, we will review the various coordinate systems of the Hotine hierarchy indicating their various defects and the ideas motivating their construction. The details involved in their construction are complicated and we will merely cite the principal properties which are relevant to our critique. The primary reference is Part II of Hotine's treatise and some results will be cited in the problems at the end of this chapter.

Before proceeding to the individual members of $\mathfrak{H}$, let us briefly review some basic equations and general equations which hold for all these coordinate systems. Given the covariant components of the Hotine 3-leg vectors $\{\boldsymbol{\lambda}, \boldsymbol{\mu}, \boldsymbol{\nu}\}$ we have

$$g_{rs} = \lambda_r \lambda_s + \mu_r \mu_s + \nu_r \nu_s, \tag{3.3}$$

so the line element ds^2 of $\mathbf{E}_3$, i.e.

$$ds^2 = g_{rs} dx^r dx^s \tag{3.4}$$

becomes

$$ds^2 = (\theta_1)^2 + (\theta_2)^2 + (\theta_3)^2 , \tag{3.5}$$

i.e. (3.4), with the usual definition of the Pfaffian forms

$$\theta_1 := \lambda_r dx^r, \quad \theta_2 := \mu_r dx^r, \quad \theta_3 := \nu_r dx^r. \tag{3.6}$$

Note that (3.5) does not imply that (3.4) is diagonal, viz. that the components g_{12}, g_{13} and g_{23} of g_{rs} automatically vanish.

Indeed, since the metric is based on the geometry of the family $\boldsymbol{\Sigma}$ of N-surfaces and its normal congruence $\boldsymbol{\Gamma}$, the explicit construction of the curvilinear coordinate system x^r makes use of the contravariant components of the 3-leg vectors $\{\boldsymbol{\lambda}, \boldsymbol{\mu}, \boldsymbol{\nu}\}$ which are regarded as tangent vectors to the congruences $\boldsymbol{\Gamma}_1$, $\boldsymbol{\Gamma}_2$, $\boldsymbol{\Gamma}_3 :=\boldsymbol{\Gamma}$. Hence, we have

$$\lambda^r := x^r_{/1}, \quad \mu^r := x^r_{/2}, \quad \nu^r := x^r_{/3} \tag{3.7}$$

where "/" denotes leg derivatives with respect to the Hotine 3-leg. The systems $\mathfrak{H}_0$, $\mathfrak{H}_1$ and $\mathfrak{H}_3$ are subject to the Marussi Ansatz

$$(\mathcal{A}) \qquad x^3 := N \tag{3.8}$$

so by virtue of the leg version of $(\mathcal{B})$, i.e. $\{\mathcal{B}\}$ of SECTION V-4, we have

$$\lambda^r = \left(x^1_{/1}, x^2_{/1}, 0\right), \tag{3.9}$$

$$\mu^r = \left(x^1_{/2}, x^2_{/2}, 0\right), \tag{3.10}$$

$$\nu^r = \left(x^1_{/3}, x^2_{/3}, n\right). \tag{3.11}$$

Now we consider the first system of the Hotine hierarchy $\mathfrak{H}$.

$$\mathfrak{H}_0 : x^r = (\omega, \phi, N).$$

Here ω is the longitude $0 < \omega < 2\pi$, and ϕ is the latitude $-\pi/2 < \phi < \pi/2$. The construction of this system involves three basic systems of equations which we call the *geometrical*, *primary*, and *consistency equations* respectively.

The *geometrical equations* involve the expression of the Hotine vectorial 3-leg $\{\boldsymbol{\lambda}, \boldsymbol{\mu}, \boldsymbol{\nu}\}$ in terms of a (constant) Cartesian 3-leg {A, B, C} fixed at the center of the Earth with **C** being aligned along the terrestrial axis of rotation. These were elegantly derived by Hotine in his treatise [pages 70-72] and include explicit definitions of the angles ω and ϕ (see PROBLEM X.1).

The *primary equations* then relate the various components of the 3-leg vectors to the leg coefficients/parameters. This involves obtaining covariant derivatives of the 3-leg vectors (see

PROBLEM X.2), and ultimately leads to the following explicit versions of equations (3.9)-(3.11):

$$\begin{aligned}\lambda^r &= \left(\omega_{/1}, \phi_{/1}, 0\right) = (-k_1 \sec\phi, -t_1, 0)\,,\\ \mu^r &= \left(\omega_{/2}, \phi_{/2}, 0\right) = (-t_1 \sec\phi, -k_2, 0)\,,\\ \nu^r &= \left(\omega_{/3}, \phi_{/3}, n\right) = (\gamma_1 \sec\phi, \gamma_2, n)\,,\end{aligned} \tag{3.12}$$

and the corresponding covariant components may be readily computed via PROBLEM X.3. Then the primary equations are the triples of the leg derivatives of the coordinates', i.e.

$$\begin{aligned}(\omega):&\quad \left(\omega_{/1}, \omega_{/2}, \omega_{/3}\right) = (-k_1 \sec\phi, -t_1 \sec\phi, \gamma_1 \sec\phi)\,,\\ (\phi):&\quad \left(\phi_{/1}, \phi_{/2}, \phi_{/3}\right) = (-t_1, -k_2, \gamma_2)\,,\\ (N):&\quad \left(N_{/1}, N_{/2}, N_{/3}\right) = (0, 0, n)\,.\end{aligned} \tag{3.13}$$

As noted in PROBLEM X.4, Hotine's analysis contains a curious gap: he failed to notice that his equations furnish *two* expressions for the gradient of ω. In other words in addition to the above (ω), we also have a second triple:

$$(\omega^*) : \left(\omega_{/1}, \omega_{/2}, \omega_{/3}\right) = (\sigma_1 \csc\phi, \sigma_2 \csc\phi, \varepsilon_3 \csc\phi)\,. \tag{3.14}$$

This curious feature of $\mathfrak{H}_0$ was discovered in ZUND (1990) and may be called the *ω-degeneracy*. Since (ω) and (ω^*) must be equal there are obviously algebraic identities relating the leg coefficients k_1, t_1, γ_1 and σ_1, σ_2, ε_3 respectively; i.e. these coefficients are *ϕ-related* (see PROBLEM X.5). The ω, ω^*, and ϕ primary equations may be written in several equivalent forms, i.e. the *gradient Pfaffian form*: (ω_r), (ω^*_r), (ϕ_r); and the *Pfaffian form*: $(d\omega)$, $(d\omega^*)$, $(d\phi)$ as exhibited in PROBLEM X.6. The most useful form is the full set of primary equations given by the matrix equations:

$$\left\| \begin{matrix} d\omega \\ d\phi \\ dN \end{matrix} \right\| = \mathcal{M} \left\| \begin{matrix} \theta_1 \\ \theta_2 \\ \theta_3 \end{matrix} \right\|, \tag{3.15}$$

and

$$\begin{Vmatrix} \theta_1 \\ \theta_2 \\ \theta_3 \end{Vmatrix} = \mathcal{M}^{-1} \begin{Vmatrix} d\omega \\ d\phi \\ dN \end{Vmatrix}, \tag{3.16}$$

where the matrices $\mathcal{M}$ and $\mathcal{M}^{-1}$ are given in PROBLEM X.7.

Equation (3.15) also occurs in GRAFAREND (1975) who regards it as a matrix of integrating factors, and as "... one of the most fundamental formulae of geodetic science ... ." His reasoning is as follows: in this equation one *cannot* take

$$\theta_1 := dX, \quad \theta_2 := dY, \quad \theta_3 := dZ; \tag{3.17}$$

where X, Y, Z are new independent variables, viz. if this were true, then (3.15) would be a transformation between two holonomic systems. By using (3.3) and (3.5) one may obtain the resulting ds^2 for $\mathfrak{H}_0$. It is complicated and given in PROBLEM X.8. Obviously Hotine's treatment of the primary equations was incomplete due to his decidedly rudimentary use of the leg calculus, and his failure to note the ω-degeneracy, i.e. he did not have (ω^*), or (ω_r^*), or ($d\omega^*$).

The final step in the construction of $\mathfrak{H}_0$ involves consideration of the *consistency equations.* These involve the Hotine-Marussi equations $\{\mathcal{HM}\}$ discussed in SECTION VI-9, and for $\mathfrak{H}_0$ they couple the (ω, ϕ) coordinates to the leg coefficients and their leg derivatives. These were partially given for the ω and ϕ-commutators in HOTINE (1957), but not repeated in his treatise! A more complete discussion — with a correction of an incorrect assertion in HOTINE (1957) — is given in ZUND (1990a, 1990b, 1993b). Due to their length and various interrelationships these are not given in our problems; however, what is essentially involved is working out the ω, ω^* and ϕ-commutators for the triples (ω), (ω^*) and (ϕ) exhibited above and identifying these commutators with the Lamé equations of SECTION IV-9. Without an investigation of the consistency equations one has only an incomplete specification of the leg coefficients in the $\mathfrak{H}_0$ system, and no guarantee that this system actually exists in $\mathbf{E}_3$, i.e. that it satisfies the conditions for weak existence discussed in SECTION 2.

In conclusion, we may summarize the *virtues* and *defects* of the $\mathfrak{H}_0$ system. The former include the following:

(i) $\mathfrak{H}_0$ is simple enough to be relatively easy to handle, yet complicated enough to avoid being trivial;

(ii) the coordinates have geometric/physical significance, and

(iii) the variables are measurable and at least roughly satisfy the physical conditions of Marussi of SECTION I-2.

On the other hand, relative to the latter, undoubtedly its worst feature is

(iv) the emphasis that Hotine and Marussi placed on it, with the suggestion that somehow it is the 'natural choice' and 'fairly general';

(v) it is astronomically based, reflecting a classical view of geodesy, and as such is inadequate for contemporary purposes;

(vi) $\mathfrak{H}_0$ is applicable only to a single N-surface, and cannot be extended to a neighboring equipotential surface, viz. (ω, ϕ) are "frozen" on the N-surface $\mathbf{S}$: $N =$ constant.

We will return to item **(iv)** in SECTION 5 where we propose a conjecture on the significance of $\mathfrak{H}_0$, while **(vi)** will be proven in our ensuing discussion of $\mathfrak{H}_1$.

$\mathfrak{H}_1$: Normal Coordinate System $x^r{=}(u^1, u^2, N)$.

This proposed coordinate system seems to be entirely Hotine's invention and it is mentioned only in Chapter 15 of his treatise. The basic idea was to construct a system in which the N-surfaces would form a *generalized parallel system* $\mathcal{P}'$ having *curved* orthogonal trajectories and as such $\mathfrak{H}_1$ represents Hotine's first attempt to formulate and solve the Extension Problem.

Unfortunately, Hotine's analysis is both flawed and incomplete and apparently based on a misunderstanding of what is actually involved in the construction of $\mathfrak{H}_1$. The difficulties occur almost immediately in his discussion [see page 103] when he speaks of the family $\boldsymbol{\Sigma}$ of N-surfaces being a family of *geodesic parallels* $\mathcal{G}$. This notion requires the congruence $\boldsymbol{\Gamma}$ of plumblines orthogonal to the $\boldsymbol{\Sigma}$ to be geodesics! Such an approach is interesting and physically meaningful in a curved *Riemannian space*, e.g. in a 3-dimensional $\mathbf{V}_3$, but in $\mathbf{E}_3$ it is essentially trivial since all such geodesic congruences $\boldsymbol{\Gamma}$ are linear congruences $\boldsymbol{\Lambda}$, i.e. consists of straight lines ($\chi = 0$). In other words, the $\mathcal{P}'$ sought by Hotine in his construction of $\mathfrak{H}_1$ exists in $\mathbf{E}_3$ only when it reduces to the ordinary parallel system $\mathcal{P}$ discussed in SECTION VII-5.

It is hard to believe that Hotine could have overlooked such an elementary fact, however, that seems to be the case. Apparently, he simply took over a very appealing theory from $\mathbf{V}_3$ without realizing that in $\mathbf{E}_3$ this theory is trivial. The resulting generality inherited from the situation in $\mathbf{V}_3$ then effectively obscured the resulting construction and its triviality in $\mathbf{E}_3$.

The basic steps of the construction are as follows. One seeks to specialize (3.4) in such a manner that a region of $\mathbf{E}_3$ is described by a family of coordinate surfaces $\boldsymbol{\Sigma}$ which is orthogonal to a curved congruence $\boldsymbol{\Gamma}$. Intuitively, we may regard the region as being 'swept out' by the $\boldsymbol{\Sigma}$ along $\boldsymbol{\Gamma}$ such that each point of the region has one and only one surface of $\boldsymbol{\Sigma}$ and one curve of $\boldsymbol{\Gamma}$ passing through it. Then, choosing $\boldsymbol{\Sigma}$ to be the x^3-surfaces and $\boldsymbol{\Gamma}$ the x^3-curves, one requires that the x^1 and x^2-curves be orthogonal to $\boldsymbol{\Gamma}$. Analytically this demands that the gradients of x^1 and x^2 be orthogonal to the gradient of x^3 (these gradients being computed with respect to some ambient coordinate system in $\mathbf{E}_3$). Then (3.4) reduces to

$$ds^2 = g_{\alpha\beta} dx^\alpha dx^\beta + g_{33} \left(dx^3\right)^2 \tag{3.18}$$

where $g_{\alpha 3} = 0$ and the Greek indices range over the first two spatial values 1, 2. Since the x^α-curves are by construction orthogonal to the x^3-curves, they must lie in a surface S and we may choose them to be the Gaussian parameters of $\mathbf{S}$, viz. $x^\alpha = u^\alpha$ with the Greek indices now being proper surface indices. Hence

(3.18) reduces to

$$ds^2 = a_{\alpha\beta}du^\alpha du^\beta + g_{33}\left(dx^3\right)^2 \tag{3.19}$$

where $g_{\alpha\beta} = a_{\alpha\beta}$ which describes a *singly-orthogonal system* $\mathcal{O}_1$, with each **S** of **Σ** being orthogonal to **Γ**. The covariant and contravariant components of the (spatial) metric tensor are given by

$$\|g_{rs}\| = \begin{Vmatrix} a_{\alpha\beta} & & 0 \\ & & 0 \\ 0 & 0 & g_{33} \end{Vmatrix} \tag{3.20}$$

and

$$\|g_{rs}\| = \begin{Vmatrix} a^{\alpha\beta} & & 0 \\ & & 0 \\ 0 & 0 & g^{33} \end{Vmatrix} \tag{3.21}$$

respectively where $g^{33} = 1/g_{33}$; and for each particular (constant) value of x^3, $a_{\alpha\beta}du^\alpha du^\beta$ is the first basic form I of an **S** of **Σ**. Clearly, for such a privileged choice of x^3, $\mathcal{O}_1$ as defined by (3.19), is the most general line element in $\mathbf{E}_3$ which admits a family of surfaces **Σ** orthogonal to the congruence **Γ**. The length along an x^3-curve of **Γ** is given by the line integral

$$\ell = \int \sqrt{g_{33}}dx^3$$

where the limits are taken to be the x^3-coordinates of an initial point P_0 on a *base surface* $\mathbf{S}_0$ of **Σ**, and a final point P on some prescribed surface S of **Σ**. The value $\ell = 0$ characterizes $\mathbf{S}_0$ and henceforth a zero subscript/superscript will be attached to quantities on $\mathbf{S}_0$ which depend on u^1, u^2, while quantities on **S** depend on u^1, u^2 and a non-zero value of ℓ which is now an additional parameter. Note that Hotine denotes ℓ by s, but this invites the possibility of confusion with a general non-**Γ** (!) arc-length in $\mathbf{E}_3$.

Upon adopting the Marussi Ansatz ($\mathcal{A}$), i.e. $x^3 := N$ we have

$$g_{33} = 1/n^2, \quad g^{33} = n^2$$

in (3.20) and (3.21), and Hotine's normal coordinate system line element

$$ds^2 = a_{\alpha\beta} du^\alpha du^\beta + \frac{1}{n^2} dN^2 \tag{3.22}$$

i.e. his [15.02]. Hotine's deduction of (3.22) is curious in that he obtained it by means of a conformal transformation [see page 103]! This is quite unnecessary — in fact it is a red herring since (3.22) is an immediate consequence of ($\mathcal{B}$), i.e. $N_r = n\nu_r$, and ($\mathcal{A}$) upon using $\nu_r = n^{-1}\delta_r^3$ in (3.3) and contracting with $dx^r dx^s$.

$\mathfrak{H}_2$: (x^1, x^2, N) Triply-Orthogonal System

Due to the effectiveness of orthogonal curvilinear coordinates in mathematical physics it was natural for Hotine to seek an analogous construction in differential geodesy. Having the Marussi Ansatz ($\mathcal{A}$) in mind, the obvious procedure was to choose a doubly-orthogonal curvilinear system $\mathcal{O}_2$ of surfaces

$$x^1 = \text{const}, \quad x^2 = \text{const}$$

in $\mathbf{E}_3$ and seek to *complete* it as an $\mathcal{O}_3$ by choosing ($\mathcal{A}$):

$$x^3 = N = \text{constant}$$

with $(x^1, x^2) = (u^1, u^2)$ being Gaussian parameters on the family of N-surfaces $\mathbf{\Sigma}$. If this were possible, one would automatically obtain a global — not a limiting — solution of the **EP**. Hence $\mathfrak{H}_2$ may be regarded as Hotine's second attempt to solve this problem. This construction is based on the so called

Hotine Conjecture

Any real-valued smooth function N which has a non-vanishing gradient N_r on a region $\mathcal{R}$ of $\mathbf{E}_3$ can be chosen as a member of an $\mathcal{O}_3$.

At first glance this conjecture appears very reasonable, however its 'obviousness' fades as one begins to understand what is geometrically involved in the construction of an $\mathcal{O}_3$, or an $\mathcal{O}_3^*$, e.g. recall the Dupin Theorem of SECTION V-13. It turns out that the above completion process is highly non-trivial and that the conjecture is equivalent to requiring that any real-valued smooth function N having a non-vanishing gradient is a solution of the Cayley-Darboux equation which we symbolically denote by

$$\aleph N = 0. \tag{3.23}$$

The theory of triply-orthogonal systems has a rich history and was investigated by some of the most distinguished geometers of the nineteenth century. Properly speaking it was begun by Dupin with his theorem in 1813, and it was further enriched by Lamé and his theories of curvilinear coordinate systems in $\mathbf{E}_3$ during the years 1840-1859. However, as early as 1846 Bouquet had shown that an arbitrarily chosen surface *cannot* belong to an $\mathcal{O}_3$! In 1862 Bonnet proved that the completion of an $\mathcal{O}_2$ to an $\mathcal{O}_3$ required that the function defining the new surface had to be a solution of a third order partial differential equation, but he was unable to exhibit this equation. Finally in 1872 Cayley, and in 1873 Darboux, explicitly exhibited the required equation, which we call the Cayley-Darboux equation. The subject remained a popular one for geometers for another fifty years.

The above historical sketch is essentially that given in the introduction to Chapter XI of FORSYTH (1912) which (see ZUND (1990)) was one of Hotine's favorite sourcebooks on differential geometry and cited in Chapters 6, 10 and 16 of his treatise. Indeed, Bouquet's result cited above, is sufficient to show that Hotine's Conjecture cannot be valid. However, despite this Hotine, with what can only be viewed as adamant obstinacy, stuck to his guns. He believed that the Cayley-Darboux equation (see PROBLEMS X.9 and 10) was an identity and hence trivially satisfied. The reason was that his form of the equation, which he wrote in tensor form (see [16.03] and [16.04]), was incorrect and in fact one of the Bianchi identities. This was sorted out several years ago in ZUND and MOORE (1987).

X.4 The Marussi-Hotine Approach: A Critique

The failure of the Hotine hierarchy $\mathfrak{H}$ of local coordinate systems $\mathfrak{H}_0, \mathfrak{H}_1, \cdots, \mathfrak{H}_4$ to provide more than one useful reference system for geodetic problems is undeniably disconcerting. Perhaps this suggests one reason why the methods of differential geodesy have been used little by theoretical geodesists. If the methodology has produced only *one* reasonably well-understood coordinate system then the applicability of the theory is severely limited and those situations which do not obviously fit into the (ω, ϕ, N) system can be regarded as either intractable or unnecessarily difficult. Moreover, the phrase 'reasonably well-understood' is of a somewhat dubious nature: Part II of Hotine's treatise is difficult reading, and in particular Chapter 12 is a veritable jungle of calculations and formulas. Even worse, as noted in SECTION 3, Hotine's exposition did not succeed in putting the (ω, ϕ, N) system on a firm mathematical foundation. Since Hotine's death, we know of no geodetic research which has attempted to make serious use of $\mathfrak{H}_1, \mathfrak{H}_2, \mathfrak{H}_3$ or $\mathfrak{H}_4$. This is probably understandable since the presentation of these systems in HOTINE (1969) was too sketchy and tentative to permit their immediate use.

Finally, if as we suggested in SECTION 3, $\mathfrak{H}_0$ is itself of limited practical use, is there a more general (ω, ϕ, N)-like coordinate system of the form (u, v, N) where $u^\alpha = (u^1, u^2) = (u, v)$ are Gaussian parameters on a N-surface $\mathbf{S}$ of $\mathbf{E}_3$? The investigation of $\mathfrak{H}_0$ and $\mathfrak{H}_1$ seems to indicate that if such a new system were constructed, it would likely be applicable only to a single N-surface of the family $\mathbf{\Sigma}$. Thus, by the discussion of the **EP** in CHAPTER VII, such a result would provide at most a snapshot rather than a continuous picture of the geopotential field.

One remedy to the situation is to attempt to devise a new local coordinate system. However, since the aforementioned difficulties of the Hotine hierarchy of systems is structural, such an endeavor cannot be a quick cosmetic fix of $\mathfrak{H}_0$ or $\mathfrak{H}_1$. In our opinion, a more useful tactic is to examine the Hotine hierarchy as a whole, and seek it's common features, and then attempt to assess to what extent these features can be satisfied. The obvious answer for $\mathfrak{H}_0, \cdots, \mathfrak{H}_4$ is the Marussi Ansatz $(\mathcal{A})$ which we now examine in some detail.

The meaning of $(\mathcal{A})$, and whether or not it is an integral part of the Marussi-Hotine approach to differential geodesy, are delicate issues. Indeed, it appears that the usual answers are based more on tradition than on a careful logical analysis of the situations. We will *attempt* to do better!

First, it is true that in almost all their work $(\mathcal{A})$ was employed. However, an examination of the Marussi Conditions — recall SECTION I-2; (iii)-(v) in particular — suggests that $(\mathcal{A})$ is a *choice* rather than an obligatory accoutrement of their approach. If this is so, then $(\mathcal{A})$ must necessarily involve no loss of generality in their description of the geopotential field. For example, an immediate consequence of $(\mathcal{A})$ is that the Marussi tensor N_{rs} reduces to

$$N_{rs} = -\Gamma^3_{rs} \tag{4.1}$$

which under an arbitrary regular coordinate transformation:

$$x^r \to \overline{x}^r = f^r\left(x^1, x^2, x^3\right) \tag{4.2}$$

does not have tensor character! This seems paradoxical, but having chosen $(\mathcal{A})$, viz.

$$x^3 = N, \tag{4.3}$$

and restricting (4.2) to those transformations such that

$$x^3 \to \overline{x}^3 = N \pm C \tag{4.4}$$

(where C is a constant) (4.1) is a tensor equation! The reader will be asked to verify this in PROBLEM X-1. Hence, although subject to $(\mathcal{A})$, N_{rs} fails to have tensor character under arbitrary general regular transformations (4.2), this failure is *moot* since having chosen (4.3) the only transformations which one would admit in practice are those given by (4.4). In other words, the real loss of generality *relative to their theory* is illusory, despite the apparent restrictiveness of (4.3) in a general context. It is interesting to note that (4.1) is included in [Section 79, page 82] as an unnumbered equation *before* the formal discussion of the Marussi tensor in [Section 101, page 86]. However, the meaning of (4.1) is not explained by Hotine.

Finally, we observe that under $(\mathcal{A})$ the basic gradient equation $(\mathcal{B})$, i.e.

$$N_r = n\nu_r,$$

becomes

$$\delta_r^3 = n\nu_r \Rightarrow \nu_r = (0, 0, 1/n)\,. \tag{4.5}$$

In general, the restriction that *only* $\nu_3 \neq 0$ for a vector field is non-trivial, however it is natural when one imposes the condition (4.4). Indeed (4.5) occurs in all the systems of the Hotine hierarchy $\mathfrak{H}$.

X.5 Generalized Marussi-Hotine Approach : A Proposal

In the previous section, we have seen that the usual Marussi-Hotine formulation of differential geodesy has limited applicability, and indeed may be reduced to the consideration of a single system of coordinates if our conjecture is correct. Even if this were to turn out not to be the case, the Marussi Ansatz ($\mathcal{A}$) restricts their theory to a particular viewpoint in which one is limited to trying to deduce the geometry of a *known* geopotential field N. This may be very easy, or virtually impossible, depending on the coordinate dependence of the function N. Presumably N is determined in some physical coordinate system, or at least some system in which the Laplace-like geopotential equation is reasonably manageable; however such coordinates need not be convenient for investigating the geometry of the geopotential field. In this case, one is faced with finding an appropriate geometric coordinate system, and this may be pure guesswork.

As an extreme case, let us consider the classical Laplace equation

$$\Delta N = 0 \tag{5.1}$$

describing the field of a non-rotating Earth ($\tilde{\omega} = 0$) in ordinary Cartesian coordinates $x^r = (x, y, z)$. In a feat of considerable mathematical cunning, WHITTAKER (1902), (see also Chapter XVIII of WHITTAKER and WATSON (1927)), showed that the most general solution of (5.1) could be written as a definite integral:

$$N = \int_{-\pi}^{\pi} f\,(z + ix\cos u + iy\sin u; u)\,du \tag{5.2}$$

where f is an arbitrary smooth function and u is a disposable parameter. Functionally speaking this solution is of the form

$$N = F\,(ax + by + cz) \tag{5.3}$$

where a, b, c are (complex) constants such that

$$a^2 + b^2 + c^2 = 0, \tag{5.4}$$

and has the more familiar 2-dimensional analogue

$$N = F\,(x + iy) + G\,(x - iy)\,. \tag{5.5}$$

Such solutions are undeniably elegant — or at least curious — and although Whittaker did indicate how to transform (5.2) into various orthogonal curvilinear coordinate systems, it is apparent that neither (5.2) nor (5.3) *per se* is immediately suited for geometric considerations. These coordinates are called *determinative coordinates* since they were used to determine a solution of (5.1) or more generally the modified Laplace equation VI-(2.3). Hopefully they have *some* semblance of a physical or geometric significance, but more commonly they are chosen purely for their mathematical convenience and — unless the situation has an obvious geometric symmetry, e.g. the Dirichlet or Neumann problems for a sphere etc. — the connection may be weak or nonexistent.

The challenge to be met in the Marussi-Hotine approach is to fit an appropriate coordinate system, or literally a geometry, onto a given solution of the modified Laplace equation VI-(2.3), or its equivalent gravity equation VI-(7.2). The situation is not unlike that indicated by A.N. Whitehead when he spoke of having a grin and being required to find a Cheshire cat to fit onto it! Recall that three of Hotine's systems, i.e. $\mathfrak{H}_2$, $\mathfrak{H}_4$ and $\mathfrak{H}_5$, have straight plumblines and hence are not suitable to describe nontrivial physical geodetic situations having $\chi \neq 0$.

The Marussi-Hotine approach is really a *descriptive* one in which one seeks to elucidate the geometry of a geopotential field, and consequently its admissible coordinate systems x^r, by so called *descriptive coordinates.* Obviously the situation is exacerbated by the lack of an abundant supply of such systems, e.g. $\mathfrak{H}$ consists of only $\mathfrak{H}_0$. Indeed, this is well illustrated by Hotine's discussion in Part III of his treatise when little of his analysis made serious use of the geometric methodology developed in Parts I and II. Hence, in this sense, Hotine's discussion demonstrates the limitations inherent in the Marussi-Hotine approach.

Indeed, one can go further and make the

Conjecture

> $\mathfrak{H}_0$, *or more generally the class of all local coordinate systems which are related to it by regular coordinate transformations which preserve the Marussi Ansatz* $(\mathcal{A})$, *are the only admissible coordinates in the Marussi-Hotine theory.*

By virtue of our result in SECTION 3, that $\mathfrak{H}_0$ is the only viable system in the Hotine hierarchy $\mathfrak{H}$, this conjecture seems to be reasonable — at least it comprehensively summarizes what is known about coordinate systems in the Marussi-Hotine approach. Heuristically, one can argue that having chosen $x^3 := N$ by $(\mathcal{A})$, then the remaining two coordinates x^α on the N-surface $\mathbf{S}$: $N =$ constant (when they are identified with a pair of generally curvilinear Gaussian parameters u^α) are essentially $x^1 := \omega$ and $x^2 := \phi$. The hedge 'essentially' means that — apart from questions of orientation — the pair (ω, ϕ) could be replaced by (ϕ, ω) — and the choice is naturally modulo a regular coordinate transformation

$$x^\alpha \mapsto \overline{x}^\alpha = f^\alpha\left(x^1, x^2\right).$$

The gaps in the argument are whether or not $\mathfrak{H}$ is in fact all of the possible local coordinate systems representing $(\mathcal{A})$, and, whether it is possible to generate "non-frozen" descriptive coordinate systems valid on not merely $\mathbf{S}$ but on a non-trivial family $\mathbf{\Sigma}$ of N-surfaces. Our feeling, and it is only a *conjecture* (but one reinforced by numerous futile calculations), is that $\mathfrak{H}$ exhausts such possible systems and that they are all "frozen" systems.

While it would be nice to have a proof, or a counterexample to this conjecture, we believe that it will be difficult to produce a truly convincing argument, and objectively speaking it is probably not worth the effort since the descriptive approach is *not* the only viewpoint in which differential geodesy can be developed.

In our opinion a more robust approach is not to start from a given solution of the modified Laplace equation, but rather to include it as one of the Hotine-Marussi equations $(\mathcal{HM})$ and try to solve and satisfy this system of equations subject to certain geodetic and geometric conditions. The Lamé equations $(\mathcal{L})$ would be included in $(\mathcal{HM})$ and obviously such a procedure is highly non-trivial. However, it has the advantage of seeking a geopotential which has 'built in' geodetic/geometric properties. If such a search were successful, the resulting N, or equivalently n by virtue of Hotine's theorem (SECTION VI-7), would have prescribed properties, and there would be no guesswork involved in attempting to extract the geometry from a given N. Likewise, the inability to solve for N, or n, for a given set of conditions could also be of importance in ascertaining what kind of conditions are proper. Typically, the conditions might be incompatible (as in Example A given below), or inadequate to permit solution, or a trivialization of the $(\mathcal{HM})$ system. By 'trivialization' of an equation we mean that it is not a determinative equation, i.e. an equation serving to determine a quantity or quantities, but an identity which is identically satisfied by virtue of previously determined or specified quantities.

We call such a formulation of differential geodesy the *generalized Marussi-Hotine approach*, although strictly speaking we know of no evidence that they ever seriously considered this approach *per se*. The theory is formulated in terms of the leg calculus, and the basic variables are the contravariant/covariant components of the 3-leg vectors, viz. the vectorial Hotine 3-leg $\{\boldsymbol{\lambda}, \boldsymbol{\mu}, \boldsymbol{\nu}\}$, and the full set of leg coefficients:

$$k_1, k_2, t_1, t_2, \gamma_1, \gamma_2, \sigma_1, \sigma_2, \varepsilon_3.$$

Some of these components and coefficients are specified via the geometric conditions, with the remaining ones being determined by solving or satisfying, i.e. trivializing, the system $(\mathcal{HM})$. One can also impose dynamical conditions: the most obvious one being the value of $\tilde{\omega}$, i.e. $\tilde{\omega} \neq 0$ for a rotating Earth, $\tilde{\omega} = 0$ for a non-rotating Earth (see Example B below).

Coordinates are no longer basic quantities but rather *derived* ones which occur via the specification of the components of the 3-leg vectors, and in particular by (3.7). The final step in the analysis would be to solve, or at least employ, these equations to deduce the coordinates or their geometric significance. The resulting coordinates are said to be *permissible* in that they have the properties permitted by the geodetic/geometric conditions employed to obtain N or n. Hence, we call such a procedure a *permissible approach* to the Marussi-Hotine theory (see Example C below for a rather trivial but instructive example).

We present three examples designed purely for illustrative purposes and intended to indicate the type of construction we have in mind. The first is really of a descriptive character and is a "no-go" result.

Example A

By Whittaker's example we know that (5.1) admits a solution of the form (5.3). Suppose now that F is the identity function, then

$$N = ax + by + cz \tag{5.6}$$

is a solution of (5.1) without imposition of the condition (5.4), although a, b, c are arbitrary constants. Then **S**: $N = C$ where C is a set of constants chosen over some range, defines a family of parallel planes $\mathbf{\Sigma}$ in $\mathbf{E}_3$. We then seek a *curved* congruence of curves $\mathbf{\Gamma}$ normal to $\mathbf{\Sigma}$. Obviously this is not possible. However, we wish to deduce this via the Hotine-Marussi equations. By the basic gradient equation $(\mathcal{B})$:

$$n\nu_r = (0, 0, c)\,,$$

and since we are dealing with a Cartesian coordinate system $x^r = (x, y, z)$ in $\mathbf{E}_3$ with $\boldsymbol{\nu}$ being a unit vector, we have

$$\nu_r = (0, 0, 1) = \delta_r^3,$$

with

$$n = \text{ constant } = c.$$

Then since each **S** is a plane we have $k_1 = k_2 = 0$, and consequently the set of n-equations with $n \neq 0$ yields:

$$\begin{aligned} n_{/1} &= \gamma_1 n && \Rightarrow \gamma_1 n = 0 \Rightarrow \gamma_1 = 0, \\ n_{/2} &= \gamma_2 n && \Rightarrow \gamma_2 n = 0 \Rightarrow \gamma_2 = 0, \\ n_{/3} &= (k_1 + k_2)\, n && \Rightarrow 0 = 0; \end{aligned}$$

and thus $\chi = 0$ so $\mathbf{\Gamma}$=$\mathbf{\Lambda}$. Hence, the proposed construction with $\mathbf{\Gamma} \neq \mathbf{\Lambda}$ is impossible.

Example B

This is a reworking — from the determinative viewpoint — of an example done in SECTION VI-7. We assume a normal linear congruence $\mathbf{\Lambda}$, i.e.

$$\chi = 0 : \gamma_1 = \gamma_2 = 0,$$

which satisfies property (T*) with

$$t_1 = -t_2 = 0.$$

Our problem is then to determine the family of geopotential surfaces $\mathbf{\Sigma}$ orthogonal to $\mathbf{\Lambda}$, i.e. the geopotential function N or equivalently the local gravity n. Letting $\boldsymbol{\nu}$ be the tangent to $\mathbf{\Lambda}$ with h being a height above a base surface $\mathbf{S}_0$ to $\mathbf{\Sigma}$, then

$$\nu^r = dx^r / dh.$$

Upon choosing

$$\nu^r = \delta^r_3,$$

which amounts to a particular alignment of the third Hotine 3-leg vector, we have

$$\frac{dx^3}{dh} = 1,$$

and integration yields

$$x^3 - \left(x^3\right) |_{S_0} = h.$$

This suggests identifying x^3 with a radial coordinate r, i.e.

$$x^3 := r.$$

By use of Hotine's theorem (recall again SECTION VI-7) we may consider n rather than the geopotential function N. Then the first pair of n-equations

$$\begin{aligned} n_{/1} - \gamma_1 n = 0 &\Rightarrow n_{/1} = 0, \\ n_{/2} - \gamma_2 n = 0 &\Rightarrow n_{/2} = 0, \end{aligned}$$

indicate that there is no tangential variation of n on the surfaces of $\mathbf{\Sigma}$. The third n-equation

$$n_{/3} - (k_1 + k_2)\, n = -2\tilde{\omega}^2$$

must now be considered. The values of k_1, k_2 are not known, but subject to our assumptions, two of the Lamé equations $(\mathcal{L}_{\mathrm{V}})$ and $(\mathcal{L}_{\mathrm{IX}})$ of SECTION V-9, are respectively

$$\begin{aligned} k_{1/3} - k_1^2 &= 0, \\ k_{2/3} - k_2^2 &= 0. \end{aligned}$$

Since these leg differential equations are identical, k_1 and k_2 must have the same r-dependence and modulo our assumptions $(\mathcal{L}_{\mathrm{IV}})$ and $(\mathcal{L}_{\mathrm{VII}})$ are respectively

$$\begin{aligned} -k_{1/2} + k_1\sigma_1 &= k_2\sigma_1, \\ -k_{2/1} + k_2\sigma_2 &= k_1\sigma_2. \end{aligned}$$

We have made no assumptions about the leg coefficients σ_1, σ_2, but by $(\mathcal{L}_{\mathrm{I}})$ we have

$$\sigma_{1/2} - \sigma_{2/1} - \sigma_1^2 - \sigma_2^2 = k_1 k_2 = K$$

so not both σ_1, σ_2 can be zero. Then both $(\mathcal{L}_{\mathrm{IV}})$ and $(\mathcal{L}_{\mathrm{VII}})$ are identically satisfied by choosing

$$k_1 = k_2,$$

and hence k_1 is independent of the $\boldsymbol{\mu}$-direction while k_2 is independent of the $\boldsymbol{\lambda}$-direction. The simplest choice is that these

k depend *only* on the $\boldsymbol{\nu}$-direction, viz. on the previously introduced r coordinate. Then

$$k_{/3} - k^2 = 0,$$

i.e.

$$\frac{dk}{dr} - k^2 = 0,$$

which has a solution

$$k = -\frac{1}{r}.$$

Thus, the third n-equation becomes

$$\frac{dn}{dr} + \frac{2}{r} n = -2\tilde{\omega}^2$$

which is easily seen to have the solution

$$n = -\frac{2}{3}\tilde{\omega}^2 r + \frac{n^0}{r^2}$$

where n^0 denotes a function of integration which is independent of r.

<u>Example</u> C

This is a radical reworking of the case of spherical polar coordinates $x^r = (\omega, \phi, r)$ as previously worked out in SECTION III-6.

Suppose that we are given a line element ds^2 in $\mathbf{E}_3$ having the form

$$ds^2 = (r\cos\phi)^2\, d\omega^2 + r^2 d\phi^2 + dr^2$$

with the coordinates being labelled as above, but, apart from this, *nothing* is known about their character or properties. From ds^2 one knows only that the line element is orthogonal, i.e.

$$g_{12} = g_{23} = g_{31} = 0,$$

and from its explicit character one sees that (ω, ϕ) are angular coordinates while r is not angular. Then upon employing the vectorial 3-leg $\{\boldsymbol{\lambda}, \boldsymbol{\mu}, \boldsymbol{\nu}\}$ and the canonical leg representation (3.3) and (3.6) we have

$$\begin{aligned}
\lambda_r &= (r\cos\phi, 0, 0)\,, \\
\mu_r &= (0, r, 0) \\
\nu_r &= (0, 0, 1)\,;
\end{aligned}$$

and

$$\begin{aligned}
\theta_1 &= r\cos\phi d\omega, \\
\theta_2 &= r d\phi, \\
\theta_3 &= dr.
\end{aligned}$$

The latter expressions are especially useful since by virtue of the Pfaffian representations of the coordinate differentials, i.e.

$$\begin{aligned}
d\omega &= \omega_{/1}\theta_2 + \omega_{/2}\theta_2 + \omega_{/3}\theta_3, \\
d\phi &= \phi_{/1}\theta_1 + \phi_{/2}\theta_2 + \phi_{/3}\theta_3, \\
dr &= r_{/1}\theta_1 + r_{/2}\theta_2 + r_{/3}\theta_3;
\end{aligned}$$

one readily sees that

$$\begin{aligned}
\omega_{/1} &= (r\cos\phi)^{-1}\,, \\
\phi_{/2} &= r^{-1}, \\
r_{/3} &= 1.
\end{aligned}$$

Thus, by (3.7), viz.

$$\begin{aligned}
\lambda^r &= \left(\omega_{/1}, \phi_{/1}, r_{/1}\right)\,, \\
\mu^r &= \left(\omega_{/2}, \phi_{/2}, r_{/2}\right)\,, \\
\nu^r &= \left(\omega_{/3}, \phi_{/3}, r_{/3}\right)\,;
\end{aligned}$$

we have the following contravariant components of the 3-leg vectors:

$$\begin{aligned}
\lambda^r &= (r^{-1}\sec\phi, 0, 0)\,, \\
\mu^r &= (0, r^{-1}, 0)\,, \\
\nu^r &= (0, 0, 1)\,.
\end{aligned}$$

Note that these could also have been obtained via the orthogonality properties of the vectorial 3-leg, i.e. PROBLEM X.3, or in this very simple case by inspection. We now proceed to examine the full set of leg derivatives of the coordinates, i.e.

$$\begin{array}{lll} \omega_{/1} = \frac{1}{r}\sec\phi, & \omega_{/2} = 0, & \omega_{/3} = 0 \\ \phi_{/1} = 0, & \phi_{/2} = \frac{1}{r}, & \phi_{/3} = 0 \\ r_{/1} = 0, & r_{/2} = 0, & r_{/3} = 1; \end{array}$$

as a system of leg differential equations. These derivatives are subject to the commutator rules of SECTION IV-4 which we repeat for convenience:

$$\begin{array}{ll} (\mathrm{F_I}): & F_{/1/2} - F_{/2/1} = \sigma_1 F_{/1} + \sigma_2 F_{/2} - (t_1 + t_2) F_{/3} \\ (\mathrm{F_{II}}): & F_{/3/1} - F_{/1/3} = -k_1 F_{/1} - (t_1 + \varepsilon_3) F_{/2} + \gamma_1 F_{/3} \\ (\mathrm{F_{III}}): & F_{/2/3} - F_{/3/2} = -(t_2 + \varepsilon_3) F_{/1} + k_2 F_{/2} - \gamma_2 F_{/3} \end{array}$$

where F is an arbitrary smooth function. We now apply these integrability conditions to the above (ω, ϕ, r) system of leg differential equations (considering them in the order of their simplicity):

$(r_{\mathrm{I}}) : 0 = -(t_1 + t_2) \cdot 1 \Rightarrow t_1 = -t_2,$

$(r_{\mathrm{II}}) : 0 = \gamma_1 \cdot 1 \Rightarrow \gamma_1 = 0,$

$(r_{\mathrm{III}}) : 0 = \gamma_2 \cdot 1 \Rightarrow \gamma_2 = 0;$

$(\phi_{\mathrm{I}}) : -(r^{-1})_{/1} = \sigma_2 r^{-1}$, i.e. $r_{/1} \cdot r^{-2} = \sigma_2 r^{-2} \Rightarrow \sigma_2 = 0,$

$(\phi_{\mathrm{II}}) : 0 = -(t_1 + \varepsilon_3) \cdot r^{-1} \Rightarrow t_1 + \varepsilon_3 = 0,$

$(\phi_{\mathrm{III}}) : (r^{-1})_{/3} = k_2 r^{-1}$, i.e. $-r^{-2} r_{/3} = k_2 r^{-1} \Rightarrow k_2 = -r^{-1};$

$(\omega_{\mathrm{I}}) : (r^{-2}\sec\phi)_{/2} = \sigma_1 r^{-1}\sec\phi$, i.e.

$$r^{-1}\sec\phi\tan\phi\,\phi_{/2} = \sigma_1 r^{-1}\sec\phi \Rightarrow \sigma_1 = r^{-1}\tan\phi,$$

$(\omega_{\mathrm{II}}) : -(r^{-1}\sec\phi)_{/3} = -k_1 r^{-1}\sec\phi$, i.e.

$$r^{-2} r_{/3}\sec\phi = -k_1 r^{-1}\sec\phi \Rightarrow k_1 = -r^{-1},$$

$(\omega_{\mathrm{III}}) : 0 = -(t_2 + \varepsilon_3) r^{-1}\sec\phi \Rightarrow t_2 + \varepsilon_3 = 0.$

Hence, we have the following values for the leg coefficients:

$$k_1 = k_2 = r^{-1}, \quad \sigma_1 = r^{-1} \tan \phi,$$

$$\gamma_1 = \gamma_2 = t_1 = \varepsilon_3 = \sigma_2 = 0.$$

Note that the above analysis could also have been done without recourse to the commutator rules, by using the Cartan calculus and computing $d\,(d\omega) = d^2\omega = 0$, $d\,(d\phi) = d^2\phi = 0$, and $d\,(dr) = d^2 r = 0$.

These considerations now lead to *two* important conclusions. First, since the values of the leg coefficients were *determined* by the commutators, the commutators are identically satisfied and consequently the system of Lamé equations ($\mathcal{L}$) are automatically satisfied. Thus, there is no ambiguity about whether the coordinate system $x^r = (\omega, \phi, r)$ exists and is properly defined in $\mathbf{E}_3$. Second, an analysis of the (ω, ϕ, r)-system of leg differential equations reveals the following properties:

(a) since $r_{/3} = 1$ with $r_{/1} = r_{/2} = 0$ then r is aligned along $\boldsymbol{\nu}$, and since $\boldsymbol{\nu}$ is a constant vector r is a linear coordinate;

(b) by $\phi_{/2} = r^{-1}$ it follows from the above r-differential equations that $(r\phi)_{/2} = 1$, and hence $r\phi$ is aligned along $\boldsymbol{\mu}$;

(c) from $\omega_{/1} = r^{-1} \sec \phi$, and the r and ϕ differential equations, we have $(r\omega \cos \phi)_{/1} = 1$ so that $r\omega \cos \phi$ is aligned along $\boldsymbol{\lambda}$.

Properties **(a)**, **(b)** and **(c)** explicitly indicate the nature of the x^r and how they are related to the 3-leg vectors. While they stop just short of identifying (ω, ϕ, r) as being the familiar spherical polar coordinates, these properties obviously duplicate those shared with spherical polar coordinates. The missing link being the *geometrical equations* which could now be deduced from the properties **(a)**-**(c)**. Note, our investigation has essentially been based on the *primary equations*, and when their integrability conditions are imposed, then the *consistency equations* are trivialized.

A major problem confronting the generalized Marussi-Hotine approach is to construct a solution of the system $(\mathcal{HM})$. A primary issue is how to specify meaningful, i.e. workable, geodetic and geometric conditions. We will conclude this section with a discussion of such conditions, omitting from consideration the important topic of boundary conditions, since these are best handled within the theory of geodetic boundary value problems.

In this regard it is useful to supplement the eight Marussi Conditions **(i)**-**(viii)** of SECTION I-2 by a pair of conditions which seem especially relevant to the question of handling the system $(\mathcal{HM})$. These are

(ix) the equipotential surfaces are general convex surfaces;

(x) general smoothness hypotheses on the equipotential surfaces and the geometric quantities associated with them.

We will discuss these new conditions in some detail, but it should be kept in mind that like the earlier conditions **(i)**-**(viii)**, these should be merely regarded as *suggestions* or *working hypotheses*, rather than a rigid system of axioms as in elementary geometry. In this sense **(ix)** and **(x)** should be taken as *indications* of what is involved in satisfying the system $(\mathcal{HM})$, and nothing more. Our discussion concludes with some general comments about the theory of systems of partial differential equations.

Condition **(ix)** is highly non-trivial, and not easy to work with. The first difficulty is that the notion of convexity is complicated and has been used in different senses in the literature (see in particular ALEKSANDROV and ZALGALLER (1967) and the treatise of POGORELOV (1973)). For our purposes we will take the term 'convex surface' to be synonymous with a surface **S** which is

1) a 2-dimensional smooth oriented Riemannian manifold,

2) simply-connected, and

3) has strictly positive Gauss curvature K.

Then, roughly speaking, it can be shown that such an **S** can always be imbedded in $\mathbf{E}_3$. This does not quite guarantee the requirement of **(ii)** of SECTION I-2 that the local imbedding is automatically isometric, but it is a significant step in this direction. Likewise, since the **EP** of CHAPTER VII is not necessarily solvable we speak only of an individual surface **S** and not the family $\boldsymbol{\Sigma}$. Of course when the hypotheses of the Fundamental theorem (Pizzetti's Theorem) are satisfied then **S** is compact, and can be parametrized or covered by a finite number of coordinate neighborhoods (recall SECTION VII-4). Relative to **(ix)** two caveats should be kept in mind: first, neither the topographic surface of the Earth, nor the geoid, are convex surfaces (for the former this is obvious, but less obvious for the latter — see the excellent discussion in Part II of HOPFNER (1949)); second the adjective 'convex' need not be always employed in the geodetic literature in any more than a rough intuitive sense. Hence, in a strict mathematical sense, **(ix)** is meant to apply to 'smoothed out' equipotential surfaces lying above the geoid but satisfying the Pizzetti inequality VII-(3.1).

In contrast to **(ix)**, condition **(x)** is relatively straightforward. Essentially it involves inspection of the number of derivatives — each assumed to be continuous — required for the quantities appearing in the analysis to be meaningful. Up to now we have avoided such considerations by employing the adjective 'smooth', but now more detail is required. We denote by $\mathbf{C}^k\,(\mathcal{R})$ the space, or class, of continuous functions possessing continuous derivatives up to order k (an integer!) on the domain (i.e. open connected subset) $\mathcal{R}$ of $\mathbf{E}_3$. Usually $\mathcal{R}$ will be omitted but it must be understood that this omission does not suggest that $\mathcal{R}$ is identical with $\mathbf{E}_3$. Likewise, $\mathbf{C}^\omega\,(\mathcal{R})$ is the space, or class of real-analytic functions on $\mathcal{R}$, i.e. those functions which can be expanded in a convergent power series about every point of $\mathcal{R}$. To denote inclusion of functions in these classes we write $F \in \mathbf{C}^k$ etc.

Then, by examination of the number of derivatives appearing in our equations, and taking $\mathbf{S} \in \mathbf{C}^k$ to mean that the parametrized coordinates $x^r\,(u^\alpha)$ are $\mathbf{C}^k$ with respect to the

Gaussian parameters u^α, clearly $x^r_\alpha \in \mathbf{C}^{k-1}$. Consequently,

$$a_{\alpha\beta} \in \mathrm{C}^{k-1}, \tag{5.7}$$

$$b_{\alpha\beta}, c_{\alpha\beta}, K, H \in \mathrm{C}^{k-2}, \tag{5.8}$$

and

$$g_{rs} \in \mathrm{C}^{k-1} \tag{5.9}$$

(recall our construction of Gaussian differential geometry in SECTION V-2). These relationships yield the following results in the leg calculus:

$$\lambda_r, \mu_r, \nu_r, \lambda^r, \mu^r, \nu^r \in \mathrm{C}^{k-1}, \tag{5.10}$$

and

$$k_1, k_2, t_1, t_2, \gamma_1, \gamma_2, \sigma_1, \sigma_2, \varepsilon_3 \in \mathrm{C}^{k-2}. \tag{5.11}$$

The $(\mathcal{L})$ equations involve derivatives of the leg coefficients (5.11) and thus we must have at least $k-2=1$, i.e. $k=3$. Apart from these naive counting procedures the integer k may be decreased or increased according to special circumstances occurring in existence and uniqueness proofs. By an ingenious argument, where the Gauss and Codazzi equations are replaced by integral equations, one can take $k=2$, or even replace $\mathbf{C}^k$ by the Hölder class $\mathbf{C}^{k,\ell}$ for $k \geq 0$ and $0 < \ell < 1$ in which the kth order partial derivatives satisfy a uniform Hölder condition of order ℓ (see references in ZUND (1992)). For our purposes a more useful value is $k=4$ which was obtained by NIRENBERG (1953) in his solution of the Weyl Problem (see WEYL (1916)), which is concerned with the imbedding of a compact general convex surface $\mathbf{S}$ in $\mathbf{E}_3$. Hence, although the choice $k=4$ may seem extravagant, we know that it is workable and insures that $(\mathbf{x})$ is feasible.

Finally, let us consider the system $(\mathcal{HM})$ from the viewpoint of the theory of partial differential equations. Unfortunately, the contemporary results of this theory are not well-known outside the mathematical community and within this community the modern theory is regarded as being one of the most difficult in mathematics. The problem is that most physical scientists confuse this subject with the relatively simple and straightforward techniques they learned in courses on the partial differential equations of mathematical physics, e.g. boundary value problems and Fourier series. While such material is important and undeniably useful in physical geodesy, it is essentially focused on formal calculational methods and consequently gives the practitioner a wholly inadequate and misleading impression about the general solvability of partial differential equations.

Theoretically speaking, and aside from clever techniques developed by Lagrange, Monge, Jacobi *et al* at the end of the eighteenth and first half of the nineteenth centuries, the theory actually began with the celebrated Cauchy-Kowalewski theorem. This result was proved by Cauchy in 1842, and in greater generality by Kowalewski in her thesis of 1874 (for references see HADAMARD (1923)), and concerns the existence and uniqueness of solutions of a system of partial differential equations within the class of real-analytic functions $\mathbf{C}^\omega$. For awhile, the restriction to $\mathbf{C}^\omega$ functions was regarded merely as a *mathematical convenience* which was adequate for the needs of classical physics. However, as physical problems became more reliable it became apparent that the $\mathbf{C}^\omega$ requirement was a major impediment — particularly as all attempts to extend the theorem to $\mathbf{C}^k$ functions (or $\mathbf{C}^{k,\ell}$ functions) failed. Nevertheless, the appearance of such unexpected solutions as (5.2) of (5.1) did not dampen the expectation that, given sufficient cunning, most partial differential equations were in principle solvable. However, despite its ingenuity, Whittaker's formal analysis sidestepped the issue of whether such solutions were really useful, i.e. 'reasonable' in the sense that physical boundary conditions could be imposed, or merely mathematical curiosities. In 1917, Hadamard produced a solution of the 2-dimensional Laplace equation which exhibited a discontinuous dependence on the boundary conditions and hence was not a physically meaningful solution (see Hadamard's book *loc cit supra*, or the textbook of GARABEDIAN (1964)). This example dramatically indicated that formal calculational techniques — notwithstanding the deviousness of their construction — need not lead to solutions which could be fit with so-called reasonable boundary conditions. It thus revealed that the appropriate selection of the boundary conditions is as important as the choice of solution technique.

Despite these two warning signs, the naive hope persisted that solvability would be assured as long as one considered only 'simple' equations with 'simple' boundary conditions. This hope vanished when LEWY (1957) constructed an embarrassingly simple example of a first order partial differential equation which (with no boundary conditions imposed) had no solutions. The full impact and implications of Lewy's example became inescapable only when JACOBOWITZ and TREVES (1983) proved that his result — in a suitably specified sense — is not an unusual case. In other words, the $\mathbf{C}^{\omega}$ restriction in the Cauchy-Kowalewski theorem is structural and the *local solvability* of a partial differential equation is *atypical* rather than *typical.* Jacobowitz and Treves emphasize this by introducing the notion of an *aberrant* partial differential operator L. For such an operator the *only* $\mathbf{C}^k$-solution F of the differential equation

$$LF = 0 \tag{5.12}$$

is the constant solution, and they proved that such operators/ equations are by no means unusual or pathological.

The relevance of these somewhat esoteric-appearing results to the $(\mathcal{HM})$ system is that upon specifying certain geodetic/geometric conditions as in Example B we expect that in the permissible approach some of the equations will be:

1) identically satisfied, i.e. they reduce to identities which lead to no new results;

2) will simplify into differential equations which upon a suitable choice of the contravariant components of a leg vector can be integrated in closed form;

3) will reduce to algebraic equations which may either be identically satisfied or lead to expressions for some unknown leg coefficients as in the commutators in Example C;

4) reduce to differential equations which, when not readily solvable, are satisfied by choosing the involved functions to be constant functions. The Jacobowitz-Treves result suggest that, contrary to the classical viewpoint, this 'specialization' may not be a specialization but the *only* solution.

This is the procedure we expect to be used in the generalized Marussi-Hotine approach to geodesy. It is a miniature Euclidean version of the Newman-Penrose method of spin coefficients, which has proven so useful in solving the Einstein equations in general relativity. Since the $(\mathcal{HM})$ system is significantly simpler, there is really no question that our scheme is workable provided that theoretical geodesists learn it and supply appropriate boundary conditions on N or n and the relevant non-zero leg coefficients.

X.6 Conclusion

In this monograph we have sought to work out the foundations of differential geometry as pioneered by Antonio Marussi and Martin Hotine. Our task has essentially been focused on two separate goals: first, to seek to enunciate the fundamental ideas underlying their work; and second, to complete that part of their work which seems to have a valid conceptual basis and to reject those parts of it which are untenable on various grounds. As a consequence, all the results derived in CHAPTERS I-IX are independent of the Marussi Hypothesis i.e. the general existence of local (holonomic) coordinate systems, and the Marussi Ansatz $(\mathcal{A})$ in particular.

In this regard the leg calculus, as realized by the congruence calculus of Ricci *and* the exterior calculus of Cartan, has been shown to be an appropriate formalism in that it can be employed either in the usual Marussi-Hotine approach, or more generally in the generalized Marussi-Hotine approach suggested in SECTION 5. In other words the leg calculus, as developed in CHAPTERS IV and V, presupposes the use of neither descriptive nor permissible local coordinate systems. Indeed, it assumes only the existence of an unspecified ambient coordinate system in $\mathbf{E}_3$, and may be considered to be a "coordinate-free" formalism unlike the classic tensor calculus. Our discussion in CHAPTER I was intended to disturb the reader's naive belief in the infallibility and universal utility of local coordinate systems. Our conclusion is that rather than being a primary notion, coordinates — in particular those which possess some semblance of geometrical/physical significance — are a secondary, viz. a derived, notion. Apart from Cartesian coordinates, which are the natural candidate for an ambient system in $\mathbf{E}_3$, such coordinates do not automatically occur but must be *constructed* to fit the geometrical/physical situation. Regardless of whether one employs the original descriptive approach or the new permissible approach this construction is a delicate process which requires a careful analysis of the full set of Hotine-Marussi equations (recall CHAPTER VI). This set of equations is non-trivial, and the success or failure in satisfying/solving them crucially depends on imposition of proper geometrical and physical conditions. These may be more or less obvious conditions: e.g. curved plumblines, convex equipotential surfaces, etc.; or more subtle conditions related to various properties of the geometry of $\mathbf{\Gamma}$ or $\mathbf{\Sigma}$.

PROBLEMS FOR CHAPTER X

X.1 Derive Hotine's geometric equations [12.008], i.e.

$$\begin{aligned}
\lambda_r &= -A_r \sin\omega + B_r \cos\omega,\\
\mu_r &= -A_r \cos\omega \sin\phi - B_r \sin\omega \sin\phi + C_r \cos\phi,\\
\nu_r &= A_r \cos\omega \cos\phi + B_r \sin\omega \cos\phi + C_r \sin\phi;
\end{aligned}$$

and his inverse geometric equations [12.009], i.e.

$$\begin{aligned}
A_r &= -\lambda_r \sin\omega - \mu_r \cos\omega \sin\phi + \nu_r \cos\omega \cos\phi,\\
B_r &= \lambda_r \cos\omega - \mu_r \sin\omega \sin\phi + \nu_r \sin\omega \cos\phi,\\
C_r &= \mu_r \cos\phi + \nu_r \sin\phi.
\end{aligned}$$

ZUND (1993b)

X.2 Derive Hotine's equations [12.014-.016],

$$\begin{aligned}\lambda_{rs} &= (\mu_r \sin\phi - \nu_r \cos\phi)\,\omega_s,\\ \mu_{rs} &= -\sin\phi\lambda_r\omega_s - \nu_r\phi_s,\\ \nu_{rs} &= \cos\phi\lambda_r\omega_s + \mu_r\phi_s.\end{aligned}$$

Hint: From the inverse geometric equations, compute $A_{rs} = 0$, $B_{rs} = 0$ and $C_{rs} = 0$.

ZUND (1993b)

X.3 Show that equation II-(2.12) is equivalent to the matrix equation

$$\left\|\begin{matrix}\lambda^1 & \lambda^2 & \lambda^3\\ \mu^1 & \mu^2 & \mu^3\\ \nu^1 & \nu^2 & \nu^3\end{matrix}\right\| \cdot \left\|\begin{matrix}\lambda_1 & \mu_1 & \nu_1\\ \lambda_2 & \mu_2 & \nu_2\\ \lambda_3 & \mu_3 & \nu_3\end{matrix}\right\| = \left\|\begin{matrix}1 & 0 & 0\\ 0 & 1 & 0\\ 0 & 0 & 1\end{matrix}\right\|$$

and provides a direct method for computing the covariant components of the leg vectors when the contravariant components are known.

ZUND (1993b)

X.4 In his treatise on [page 74] Hotine noted that

$$\begin{aligned}\nu_{rs}\lambda^r &= \cos\phi\omega_s = -\lambda_{rs}\nu^r,\\ \mu_{rs}\nu^r &= -\phi_s = -\nu_{rs}\mu^r,\end{aligned}$$

i.e. [12.026] and [12.025], but not that

$$\lambda_{rs}\mu^r = \sin\phi\omega_s = -\mu_{rs}\lambda^r.$$

Explain this oversight, which leads to the ω-degeneracy.

ZUND (1990b)

X.5 By virtue of the ω-degeneracy, viz. (ω) and (ω^*) show that one has the *linear identities*:

$$\begin{aligned}\sigma_1 &= -k_1 \tan\phi,\\ \sigma_2 &= -t_1 \tan\phi,\\ \varepsilon_3 &= \gamma_1 \tan\phi;\end{aligned}$$

and hence the *quadratic identities*:

$$\begin{aligned}\sigma_1 t_1 &= k_1 \sigma_2,\\ \sigma_1\gamma_1 &= -k_1\varepsilon_3,\\ \sigma_2\gamma_2 &= -t_1\varepsilon_3.\end{aligned}$$

ZUND (1990)

X.6 Show that equivalent versions of primary equations (ω), (ω^*), (ϕ), viz. (3.13) and (3.14), are given by the gradient expressions:

$$\begin{aligned}(\omega):&\quad \omega_r = (-k_1\lambda_r t_1\mu_r + \gamma_1\nu_r)\sec\phi,\\ (\omega^*):&\quad \omega_r = (\sigma_1\lambda_r + \sigma_2\mu_r + \varepsilon_3\nu_r)\cos\phi,\\ (\phi):&\quad \phi_r = (-t_1\lambda_r - k_2\mu_r + \gamma_2\nu_r);\end{aligned}$$

and the Pfaffian expressions:

$$\begin{aligned}(d\omega):&\quad d\omega = (-k_1\theta_1 - t_1\theta_2 + \gamma_1\theta_3)\sec\phi,\\ (d\omega^*):&\quad d\omega = (\sigma_1\theta_1 + \sigma_2\theta_2 + \varepsilon_3\theta_3)\cos\phi,\\ (d\phi):&\quad d\phi = -t_1\theta_1 - k_2\theta_2 + \gamma_2\theta_3.\end{aligned}$$

HOTINE (1969) and ZUND (1990b)

X.7 Show that the matrices $\mathcal{M}$ and $\mathcal{M}^{-1}$ in (3.15) and (3.16) are given by

$$\mathcal{M} = \left\| \begin{matrix} -k_1 \sec\phi, & -t_1 \sec\phi, & \gamma_1 \sec\phi \\ -t_1, & -k_2, & \gamma_2 \\ 0, & 0, & n \end{matrix} \right\|,$$

$$\mathcal{M}^{-1} = \left\| \begin{matrix} -\dfrac{-k_2\cos\phi}{K}, & \dfrac{t_1}{K}, & \dfrac{k_2\gamma_1 - t_1\gamma_2}{nK} \\ \dfrac{t_1\cos\phi}{K}, & -\dfrac{k_1}{K}, & \dfrac{k_1\gamma_2 - t_1\gamma_1}{nK} \\ 0, & 0, & n \end{matrix} \right\|.$$

GRAFAREND (1975) and ZUND (1993b)

X.8 By using (3.12) and the results of PROBLEM X.3 for computing the covariant components of the Hotine 3-leg vectors use II-(2.13) to calculate the metric tensor in the (ω, ϕ, N) system

$$\begin{aligned}
K^2 g_{11} &= (k_2^2 + t_1^2)\cos^2\phi = K^2 a_{11} \\
K^2 g_{12} &= -2Ht_1\cos\phi = K^2 a_{12} \\
K^2 g_{22} &= (k_1^2 + t_1^2) = K^2 a_{22} \\
K^2 g_{13} &= -\left[\gamma_1 (k_2^2 + t_1^2) - 2H\gamma_2 t_1\right] / (n\sec\phi) \\
K^2 g_{23} &= -\left[\gamma_2 (k_1^2 + t_1^2) - 2H\gamma_1 t_1\right] / n \\
K^2 g_{33} &= \left[\gamma_1^2 (k_2^2 + t_1^2) + \gamma_2^2 (k_1^2 + t_1^2) - 4Ht_1\gamma_1\gamma_2 + K^2\right] / n^2 \\
g &= \det \|g_{rs}\| = \cos^2\phi / (n^2 K^2).
\end{aligned}$$

HOTINE (1969)

X.9 For an arbitrary smooth F, show that the Cayley-Darboux equation: $\aleph F = 0$ in Cartesian tensor notation is equivalent to

$$(*) \qquad \varepsilon_{ijk}\varepsilon_{\ell mn}\varepsilon_{pqr}F_{i\ell}\left(F_s F_{jps} - 2F_{js}F_{ps}\right)\delta_{mq}F_k F_n F_r = 0.$$

ZUND and MOORE (1987)

X.10 Show that an explicit expansion of (∗) in PROBLEM X.9 is

$$\begin{aligned}
&\textstyle\sum_{a,\beta,\gamma}\{(F_s F_{\alpha\alpha s} - 2F_{\alpha s}F_{\alpha s})\,[F_\beta^3 F_{\gamma\alpha} + F_\beta F_\gamma^2 F_{\gamma\alpha} \\
&\quad + F_\alpha F_\gamma^2 F_{\beta\alpha} - \left(F_\gamma^3 F_{\alpha\beta} + F_\beta^2 F_\gamma F_{\alpha\beta} + F_\alpha F_\beta F_{\beta\gamma}\right) \\
&\quad + F_\alpha F_\beta F_\gamma\,(F_{\beta\beta} - F_{\gamma\gamma})] \\
&\quad + (F_s F_{\alpha\beta s} - 2F_{\alpha s}F_{\beta s})\,[(F_{\gamma\gamma} - F_{\beta\beta})\,F_\alpha^2 \\
&\quad + (F_{\alpha\alpha} - F_{\gamma\gamma})\,F_\beta^2 + (F_{\alpha\alpha} - F_{\beta\beta})\,F_\gamma^2 \\
&\quad + 2\left(\left(F_\alpha^2 + F_\gamma^2\right) F_\beta F_{\beta\gamma} - \left(F_\beta^2 + F_\gamma^2\right) F_\alpha F_{\gamma\alpha}\right)]\} = 0
\end{aligned}$$

where the summation sign denotes a cyclic sum over α, β, $\gamma = 1, 2, 3$. Thus the Cartesian version of $\aleph F = 0$ involves 324 terms!

ZUND and MOORE (1987)

Bibliography

The following bibliography — although rather lengthy — is by no means complete and contains only those items which are concerned with those aspects of the Marussi-Hotine approach to differential geodesy discussed in the text. Complete bibliographies of Marussi and Hotine can be found in their monographs MARUSSI (1985) and HOTINE (1991) respectively. Note that in order to indicate the priority and vintage of the following items, we will usually cite the original edition of a publication with English translations — if any — being duly noted.

A.D. ALEKSANDROV and V.A. ZALGALLER (1967)
Intrinsic geometry of surfaces, English of the 1962 Russian edition, Volume 15 *Translation of Mathematical Monographs*, American Mathematical Society, Providence.

E.J. ATZEMA (1993)
The structure of systems of lines in 19th century geometrical optics: Malus' theorem and the description of the infinitely thin pencil, Ph.D. thesis, Universiteit Utrecht, 204 pp.

E. BELTRAMI (1864, 1865)
Ricerche di analisi applicata alla geometria, *Giornale di Matematiche* 2: 267-282, 297-282, 297-306, 331-339, 355-375; 3: 15-22, 33-41, 82-91, 228-240, 311-314.

L. BIANCHI (1927)
Lezioni di geometria differenziale, Volume I - Parte I, 3rd edition, N. Zanichelli, Bologna.

W. BLASCHKE (1921)
Vorlesungen über Differentialgeometrie - I: Elementare Differentialgeometrie, Springer-Verlag, Berlin.

H. BLASCHKE (1928)
Eine Verallgemeinerung der Theorie der konfokalen F_2, *Mathematische Zeitscrift* 27, 653-668.

W. BLASCHKE and K. LEICHTWEISS (1970)
Elementare Differentialgeometrie, Springer-Verlag, Berlin (this is the 5th edition of BLASCHKE (1921) completely rewritten by the second author).

F. BOCCHIO (1970)
Su alcune applicazioni di interesse geodetico delle connessioni non simmetriche, *Rendiconti della classe di Scienze Fisiche, matematiche e naturali, Accademia Nazionale dei Lincei (Roma), Ser.* VIII, 48, 343-351.

F. BOCCHIO (1974)
From differential geodesy to differential geophysics, *Geophysics Journal Royal Astronomical Society* 39, 1-10.

F. BOCCHIO (1975)
The holonomity problem in geophysics, *Bollettino di Geodesia e Scienze Affini*, anno XXXIV, 453-459.

F. BOCCHIO (1978)
Some of Marussi's contributions in the field of two-dimensional and three-dimensional representation, *Bollettino di Geodesia e Scienze Affini*, anno XXXVII, 441-450.

F. BOCCHIO (1981)
An inverse geodetic singularity problem, *Geophysics Journal Royal Astronomical Society* 67, 181-187.

F. BOCCHIO (1982)
Geodetic singularities, *Reviews of Geophysics and Space Physics* 20, 399-409 = 1-11 in *Advances in Geodesy*, edited by E. W. Grafarend and R. H. Rapp, American Geophysical Union, Washington, D.C. (1984).

A. BODE and E.W. GRAFAREND (1982)
The telluroid mapping based on a normal gravity potential including the centrifugal term, *Bollettino di Geodesia e Scienze Affini*, anno XLI, 22-56.

N. BOURBAKI (1948)
Éléments de mathematique, Livre II Algèbre, Chapitre 3, Hermann, Paris.

H. BRUNS (1878)
Die Figur der Erde, *Publication des Königl. Preussischen Geodätischen Institutes*, Druck und Verlag P. Stankiewicz, Berlin, 49 pp.

E. CARTAN (1928)
Leçons sur la géométrie des espaces de Riemann, Gauthier-Villars, Paris; enlarged second edition *ibid* (1946).

H. CARTAN (1967a)
Calcul différentiel, Hermann, Paris = *Differential calculus*, Houghton-Mifflin Co., Boston (1971).

H. CARTAN (1967b)
Formes différentiel, Hermann, Paris = *Differential forms*, Houghton-Mifflin Co., Boston (1970).

F. CASORATI (1890)
Mesure de la courbure des surfaces suivant l'idée commune — ses rapports avec les measures de courbore Gaussienne et mayenne, *Acta Mathematica* 14, 95-110.

A. CAYLEY (1860a)
Sur la courbe parellèle à l'ellipse, *Annali di Matematica pura ed applicata*, 3, 311-316 = 152-157, paper #253 of Volume IV of CAYLEY (1889/1897).

A. CAYLEY (1860b)
Sur la surface parallèle à l'ellipsoid, *Annali di Matematica pura ed applicata*, 3, 345-352 = 158-165, paper #254 of Volume IV of CAYLEY (1889/1897).

A. CAYLEY (1888)
On systems of rays, *Messenger of Mathematics* 17, 73-78 = 571-575, paper #876 of Volume XII of CAYLEY (1889/1897).

A. CAYLEY (1889/1897)
Collected mathematical papers, thirteen volumes, Cambridge University Press, Cambridge.

G.A. CHEBOTAREV (1964)
Gravitational spheres of the major planets, moon and Sun, *Soviet Astronomy* 7, 618-622, translated from *Astronomicheskii Zhurnal* 40, (1963), 812-818.

Y. CHOQUET-BRUHAT (1968)
Géométrie différentielle et systèmes extérieurs, Dunod, Paris.

B.H. CHOVITZ (1969)
Hotine's 'Mathematical geodesy', *Proceedings IV Symposium on Mathematical Geodesy*, Trieste = 159-172 of HOTINE (1991).

B.H. CHOVITZ (1972)
Generalized three-dimensional conformal transformations, *Bulletin Geodesique* 104, 159-163.

B.H. CHOVITZ (1982)
The influence of Hotine's Mathematical geodesy, *Bollettino di Geodesia e Scienze Affini*, anno XLI, 57-64.

G. DARBOUX (1915)
Lecons sur la théorie générale des surfaces et applications géométriques du calcul infinitesimal, Deuxième partie, 2nd edition, Gauthier-Villars, Paris.

R. DEHEUVELS (1993)
Tenseurs et spineurs, Presses Universitaires de France, Paris.

P. DUHEM (1914)
La théorie physique: son objet, sa structure, M. Rivière & Cie, Paris = *The aim and structure of physical theory*, (translated by P.P. Wiener), Princeton University Press, Princeton (1954); paperback reprint *ibid* (1991).

A. EINSTEIN and B. KAUFMAN (1955)
A new form of the general relativistic field equations, *Annals of Mathematics* 62, 128-138.

L.P. EISENHART (1926)
Riemannian geometry, Princeton University Press, Princeton; enlarged edition *ibid* (1949).

L.P. EISENHART (1940)
An introduction to differential geometry with the use of the tensor calculus, Princeton University Press, Princeton, also corrected edition 1947.

W.H. FLEMING (1977)
Functions of several variables, second edition, Springer-Verlag, New York.

A.R. FORSYTH (1912)
Lectures on the differential geometry of curves and surfaces, Cambridge University Press, Cambridge.

P.R. GARABEDIAN (1964)
Partial differential equations, J. Wiley & Sons, New York.

I.M. GEL'FAND (1948)
Lectures on linear algebra (in Russian), Nauka, Moscow: four editions; (translation of second edition by A. Shenitzer) Interscience Publishers Inc., New York (1961).

E.W. GRAFAREND (1971)
The object of anholonomity and a generalized Riemannian geometry for geodesy, *Bollettino di Geofisica Teorica ed Applicata*, 13, 241-253.

E.W. GRAFAREND (1973)
Le theoreme de conservation de la courbure et la torsion or attempts at a unified theory of geodesy, *Bulletin Géodésique*, 109, 237-260.

E.W. GRAFAREND (1975)
Three-dimensional geodesy - I: the holonomity problem, *Zeitschrift für Vermessungswesen*, 269-281.

E.W. GRAFAREND (1977)
Geodäsie - Gauszsche oder Cartansche Flächengeometrie?, *Allgemeine Vermessungs-Nachrichten* 4, 139-150.

E.W. GRAFAREND (1978)
Marussian geodesy, *Bollettino di Geodesia e Scienze Affini,* anno XXXVII, 209-247.

E.W. GRAFAREND (1994)
The optimal universal transverse Mercator projection, in press *Manuscripta Geodaetica.*

N. GROSSMAN (1974)
Holonomic measurables in geodesy, *Journal Geophysical Research* 79, 689-694.

N. GROSSMAN (1978)
Is the geoid a trapped surface?, *Bollettino di Geodesia e Scienze Affini*, anno XXXIV, 173-183.

N. GROSSMAN (1979)
The nature of space near the Earth, *Bollettino di Geodesia e Science Affini,* anno XXXV, 413-424.

K.P. GROTEMEYER (1969)
Analytische Geometrie, W. de Gruyter & Co., Berlin.

V. GUILLEMIN and A. POLLACK (1974)
Differential topology, Prentice-Hall Inc., Englewood Cliffs.

J. HADAMARD (1923)
Lectures on Cauchy's problem in linear partial differential equations, Yale University Press, New Haven; reprint by Dover Publications Inc., New York (1952).

P.R. HALMOS (1958)
Finite dimensional vector spaces, second edition, D. Van Nostrand Co. Inc., Princeton; reprint Springer-Verlag, New York (1987).

SIR W.R. HAMILTON (1828)
Essay on the theory of systems of rays, *Transactions of the Royal Irish Academy* 15, 69-174 =1-106 paper I of HAMILTON (1931).

SIR W.R. HAMILTON (1931)
The mathematical papers of Sir William Rowan Hamilton: Volume I, Geometrical optics, edited by A.W. Conway and J.L. Synge, Cambridge University Press, Cambridge (see Note 2: *History of the theorem of Malus*, 463-464).

T.L. HANKINS (1980)
Sir William Rowan Hamilton, Johns Hopkins University Press, Baltimore.

M. HERZBERGER (1958)
Modern geometrical optics, Interscience Publishers Inc., New York.

P. HOLOTA (1989)
Boundary value problems and invariants of the gravitational tensor in satellite gradiometry, in *Theory of Satellite geodesy and gravity field determination*, edited by F. Sansò and R. Rummel, *Lecture Notes in Earth Sciences* 25, Springer-Verlag, Berlin, 447-458.

F. HOPFNER (1929)
Physikalische Geodäsie, Akademische Verlagsgessellschaft, Leipzig.

F. HOPFNER (1949)
Grundlagen der höheren Geodäsie, Springer-Verlag, Wien.

M. HOTINE (1967/1947)
The orthomorphic projection of the spheroid, *Empire Survey Review*, 8, No. 62, 300-311 (1946); 9, No. 63, 25-35

(1947); 9, No. 64, 52-70 (1947); 9, No. 65, 112-123 (1947); 9, No. 66, 157-166 (1947).

M. HOTINE (1957a)
Metrical properties of the earth's gravitational field, report to I.A.G. Toronto Assembly = 33-64 of HOTINE (1991).

M. HOTINE (1957b)
Geodesic coordinate systems, report to I.A.G. Toronto Assembly = 65-89 of HOTINE (1991).

M. HOTINE (1959)
A primer on non-classical geodesy, report to Venice Symposium = 91-130 of HOTINE (1991).

M. HOTINE (1965)
Trends in mathematical geodesy, *Bollettino di Geodesia e Science Affini*, anno XXIV, 607-622 = 5-21 of HOTINE (1991).

M. HOTINE (1966a)
Geodetic applications of conformal transformations, *Bulletin Géodésique* 80, 123-140.

M. HOTINE (1966b)
Triply orthogonal coordinate systems, *Bulletin Géodésique* 81, 195-222.

M. HOTINE (1966c)
Note by the writer of the paper 'On Triply orthogonal coordinate systems', *Bulletin Géodésique* 81, 223-224.

M. HOTINE (1969)
Mathematical geodesy, U.S. Department of Commerce, Washington, D.C.

M. HOTINE (1991)
Differential geodesy (edited with commentary by J.D. Zund), Springer-Verlag, Berlin.

H. JACOBOWITZ and F. TREVES (1983)
Nowhere solvable homogeneous partial differential equations, *Bulletin American Math Society* 8 (3), 467-469.

K. JUNG (1956)
Figur der Erde, *Handbuch der Physik*, Band 47, 534-639.

G. KOENIGS (1895)
La géométrie réglée et ses applications, Gauthier-Villars, paris.

E.E. KUMMER (1860)
Allgemeine Theorie der gradlinigen Strahlensysteme, *Journal für die reine und angewandte Mathematik* 57, 189-230.

K. LAMBECK (1988)
Geophysical geodesy: the slow deformations of the Earth, Clarendon Press, Oxford.

W.D. LAMBERT (1921)
Some mechanical curiosities connected with the Earth's field of force, *American Journal of Science*, Fifth Series, 2, 129-158.

C. LANCZOS (1949)
The variational principles of mechanics, University of Toronto Press, Toronto.

P.S. LAPLACE (1839)
Celestial mechanics, Volume IV, (translated by N. Bowditch), Boston; reprinted by Chelsea Publishing Company Inc., New York (1966).

T. LEVI-CIVITÀ (1925)
Lezioni di calcolo differenziale assoluto, Stock, Roma; = *The absolute differential calculus: calculus of tensors*, edited by E. Persico, translation by M. Long, Blackie & Son Limited, London (1926); reprint by Dover Publications, New York (1989).

T. LEVI-CIVITÀ and U. AMALDI (1931)
Calcolo differenziale assoluto, *Enciclopedia Italiana*, XII, Instituto della Enciclopedia Italiana, Roma, 796-798.

H. LEWY (1957)
An example of a smooth linear partial differential equation without solution, *Annals of Mathematics* 66, 155-158.

A. LICHNEROWICZ (1947)
Algèbre et analyse linéaires, Masson & Cie, Paris; = *Linear algebra and analysis*, (translated by A. Johnson), Holden-Day, San Francisco, (1967).

A. LICHNEROWICZ (1950)
Elements de calcul tensoriel, Libraire A. Colin, Paris; = *Elements of tensor calculus*, (translated by J.W. Leech and D.J. Newman), Metheun & Co. Limited, London.

A. LICHNEROWICZ (1955)
Théorie globale des connexions et des groupes d'holonomie, Cremonese, Rome.

E. LIVIERATOS (1976)
On the geodetic singularity problem, *Manuscripta Geodaetica* 1, 269-292.

A.J. McCONNELL (1931)
Applications of the absolute differential calculus, Blackie & Son Limited, London = *Applications of tensor analysis*, Dover Publications, New York (1957).

E.L. MALUS (1808)
Mémoire sur l'optique, *Journal de l'École polytechnique*, 7 (14e Cahier), 1-44.

E.L. MALUS (1811)
Traité d'optique, *Mémoires présentés a l'Institut des sciences par divers savants* 2, 214-302.

A. MARUSSI (1947)
Sulla struttura locale del geoide, e sui mezzi geometrici e mecanici atti a determinarla, *Annali Triestini* 17 (Ser. IV), 1-31.

A. MARUSSI (1949)
Fondements de géométrie différentielle absolue du champ potentiel terrestre, *Bulletin Géodésique* 14, 411-439.

A. MARUSSI (1951a)
Les principes de la géodésie intrinsèque, *Bulletin Geodesique* 19, 68-76.

A. MARUSSI (1951b)
Fondamenti di geodesia intrinseca, *Publicazioni della Commissione Geodetica Italiana*, Ser. III, 7, 1-47 = 13-58 of MARUSSI (1985).

A. MARUSSI (1951c)
Su alcune proprietà integrali delle rappresentazioni conformi di superfici su superfici, *Rendiconti della classe di Scienze fisiche, matematiche e naturali, Accademia Nazionale dei Lincei* (Roma), Ser. VIII, 10, 307-310 = 149-152 in MARUSSI (1985).

A. MARUSSI (1953)
Un'analogia fra le leggi della propagazione delle luce in mezzi rifrangenti continui isotropi, e le rappresentazioni conformi, *Atti Fondazione Giorgio Ronchi dell'Instituto Nazionale di Ottica*, anno VIII, 2-7 = 169-172 in MARUSSI (1985).

A. MARUSSI (1957)
La coordination des systèmes géodésiques, *Bulletin Geodesique* 43, 16-19.

A. MARUSSI (1959)
Dalla geodesia classica alla geodesia in tre dimensioni, *Bollettino di Geodesia e Scienze Affini*, anno XVIII, 485-495 = 3-12 in MARUSSI (1985).

A. MARUSSI (1979)
The tidal field of a planet and the related intrinsic reference systems, *Geophysics Journal Royal Astronomical Society* 56, 409-417 = 75-83 of MARUSSI (1985).

A. MARUSSI (1985)
Intrinsic geodesy, (translated by W.I. Reilly), Springer-Verlag, Berlin.

A. MARUSSI (1988)
Intrinsic geodesy (a revised and edited version of his 1952 lectures by J.D. Zund), *Report No.* 390, Department of Geodetic Science and Surveying, The Ohio State University, Columbus.

J. MILNOR (1965)
Topology from the differentiable viewpoint, based on notes by D.W. Weaver, The University Press of Virginia, Charlottesville.

G. MONGE (1809)
Application de l'analyse a la géométrie, 4th edition, M.V. Bernard, Paris; 5th edition, edited by J. Liouville with supplementary notes and memoir by Gauss Bachelier, Paris (1850).

J.R. MUNKRES (1975)
Topology, a first course, Prentice-Hall, Englewood Cliffs.

J.R. MUNKRES (1991)
Analysis on manifolds, Addison-Wesley Publishing Company, Redwood City.

I. NEWTON (1934)
Sir Isaac Newton's 'Mathematical principles of natural philosophy and his system of the world', revised version by F. Cajori of the 1729 English edition of A. Motte, The University of California Press, Berkeley.

L. NIRENBERG (1953)
The Weyl and Minkowski problems in differential geometry in the large, *Communications Pure and Applied Mathematics*, 6, 337-394.

J.C.C. NITSCHE (1975)
Vorlesungen über Minimalflächen, Springer-Verlag, Berlin.

J.M.H. OLMSTED (1947)
Solid analytic geometry, Appleton-Century-Crofts, Inc., New York.

P. PIZZETTI (1901)
Un principio fondamentale nello studio delle studio delle superfici di livello terrestri, *Rendiconti della Reale Accademia dei Lincei (Roma)*, 10, 35-39.

P. PIZZETTI (1906)
Höhere Geodäsie, *Encyklopädie der Mathematischen Wissenschaften Band VI, Erster Teil: Geodäsie und Geophysik*, 125-239, B.G. Teubner, Leipzig.

P. PIZZETTI (1913)
Principii della teoria mecannica della figura dei pianeti, E. Spoerri, Pisa.

J. PLÜCKER (1868/69)
Neuen Geometrie des Raumes, gegründet auf die Betrachtung der geraden Linie als Raumelement, Teil I & II, B.G. Teubner, Leipzig.

A.V. POGORELOV (1955)
Lectures on differential geometry (in Russian) "Nauka", Moscow, six editions 1955-1974; = *Differential geometry* (translated by L.F. Boron), P. Noordhoff, Groningen, 1959.

A.V. POGORELOV (1973)
Extrinsic geometry of convex surfaces, English translation of 1969 Russian edition, *American Mathematical Society*, Volume 31, *Translation of Mathematical Monographs*, Providence.

G. RICCI (1895)
De sistemi di congruenze ortogonali in una varietà qualunque, *Memoria Accademia Lincei*, (5) 2, 276-322 = 1-62, Volume II of RICCI (1956/1957).

G. RICCI (1898)
Lezioni sulla teoria delle superficie, F. Drucker, Verona-Padova.

G. RICCI and T. LEVI-CIVITÀ (1901)
Methodes de calcul differentiel absolu et leurs applications, *Matehmatische Annalen* 54, 125-201.

G. RICCI (1956/1957)
Opere, two volumes, Edizioni Cremonese, Roma.

S. ROBERTS (1872)
On the parallel surfaces of conicoids and conics, *Proceedings of the London Mathematical Society*, *Ser.* 1, 4, 57-91.

S. ROBERTS (1873)
On parallel surfaces, *Proceedings of the London Mathematical Society*, *Ser.* 1, 4, 218-235.

R. RUMMEL (1985)
Satellite gradiometry: a promising concept for 25 years, Die Arbeiten des Sonderforschungsbereiches 78 Satellitengeodasie der Technischen Universität München 1984 und 1985, vorgelegt von M. Schneider, *Bayerischen Kommission für die Internationale Erdmessung*, N. 48, 123-139.

R. RUMMEL (1986)
Satellite gradiometry, in *Mathematical and numerical techniques in physical geodesy*, edited by H. Sünkel, *Lecture notes in Earth sciences*, 7, Springer-Verlag, Berlin, 317-363.

R. RUMMEL and M. VAN GELDEREN (1992)
Spectral analysis of the full gravity tensor, *Geophysics Journal International* 111, 159-169.

G. SALMON (1879)
A treatise on conic sections containing an account of some of the most important modern algebraic and geometric methods, 6th edition, Longmans, Green & Company, London.

J.A. SCHOUTEN (1951)
Tensor analysis for physicists, Oxford at the Clarendon Press, Oxford, 2nd edition *ibid* (1954).

J.A. SCHOUTEN (1954)
Ricci-calculus: an introduction to tensor analysis and its geometrical applications, second edition, Springer-Verlag, Berlin.

P. STÄCKEL (1891)
Über die integration der Hamilton-Jacobischen differentialgleichung mittelst separation der variablen, *Habilitationschrift,* Halle.

P. STÄCKEL (1893)
Über die Bewegung eines Punktes in einer n-fachen Mannigfaltigkeit, *Mathematische Annalen* 42, 537-563.

J.L. SYNGE (1937)
Geometrical optics, Cambridge University Press, Cambridge, 110 pages.

T.Y. THOMAS (1965)
Concepts from tensor analysis and differential geometry, Academic Press, New York.

E. VESSIOT (1906)
Leçons de géométrie supérieure, Libraire Scientifique, J. Hermann, Paris.

B.L. VAN DER WAERDEN (1966)
Algebra, Erster Teil; 7th edition of *Moderne Algebra* - I originally published in 1930; Springer-Verlag, Berlin.

C.E. WEATHERBURN (1938)
An introduction to Riemannian geometry and the tensor calculus, Cambridge University Press, Cambridge.

H. WEYL (1916)
Über die Bestimmung einer geschlossen konvexen Fläche durch ihr Linienelement, *Viertel jahrsschrift der Naturforschenden Gesellschaft in Zürich* 61, 40-72 = 148-178 (mit Nachtrag by author in 1955) *Selecta Hermann Weyl*, Birkhäuser Verlag Basel (1956).

H. WEYL (1918)
Raum-Zeit-Materie: Vorlesungen über allgemeine Relativitätstheorie, five editions (1918-1924) Springer-Verlag, Berlin; reprint *ibid* (1970).

H. WEYL (1928)
Gruppentheorie und Quantenmechanik, Verlag S. Hirzel, Leipzig = *Theory of groups and quantum mechanics*, (second edition translation by H.P. Robertson), E.P. Dutton and Co., Inc., New York (1932); reprint by Dover Publications Inc., New York (1950).

E.T. WHITTAKER (1902)
On the partial differential equations of mathematical physics, *Mathematische Annalen* 57, 333-355.

E.T. WHITTAKER (1904)
A treatise on the analytical dynamics of particles and rigid bodies with an introduction to the problem of three bodies, four editions 1904-1937, Cambridge University Press, Cambridge.

E.T. WHITTAKER and G.N. WATSON (1927)
A course of modern analysis: an introduction to the general theory of infinite processes and of analytic functions; with an account of the principal transcendental functions, 4th edition, Cambridge University Press, Cambridge.

J.M. WILKES and J.D. ZUND (1982)
Group-theoretical approach to the Schwarzschild solution, *American Journal of Physics* 50, 25-27.

T.J. WILLMORE (1959)
An introduction to differential geometry, Oxford at the Clavendon Press, London.

J.D. ZUND (1988a)
Tensorial methods in classical differential geometry - I: Basic principles, *Tensor NS* 47, 74-82.

J.D. ZUND (1988b)
Tensorial methods in classical differential geometry - II: Basic surface tensors, *Tensor NS* 47, 83-92.

J.D. ZUND (1989)
Differential geodesy of the Eötvös torsion balance, *Manuscripta Geodaetica* 14, 13-18.

J.D. ZUND (1990a)
An essay on the mathematical foundations of the Marussi-Hotine approach to geodesy, *Bollettino di Geodesia e Scienze Affini*, anno XLIX, 133-179.

J.D. ZUND (1990b)
The assertion of Hotine on the integrability conditions in his general (ω, ϕ, N) coordinate system, *Manuscripta Geodaetica* 15, 362-372.

J.D. ZUND (1990c)
The Hotine problem in differential geodesy, *Manuscripta Geodaetica* 15, 373-382.

J.D. ZUND (1992)
The mathematical foundations of the Hotine-Marussi approach to geodesy, *Bollettino di Geodesia e Scienze Affini*, anno LI, 125-138.

J.D. ZUND (1993a)
The Bruns curvature equations and tangential differential in differential geodesy, *Bollettino di Geodesia e Scienze Affini*, anno LII, 191-205.

J.D. ZUND (1993b)
Hotine's (ω, ϕ, N) coordinate system, *Phillips Laboratory Research Report*, PL-TR-93-2174, 23 pages.

J.D. ZUND and W. MOORE (1987)
Hotine's conjecture in differential geodesy, *Bulletin Géodésique* 61, 209-222.

J.D. ZUND and J.M. WILKES (1988)
The significance and generalization of two transformation formulas in Hotine's Mathematical geodesy, *Bollettino di Geodesia e Scienze Affini*, anno XLVII, 77-85.

Index

A

abstract components, 32
abstract Euclidean space, 183
abstract indices, 30–31
adjoint of a matrix, 257
Aleksandrov and Zalgaller, 333
algebraic Bianchi identity, 69, 90, 97
algebraic invariants, 256
Amaldi, 17
ambient (local) coordinate system, 98
ambient coordinates, 108
angular velocity of the Earth, 198, 217, 225, 267
anholonomic coefficients, 58, 96
anholonomic coordinate system, 131
anholonomity matrices, 138
Appell's equations, 306
arc length, 22
Atzema, 250

B

base surface, 233, 317
basic equation, 115
basic equations of the leg calculus, 162
basic gradient equation, 200, 213, 218
basic system, 256
basis, 33
basis of **V**, 28
Beltrami, 244, 246, 251
Beltrami condition, 246
Beltrami first order differential operators, 66
Beltrami's form, 245
Bianchi, 224
bilinear functional, 37
binormal, 175
Blaschke, 236
Blaschke and Leichtweiss, 166
Bocchio, 138, 271, 273
Bonnet, 156, 177, 319
Bonnet formula, 179
Bouquet, 319
Bourbaki, 26
Bozzi Zadro, 273
Bruns, 10, 206
Bruns curvature equation, 211–212
Bruns equation, 213, 216, 218
Bruns-Helmert conception of geodesy, 9
Burali-Forti, 303

C

coordinate method, 1
co-latitude, 2
coordinate method, 2
conformal transformations, 273
canonical congruence, 191
canonical congruences, 55, 114
canonical enumeration scheme,

52, 58
canonical forms for the Riemann tensor, 224
canonical identification of 3-leg vectors, 46
canonical identification scheme, 187
Caputo, 273
Carminelli, 273
Cartan, 23–24, 26, 86
Cartan calculus, 23–25, 278, 283
Cartan coefficients, 87–88, 99
Cartesian basis, 32
Cartesian coordinate system, 304
Casorati, 224
Casorati curvature, 224
Cauchy, 183, 336
Cauchy Problem, 307
Cauchy-Kowalewski theorem, 336–337
Cayley, 238, 254, 319
Cayley-Darboux equation, 111, 114, 319, 342
central quadric, 257
Cesàro, 11
Chebotarev, 253
Choquet-Bruhat, 122
Chovitz, 26, 274, 295, 301, 310
Christoffel, 16
Christoffel elimination scheme, 96
Christoffel symbols, 94, 301
classification of central quadrics, 258
co-differential δ, 122
Codazzi equations, 174, 219, 280, 289
commutation formulae, 60
commutators, 60, 92
compatibility equation, 223
complete integrability, 101, 114
completely integrable Pfaffian form, 245
conformal covariant derivative, 301
conformal curve theory, 290
conformal Frenet equations, 290
conformal function, 274
conformal Ricci coefficients, 283
conformal structural equations, 278
conformal surface theory, 285
conformal transformation, 274, 318
congruence, 22
congruence-forming property of a system of surfaces, 238–239
congruences of curves, 20, 22, 50
conjugate expression, 120, 125
conjugate expressions, 113
connection 1-form, 87
connection 1-forms, 172
consistency equations, 312, 314, 332
convergence of a congruence, 235

coordinate-based indices, 46
cotangent space, 24
covariant leg derivatives, 136
covector, 29
covectors, 24
cross product, 75
curvature 2-form, 87
curvature parameters, 111, 210
curvature vectors, 154, 162
curvatures of a congruence, 54
curvatures of congruences, 113

D

Darboux, 20, 249, 319
Darboux's approach, 249
Deheuvels, 39
descriptive coordinates, 324
determinant, 18
determining equation, 223
developmental equations, 251
differential Bianchi identity, 69
differential of a function, 76
differential or non-integrable constraints, 129
differentiation, 22
dimension of **V**, 27
Dirac, 15
directional derivative, 106
directional differentiation, 21
Dirichlet problem, 323
discriminant, 259–260
divergence of a congruence, 235
dual, 43
dual basis, 29, 33
dual space $\mathbf{V}^*$, 28
dualizors, 152
Duhem, 14–15, 44
Dupin, 185, 319
Dupin Theorem, 192, 318

E

Euclid, 1
Eötvös homography, 255, 268
Eötvös tensor, 206, 255
Eötvös torsion balance, 225
Eigenstructure theorem for the Marussi tensor, 264
eigenvalue of a tensor, 258
eigenvector of a tensor, 258
Einstein, 10–11, 13, 16–18
Einstein and Kaufmann, 18
Eisenhart, 45, 181–182, 234
ellipsoid, 259–260
elliptic point, 184–185
equations of Codazzi ($\mathcal{C}$), 145
equations of Gauss ($\mathcal{G}$), 145
Euclidean space, 45
Euler, 74, 273
Euler condition, 242, 250
Euler formula, 182, 185
Euler tensor, 157, 160, 182
Euler's condition, 241
Euler's form, 240
exact 1-form, 104
exact differential equation, 100
Extension Problem, 199, 227, 232, 315
exterior differential 1-form, 78
exterior differentiation, 78
exterior power of a vector space, 75
exterior product, 74

F

Fermat's Principle, 244
Ferrers, 306
first basic form I of a surface, 141–142
first Bruns curvature equation, 212
first Cartan structural equations, 172
flat/planar point, 184–185
focal planes, 249
focal points, 249
focal surfaces, 249
formula of Liouville, 173
Forsyth, 319
fourth basic form of a surface, 161
fourth basic tensor, 225
Frenet 3-leg, 177–178, 292
Frenet equations, 175–176, 179, 181
Frenet vectorial 3-leg, 175
functional, 18
functional determinant, 77
fundamental equation of differential geodesy, 206
fundamental principle of conformal geometry, 275
Fundamental Theorem of Differential Geodesy, 227, 229, 231
Fundamental Theorem of Riemannian Geometry, 94–95

G

Garabedian, 336
Gauss, 145, 148, 173, 273
Gauss and Codazzi equations, 335
Gauss curvature, 158, 184, 190, 193, 212, 224
Gauss equation, 173, 219, 280
Gauss formula, 145–146, 153, 169
Gauss operator, 142, 148–149, 155, 189
Gaussian curvature, 146, 282
Gaussian differential geometry, 11–12, 15, 304
Gaussian representation of a surface, 141, 192
Gel'fand, 39
general classification of proper quadrics, 260
general eigenvalue problem, 271
general relativity, 3, 18
generalized Burali-Forti homography, 255
generalized conformal transformations, 274, 295–296
generalized Marussi-Hotine approach, 322, 325, 333, 338
generalized Marussi-Hotine theory, 304
generalized parallel system, 315
generate, 28
geodesic congruence, 54
geodesic congruences, 113
geodesic curvatures, 154, 181

geodesic parallels, 316
geodesic torsion, 156–157, 181
geodetic singularity problem, 267–268
geometric equations, 339
geometric optics, 237, 243–244, 273
geometrical equations, 312, 332
geopotential function, 198
Germain curvature, 148, 158, 188, 191, 212, 282
Gibbs, 9
global problem, 233
Grafarend, 10, 40, 83, 127, 131, 138, 271, 305, 314, 342
Grafarend anholonomity matrices, 92
Grafarend matrices, 132
Gram-Schmidt orthogonalization process, 84
gravitational sphere of a planet, 253
gravity tensor, 255
Grossman, 10, 12–13, 40, 83, 127, 271, 305
Grotemeyer, 260
Guillemin and Pollack, 267

H

Hilbert, 1
Hölder class, 335
Hölder condition, 335
Hadamard, 307, 336
Halmos, 39
Hamilton, 237–238, 248
Hankins, 238
Heisenberg, 13
Helmert, 10
Herzberger, 243
Hilbert, 143
Hill gravitational sphere, 253
Hodge-deRham-Laplacian, 122
holonomic, 129
holonomic leg system, 131
holonomic systems, 306, 314
Holonomity Theorem, 130
Holota, 271
homographic calculus, 7, 255, 303
homographic equations, 16
Hopfner, 253, 334
Hotine, 5–8, 11, 14–15, 25–26, 40, 83, 111, 126–127, 149, 153, 170, 201, 205–209, 214–215, 222, 225, 233, 255, 273, 295, 300, 303, 310–311, 314, 320, 338, 341–342
Hotine 2-leg, 202, 222
Hotine 3-leg, 129, 202, 211, 216, 256, 259, 262, 266
Hotine Conjecture, 318
Hotine hierarchy, 319–320, 324
Hotine leg calculus, 203
Hotine Problem, 127, 304, 308
Hotine transformation formulas, 25, 43, 67, 108
Hotine's hierarchy of local coordinate systems, 308
Hotine's theorem, 213, 325, 328
Hotine-Marussi equations, 218, 223, 314, 325–326, 339

hyperbolic point, 184–185
hyperboloid of one-sheet, 259, 261
hyperboloid of two-sheets, 259, 261

I

I-invariants, 257, 270
indicatrix of curvature, 185
integrability conditions, 116, 308
integrable constraints, 129
integrable differential equations, 74
integrating factor, 74, 101
interpretations of exterior differentiation, 80–81
intrinsic coordinates, 126
intrinsic geodesy, 5–6
intrinsic geometry, 11
inverse geometric equations, 339–340
inverse tensor, 194
isomorphic, 29
Ivory Property, 235–236

J

J-invariants, 257, 259–260, 262, 268–270
Jacobi, 61, 336
Jacobi identity, 62, 131
Jacobowitz, 337
Jacobowitz-Treves, 338
Jordan, 10
Jung, 253

K

K-inequality, 263, 267–268
Koenigs, 248–249
Kowalewski, 336
Kummer, 246, 251
Kummer's (linear) ray systems, 246
Kummer's equation, 247

L

longitude, 2
leg space, 105
leg coefficients, 139
Lagrange, 129, 273, 336
Lagrange's equations, 128, 306
Lamé, 319
Lamé equations, 67–68, 72, 88, 90–91, 102, 123–125, 132–135, 172–173, 218, 221, 314, 325, 328, 332
Lamé identities, 125, 134
Lambeck, 198, 252
Lambert, 253, 273
Lanczos, 306
Laplace, 253
Laplace equation, 197
Laplacian of a function, 65
Laplacian of a scalar function, 121–122, 136
Laplacian of local gravity (2-dimensions), 221
Laplacian of the local gravity (3-dimensions), 220, 225
Laplacian operator in 3-dimensions, 198
leg calculus, 8, 40
leg coefficients, 110, 163

leg coefficients of a tensor, 50
leg components, 109
leg derivative, 106, 109
leg expression for the Kronecker delta, 48
leg indices, 46
leg representations of the metric tensor, 48
Levi-Cività, 17, 25, 45, 274
Lewy, 337
Lichnerowicz, 39
Lie bracket, 118, 171
Lie bracket of a pair of leg vectors, 60
limiting problem, 233
line geometry approach, 248
linear congruence of curves, 236
linear functional, 28, 38
linear identities, 341
linear transformation, 28
lines of curvature, 166–167
Lipschitz, 16
Livieratos, 271
local gravity, 200, 213–214, 327
local gravity vector, 206
local problem, 233
local solvability, 337

M

m-inequality, 228, 230
Marussi tensor, 255–256, 261, 263, 268, 270
Madelung, 236
Malus, 237, 248
Malus cone, 249
Malus-Dupin theorem, 237, 243
Marcolongo, 303
Marussi, 5–12, 14–15, 25, 40, 83, 126–127, 201, 206, 224, 255, 273, 300, 303, 305, 338
Marussi Ansatz, 200, 317–318, 320, 322, 338
Marussi Condition, 83
Marussi Conditions, 12–14, 305, 308, 321, 333
Marussi Hypothesis, 126–127, 304, 338
Marussi quadric, 263, 267, 269
Marussi tensor, 206–207, 209, 213
Marussi-Hotine approach, 303, 319–320, 323–324, 338
Marussi-Hotine theory, 326
Mathematical Geodesy, 8
matrix of integrating factors, 314
Maxwell's electrodynamics, 14, 44, 18
McConnell, 7, 141, 163, 176–177, 179, 181–182
method of rotation coefficients, 20
Meusnier, 156
Meusnier formulas, 179
Milnor, 267
minimal surfaces, 167
modified Laplace equation, 198, 201, 213, 216, 232, 323, 325

Monge, 237, 336
Mongean differential geometry, 238, 240, 304
Munkres, 231

N

n, 178
Newtonian potential, 197
N-integrability conditions, 203–205, 208, 210
n-integrability conditions, 216
N-integrability conditions, 218
n-integrability conditions, 218
N-integrability conditions, 223
n-integrability conditions, 224
Neumann problem, 323
Newman-Penrose method, 338
Newton, 13
Newtonian dynamics, 12, 18
Nirenberg, 335
Nitsche, 167
non-holonomic, 11, 129
non-holonomic coordinate system, 306
non-holonomic leg system, 131
non-holonomic reference systems, 305–306
normal congruence, 51, 55, 202
normal congruence of curves, 113
normal congruences, 55
normal coordinate system, 315
normal curvatures, 156–157, 188
normal linear congruence of curves, 244
normal-system, 308

O

(ω, ϕ, N)-system, 308
(ω, ϕ, h)-system, 308
ω-degeneracy, 313, 340–341
objects of anholomity/nonholonomity, 58
Olmsted, 270
operator, 28, 121
ordinary integrability, 101
orthonormality hypothesis, 83
osculating paraboloid, 184

P

Peano, 1
Pasteur, 4
Pfaffians, 73
Pizzetti Theorem, 227
parabolic point, 184–185
paraboloid, 265
parallel curve, 254
parallel family of surfaces, 235–236
parallel surface, 254
parallel systems, 244
parameter net, 167
Pensa coefficients, 53
permissible approach, 326
permutation algorithm, 111–113, 120
Pfaff's problem, 74
Pfaffian, 40
Pfaffian 3-leg, 41
Pfaffian leg, 150–151
Pfaffians, 24
Pizzetti, 227
Pizzetti , 228, 252
Pizzetti inequality, 229, 232–233, 239, 252, 334

Pizzetti Ivory, 235
Pizzetti sphere, 229
Pizzetti's first assumption, 228
Pizzetti's second assumption, 228
Pizzetti's Theorem, 334
Plücker, 248
Pogorelov, 184, 333
Poincaré property, 100, 242
Poinsot, 183
Poisson bracket, 131
Poisson equation, 197
primary congruence, 51
primary equations, 312, 332, 341
primitive, 104
principal curvature of the normal congruence Γ, 205
principal curvatures, 225
principal direction, 166–167
principal directions, 168
principal normal, 175
principal radius of curvature, 185
projection operator, 195–196
property (T), 114, 120, 140, 153, 156, 159, 162, 172, 190, 205–206, 208–209, 243, 299

Q

quadratic identities, 341
quadric cone, 259, 261
quadric surface, 183

R

Ricci, 45
radius, 3
reciprocal tensor, 194
Reconstruction Theorem, 128, 132–133
Ricci, 16–17, 20
Ricci calculus, 20, 22, 25, 283
Ricci coefficients, 51, 53, 56, 87–88, 99, 110, 133
Ricci identity/lemma, 95
Ricci-Curbastro, 16
Riemann, 16
Riemann-Christoffel curvature tensor, 94, 125, 133
Riemannian space, 45, 316
Roberts, 254
Rodrigues, 148
Rodrigues formula, 163–164
Rodrigues vectors, 164
rotation coefficients, 53
Rummel, 271
Rummel and Van Gelderen, 271

S

synthetic method, 1
spherical polar coordinates, 2
synthetic methodology, 2
symmetric (ω, ϕ, h)-system, 309
Salmon, 254
Schouten, 86, 151, 306
Schouten identities, 61, 63, 92–93, 120, 171–172, 293
Schubert, 273
Schwarzschild Problem, 3
Schwarzschild solution, 3
second basic form II of a surface, 143

second Bruns curvature equation, 223
secondary congruences, 51
secular equation, 256
self-conjugate expression, 113, 120, 125
separable system, 236
signum, 184
singly-orthogonal system, 317
singularities of a vector field, 267
singularity condition, 264
skew congruence of curves, 243
smoothness requirements, 201
Souslow's equation, 288
space $\mathbf{V}^*$, 43
space congruence Eigenstructure Theorem, 162
span, 28
special eigenvalue problem, 271
Special Relativity, 18
sphere of activity, 253
spherical image, 147
spherical polar coordinates, 19, 329
spherical representation, 147
Stäckel, 236
Stäckel Separation Property, 236
star, 121
Strong Form of the Marussi Hypothesis, 126–127, 305–306
structural constants, 62
structural equations of Cartan, 86
super anholonomity matrix, 132, 138
surface congruence Eigenstructure Theorem, 165
surface congruences, 202
surface Eigenstructure Theorem, 167
surface Frenet equations, 154
surface geodesic congruence, 165
surface-forming property of a congruence of curves, 244
Synge, 243–244
system of Lamé equations, 69
systems, 11

T

τ-roots, 257, 265, 270
tangent, 175
tangent space, 22
tangent vectors, 22–23
tangential congruences, 51, 202
Taucer, 273
tendencies of Γ_3, 113
tendency of a congruence, 54
tensor of gravity gradients, 255
tensor product, 35
tensor product of vectors, 35
tensor spaces, 26, 35
third basic form III of a surface, 147, 160
Thomas, 254
toposphere, 228–229
torsion, 175
torsion 2-form, 87, 96

total differential equation, 73, 240
trace of a matrix, 257
transformation factors, 25
Treves, 337
triply-orthogonal system of co-ordinates , 51
triply-orthogonal system of surfaces, 55, 114, 186, 235
triply-orthogonal-system, 308, 318

V

Van der Waerden, 260
vector space **V**, 27
vectorial n-leg, 45
vectorial 3-leg, 41
Vessiot, 249

W

weak existence, 307
Weak Form of the Marussi Hypothesis, 126–127, 305–306
Weatherburn, 25, 45, 54
Weingarten formula, 146
Weingarten operator, 164–165
Weyl, 15, 26, 335
Weyl Problem, 335
Whitehead, 323
Whittaker, 183, 322–323, 336
Whittaker and Watson, 322
Whittaker's example, 326
Wilkes, 299
Wilkes and Zund, 3
Willmore, 39, 143

Z

Zund, 6, 127, 202, 223, 225, 274, 303, 313–314, 319, 335, 339–342
Zund and Moore, 319, 342–343
Zund and Wilkes, 43
Zund's Conjecture, 324

MIX
Papier aus verantwortungsvollen Quellen
Paper from responsible sources
FSC® C105338

If you have any concerns about our products,
you can contact us on
ProductSafety@springernature.com

In case Publisher is established outside the EU,
the EU authorized representative is:
Springer Nature Customer Service Center GmbH
Europaplatz 3, 69115 Heidelberg, Germany

Printed by Libri Plureos GmbH
in Hamburg, Germany